Studies in Systems, Decision and Control

Volume 518

The series "Studies in Systems, Decision and Control" (SSDC) covers both new developments and advances, as well as the state of the art, in the various areas of broadly perceived systems, decision making and control–quickly, up to date and with a high quality. The intent is to cover the theory, applications, and perspectives on the state of the art and future developments relevant to systems, decision making, control, complex processes and related areas, as embedded in the fields of engineering, computer science, physics, economics, social and life sciences, as well as the paradigms and methodologies behind them. The series contains monographs, textbooks, lecture notes and edited volumes in systems, decision making and control spanning the areas of Cyber-Physical Systems, Autonomous Systems, Sensor Networks, Control Systems, Energy Systems, Automotive Systems, Biological Systems, Vehicular Networking and Connected Vehicles, Aerospace Systems, Automation, Manufacturing, Smart Grids, Nonlinear Systems, Power Systems, Robotics, Social Systems, Economic Systems and other. Of particular value to both the contributors and the readership are the short publication timeframe and the world-wide distribution and exposure which enable both a wide and rapid dissemination of research output.

Indexed by SCOPUS, DBLP, WTI Frankfurt eG, zbMATH, SCImago.

All books published in the series are submitted for consideration in Web of Science.

Mohammad Yazdi
Editor

Progressive Decision-Making Tools and Applications in Project and Operation Management

Approaches, Case Studies, Multi-criteria Decision-Making, Multi-objective Decision-Making, Decision under Uncertainty

Springer

Editor
Mohammad Yazdi
Faculty of Science and Engineering
Macquarie University
Sydney, NSW, Australia

ISSN 2198-4182 ISSN 2198-4190 (electronic)
Studies in Systems, Decision and Control
ISBN 978-3-031-51718-1 ISBN 978-3-031-51719-8 (eBook)
https://doi.org/10.1007/978-3-031-51719-8

This Springer imprint is published by the registered company Springer Nature Switzerland AG
The registered company address is: Gewerbestrasse 11, 6330 Cham, Switzerland

Paper in this product is recyclable.

To My Beloved Parents

Preface

Decision-making is a cornerstone of success in the dynamic and intricate landscape of project and operations management. This book, *Progressive Decision-making Tools and Applications in Project and Operation Management* represents a comprehensive and advanced exploration of the evolving paradigms, multifaceted challenges, and innovative methodologies shaping the contemporary decision-making landscape in these pivotal domains.

With 15 meticulously curated chapters, this volume is a compendium of cutting-edge research, offering a holistic understanding of the complexities, solutions, and transformative advancements in decision-making tools and applications. Each chapter contributes a unique facet to the mosaic of knowledge, catering to the diverse needs of scholars, practitioners, researchers, and decision-makers.

Chapter "Cutting Edge Research Topics on Operations and Project Management of Supportive Decision-Making Tools" serves as the book's foundation by addressing the formidable challenges embedded in operations and project management decision-making. It meticulously examines issues including human cognitive bias, process intricacies, problem complexity, and the multidimensional nature of contemporary challenges. Furthermore, this chapter critically assesses the frameworks and strategies proposed by experts, research centers, and academic institutions to bolster the effectiveness of decision-making tools within these domains. The chapter employs advanced statistical metadata analysis to unveil trends, frontier directions, and influential entities in operations and project management research.

Chapter "Decision-Making in Project and Operations Management" goes thought deeper into operations and project management by elucidating their interplay. It underscores how these domains are intrinsically linked and delineates the criticality of sound decision-making within both. It emphasizes the significance of informed decisions in optimizing operations management's planning, scheduling, organizing, and controlling processes. Likewise, in the context of project management, it illuminates how a meticulous approach to decision-making can spell the difference between triumph and failure. This chapter underscores managers' need to equip themselves with the requisite tools and strategies for astute decision-making.

Subsequent chapters extend this advanced exploration into diverse realms:

- Chapter "Nash Equilibrium-Based FMEA for Risk Prioritization in Hydrogen Refueling Station Design" introduces a pioneering Nash Equilibrium-Based Failure Mode and Effect Analysis (FMEA) for risk prioritization in hydrogen refuelling station design. This innovative approach leverages game theory principles and Pythagorean fuzzy uncertain linguistic variables to offer a holistic risk assessment framework, revolutionizing decision-making in hydrogen infrastructure.
- Chapter "Practice and Enlightenment of General Quality Characteristic Management of Us Military Equipment" scrutinizes the impeccable Equipment Quality Management System implemented by the United States Army, presenting strategic recommendations for enhancing the management of general quality characteristics in the military equipment sector. This chapter offers an advanced understanding of quality management in a complex operational context.
- Chapter "Bi-Objective Portfolio Optimization with Mean-CVaR Model: An Ideal and Anti-Ideal Compromise Programming Approach" ventures into financial management territory, introducing a novel bi-objective portfolio optimization model utilizing the Mean-CVaR framework. This advanced approach demonstrates how decision-making in financial portfolios can be elevated through advanced mathematical modeling and compromise programming.
- Chapter "Developing Alternative Multilinear Regression-Based Intelligence Hybrid Model" elevates the discussion by advancing quantitative structure–property relationships (QSPR) simulation for mycotoxins. It showcases the application of hybrid intelligence models, including MLR-SVM, MLR-ANN, and MLR-ANFIS, in simulating complex molecular properties, introducing a new paradigm for QSPR analysis.
- Chapter "Empowering Emergency Operations Management: A Pride Day" introduces "*Pride Day*", a visionary initiative to foster inclusivity and diversity in emergency operations management. It provides an advanced perspective on integrating diversity considerations in critical emergency response scenarios.
- Chapter "Enhancing Hydrogen Risk Management: A Modified EDAS-Based FMEA Approach" embarks on the path of innovation by introducing a modified EDAS-based FMEA approach for enhancing hydrogen risk management in steam-reforming hydrogen generation units. This pioneering methodology reshapes the risk assessment landscape in hydrogen production.
- Chapter "Multi-Criteria Decision-Making for Optimal Selection of Hydrogen Fuel Cell Types: Evaluating Sustainability and Performance Criteria" engages with the sustainable energy frontier by employing multi-criteria decision-making techniques to evaluate and rank hydrogen fuel cell types. It presents an advanced approach to selecting energy technologies, considering sustainability and performance criteria.
- Chapter "Enhancing Supplier Selection Reliability: Integrated Variable Returns to Scale-Robust Parameter R Approach" navigates the complex terrain of supplier selection, offering an integrated and advanced model that combines variable returns to scale (VRS) with robust parameter estimation. This model addresses the challenges of supplier selection in multifaceted scenarios.

- Chapter "Operation Management of Sky Magic: Consolidating Perspective of Risk and Safety" elevates the discourse by exploring operation management in the context of Sky Magic, a ground-breaking technology in aerial entertainment. It introduces advanced risk assessment frameworks, safety management systems, and real-time data analysis techniques for ensuring safe and efficient aerial performances.
- Chapter "Research on Reliability Design Analysis Application to Gas Turbines" ventures into the realm of gas turbine reliability design analysis. It evaluates existing applications in light of national military standards, offering a roadmap for the systematic and advanced application of reliability design analysis in gas turbine technology.
- Chapter "Prioritization of HSE Management Risks in the Management of Construction Projects (Non-Mechanized Tunnel) by Monte Carlo Modeling Method" discusses research conducted on health, safety, and environmental (HSE) issues in non-mechanized tunnel construction, with a focus on the Tehran Shomal Freeway project. The research identifies and prioritizes HSE indicators through expert opinions and data analysis methods. The key findings indicate that "management" and "design and implementation" risks have the most significant impact on HSE, emphasizing the importance of leadership and commitment in HSE management.
- Chapter "Enhancing Crowd Safety Resilience Through Systematic Analysis" emphasizes the growing importance of crowd safety across various domains in contemporary society and highlights the need for a systematic approach to ensure the well-being of all stakeholders. It introduces the concept of resilience as a critical aspect of crowd safety management, defined by characteristics such as disruption, adaptation, absorption, and restoration, which can evolve through continuous learning. The chapter introduces Bayesian networks as a valuable tool for assessing and improving crowd safety resilience, aiming to maintain adaptability and responsiveness in the face of evolving challenges, ultimately contributing to the overall well-being and security of everyone involved.
- Chapter "Decision-Making During Crisis in Megaprojects" examines the significance of crisis management in the context of project studies, with a particular focus on high-stake megaprojects characterized by substantial budgets, complex multi-organizational structures, and temporary nature. Crises in such projects can have a profound impact, potentially jeopardizing project success and the reputations of involved organizations. This chapter explores decision-making processes and styles within crises, offering insights and a model to aid effective crisis management in megaprojects.
- Chapter "Harnessing AI for Project Risk Management: A Paradigm Shift" provides a visionary glimpse into the future, where artificial intelligence (AI) catalyzes a paradigm shift in project risk management. This advanced discussion explores how AI transforms the landscape of risk anticipation, assessment, and mitigation in project management through predictive analytics, machine learning, and cognitive capabilities.

As the editor of this advanced and comprehensive compendium, I am honoured to present a collection of chapters that exemplify the vanguard of research, innovation, and thought leadership in decision-making tools and applications within project and operations management. This volume will inform, inspire, and challenge decision-makers to embrace advanced methodologies and technologies in their pursuit of excellence in these pivotal domains.

London, UK
September 2023

Dr. Mohammad Yazdi

Contents

Cutting Edge Research Topics on Operations and Project Management of Supportive Decision-Making Tools

Li Yang, Nankai Cheng, Rosita Moradi, and Mohammad Yazdi

Abstract In the context of operations and project management of supportive decision-making tools, this chapter addresses the challenges related to the criticality of this areas. These challenges include, human bias, process-based issues, complexity, and multi-dimensions problems, and other undesired cases. Scholars, practitioners, research centres, and academic institutes have introduced different frameworks and approaches for assessing and enhancing operations and project management techniques and tools to tackle these challenges. The primary objective of this chapter is to provide an overview of operations and project management of supportive decision-making tools and frameworks specifically applicable to different industrial sectors. These frameworks have been published over the past few decades and offer valuable insights into improving the overall performance of these operations and project management domains. Additionally, it is crucial to identify and analyse the common aspects of operations and project management, examine the existing challenges they face, and explore potential opportunities for future research directions. To accomplish this, an extensive investigation is conducted using statistical metadata analysis. this analysis explores frontier directions and development trends to uncover the current research status. Furthermore, a bibliometric study identifies this domain's most productive and influential researchers, research centres, and hotspot fields.

L. Yang
School of Media Design, College of Technology and Design, University of Economics, Ho Chi Minh City, Vietnam

N. Cheng
Department of Design, Faculty of Architecture, University of Lisbon, Lisbon, Portugal

R. Moradi
University of Wollongong, Northfields Avenue, Wollongong, NSW 2522, Australia

M. Yazdi (✉)
School of Computing, Engineering and Physical Sciences, University of the West of Scotland (UWS), London, UK
e-mail: mohammad.yazdi@mq.edu.au

Faculty of Science and Engineering, Macquarie University, Sydney, Australia

M. Yazdi (ed.), *Progressive Decision-Making Tools and Applications in Project and Operation Management*, Studies in Systems, Decision and Control 518,
https://doi.org/10.1007/978-3-031-51719-8_1

Keywords Operations management · Supportive decision-making tools · Bibliometric study

1 Introductions

Operations and project management are crucial in enabling human society's functioning and providing essential social services [1–6]. These scientific areas encompass many physically and virtually interconnected multi-dimensional elements and human skills, including leadership, risk management, stakeholder engagement, technical knowledge, team building and collaboration, adaptability, resilience, and problem-solving. They are indispensable for the smooth operation of society and the community's well-being. The significance of operations and project management becomes evident when considering the severe consequences of any damage or disruption to these systems, impacting society's health, safety, economy, and security [7–10]. Unfortunately, operations and project management are inevitably exposed to various undesired circumstances, such as scope creep, inadequate resource allocation, poor communication, inadequate planning, inadequate resource allocation, inefficient team dynamics, risk mismanagement, stakeholder dissatisfaction, inadequate quality control, insufficient monitoring and control. These undesired circumstances highlight the vulnerability of operations and project management to external factors. Overall, safeguarding operations and project management is paramount due to their vital role in sustaining society's functioning and providing essential services. Their exposure to various threats necessitates comprehensive measures to mitigate risks and enhance resilience, ensuring the stability and well-being of communities.

Recent events have highlighted the inability to accurately predict and effectively prevent all undesired circumstances [11, 12]. Consequently, managing undesired events that significantly impact operations and project management requires focused attention from scholars, practitioners, research centres, and academic institutes. These stakeholders are now directing their efforts towards developing frameworks for having innovative and intelligent operations and project management structures [13–15]. The significance of these two keywords—Project management and Operation management—within the context of can be summarized as follows:

- Project management: PMBOK 7th defined it as "*The application of knowledge, skills, tools, and techniques to project activities to meet project requirements. Project management refers to guiding the project work to deliver the intended outcomes. Project teams can achieve the outcomes using a broad range of approaches (e.g., predictive, hybrid, and adaptive)*",
- Operation management: This means that "*Operations management (OM) is the administration of business practices to create the highest level of efficiency possible within an organization. It is concerned with converting materials and labour into goods and services as efficiently as possible to maximize the profit of an organization*" [16].

Moreover, in line with the growing concern for empowering and managing operation and project tools have become critical issues for assessing, evaluating, and furthering the decision-making process. Therefore, the fundamental methods and operational techniques to measure and improve the s operations and project management framework have attracted considerable research contributions over the last few years, including but not limited to [17–20]. An example, an overview of literature reviews has been conducted on this area, such as the exponential growth of the satellite industry, the effective management of satellites has emerged as a critical concern for ground operation management [21], optimized multi-layer energy management for smart integrated-energy microgrid systems incorporating generation and demand-side flexibility is assessed [22], a comprehensive digital twin framework for smart water grids is developed [23], CFD simulations, when combined with data analysis techniques, serve as a valuable tool for enhancing the comprehension of flow structures within industrial components. This integration enables the identification of crucial design features, optimization of performance, and facilitation of operational management [24], in-depth research is conducted on the multi-domain modeling of the marine steam power system, characteristic analysis of variable working conditions, fault simulation [25], socioeconomic factors play a pivotal role as cost indicators in the suitability evaluation index system for underground space development, In conjunction with the geological environment conditions, a comprehensive assessment of underground space utilization can be achieved by considering 15 key evaluation factors. These factors encompass a range of aspects that impact the feasibility and effectiveness of utilizing underground spaces. By incorporating socioeconomic factors into the evaluation process and accounting for the geological environment, a more robust and accurate assessment of the suitability of underground space development can be attained [26].

In such cases, bibliometric analysis can be valuable for studying advanced trends in specific research areas. It offers a reliable framework for analyzing and understanding the literature. The bibliometric analysis follows a straightforward, comprehensible, replicable, logical, and intelligible procedure [27]. This approach helps to reduce biases introduced by scholars and provides a structured and objective overview of the available literature [28]. Over the past few years, bibliometric analysis has gained widespread popularity in various domains, approaches, and fields. It has proven helpful in gaining insights into the significance, properties, development trajectory, and research trends. Some areas where bibliometric analysis has been applied include but are not limited to [29–31].

This book chapter aims to provide valuable insights and frameworks for researchers, engineers, and practitioners interested in operations and project management of supportive decision-making tools improvement research. It serves as a comprehensive resource for understanding this field's current state and future prospects. Additionally, it facilitates international collaboration by highlighting opportunities for cooperation with other research centres and scholars. The chapter addresses several key questions which are crucial for advancing the literature on operations and project management of supportive decision-making tools:

1. Question 1: How has the distribution of publications related to operations and project management of supportive decision-making tools changed over time?
2. Question 2: Which journals have made the most significant contributions to research on operations and project management of supportive decision-making tools?
3. Question 3: Which institutes and scholars have been the most influential and prolific in developing improvement frameworks for operations and project management of supportive decision-making tools?
4. Question 4: What are the most highly cited references in improvement frameworks for operations and project management of supportive decision-making tools?
5. Question 5: What is the development path and emerging trends in improvement frameworks for operations and project management of supportive decision-making tools?
6. Question 6: What is the contribution of different financial sectors to improvement frameworks for operations and project management of supportive decision-making tools?

The chapter is structured as follows:

- Section 2 presents the research methodology employed to conduct the literature review.
- Section 3 provides the outcomes of the bibliometric analysis performed on the collected dataset.
- Finally, Sect. 4 offers conclusions and discusses directions for future studies.

2 Research Methodology

We are currently focused on a critical issue in our overview work, which involves assessing and reviewing existing literature on energy-critical infrastructures. The main objective of this work is to identify the limitations of the current state of the art and provide reliable and valuable insights for scholars, practitioners, research centres, and academic institutes.

There are only a limited number of comprehensive reviews approaches available. For instance, one direction utilized a five-stage framework that included question preparations, investigation locations, study evaluation and selection, analysis and synthesis, and reporting the outcomes [32, 33]. In our review work, we employed a four-stage framework to perform a bibliometric analysis of the development of energy-critical infrastructures. This framework involved three main stages: (i) defining the keywords, (ii) collecting the relevant data, and (iii) statistically analyzing the metadata and employing bibliometrics.

(i) Defining the keywords

To ensure a thorough assessment of the extensive range of research terms and achieve comprehensive coverage of literature related to energy-critical infrastructures, it is crucial to adopt a well-defined structure for selecting keywords. Considering this, a two-level structure for keyword searching is proposed as follows:

- Core keywords: "Operations Management" OR "Project Management" OR "Decision-making", AND
- Secondary keywords: "Critical" OR "Research" OR "Management".

It is important to note that the "core keywords" serve to indicate the search context, while the "secondary keywords" help in narrowing down the research scope.

(ii) Collecting the data

The Scopus database employs both "core keywords" and "secondary keywords" for searching purposes, with the following constraints considered:

Search Scope: The search includes the fields of "Topic, Title, Keywords, or Abstract."

Time Frame: The period for consideration is up until the end of September 2023.

Language: The search is limited to documents written in English.

Source Type: The search encompasses "Journal, Conference proceeding, Book series, and Book" sources.

Document Type: The selected document types include "Article, Conference paper, Review paper, Book chapter, Book, and Conference review".

Publication Stage: Both the "Final" and "Article in the press" stages of publication are considered.

The initial screening process yielded a total of 10,342 documents. These documents' abstracts, keywords, and titles were then carefully reviewed. A second screening round was conducted after this review, resulting in 1994 remaining documents. The selection criteria for this round were as follows:

Inclusion Criterion: Documents applying Operations and Project management in various application domains within decision-making purposes were retained.

Exclusion Criterion: Documents that explained the Operations and Project management solely through a simple example were excluded.

Irrelevance Criterion: Documents primarily focused on the general discussion of Operations and Project management were considered irrelevant.

Inclusion Criterion: Articles that aimed to extend Operations and Project management to address shortcomings were included.

(iii) Analyzing the metadata statistically and using bibliometrics

Figure 1 provides an overview of the publication distribution based on the period up until September 2023. The data clearly shows a gradual increase in

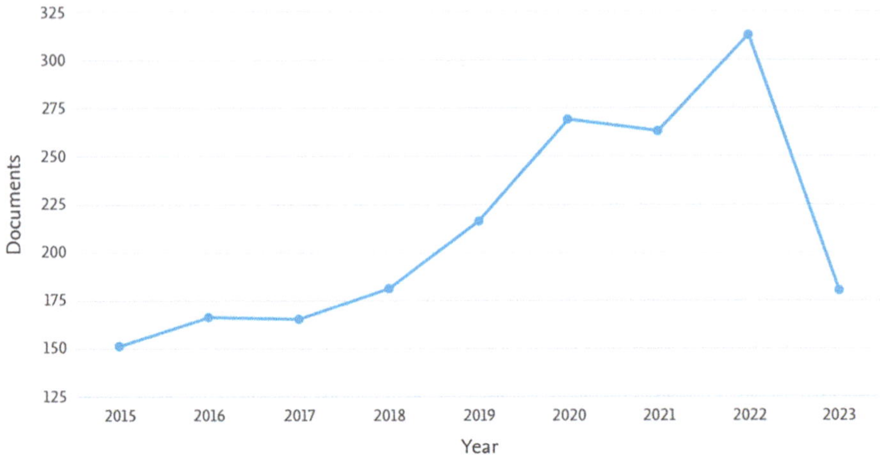

Fig. 1 The number of published documents over time

distribution over the past decade, with noticeable spikes in growth occurring in 2022 and expected to follows the same pattern.

Furthermore, during the three years from 2020, 2032, and 2022, a total of 855 papers were published, which accounts for approximately 42.3% of all the considered documents. It is essential to highlight that improving Operations and Project management has garnered significant attention from scholars, engineers, and practitioners in recent years.

Moreover, the published documents have contributed to over 26 subject areas related to improving Operations and Project management. Figure 2 showcases the distribution of these subjects. The three subjects that have received the highest priority are Engineering, with 994 documents accounting for 24.4% of the total; Business, Management and Accounting, with 714 papers representing 17.5%; and Computer Sciences, with 491 documents constituting 12.0%.

Following these subjects, Social Science has received attention with 356 documents (8.7%), Environmental Sciences with 317 papers (7.8%), and Decision Science with 299 records (7.3%).

Metadata/bibliometric analysis is an effective and dependable method for measuring and evaluating the impact of published papers. This analysis provides insights into future research by examining essential factors such as authors, citations, keywords, titles, journals, and publication years [34].

This study's bibliometric analysis focuses on the Operations and Project management improvement areas. To carry out this analysis, the researchers utilize the biblioshiny [35]. This software enables the detection, research, and visualization of trends and patterns in published scientific articles related to the areas mentioned above of infrastructure improvement.

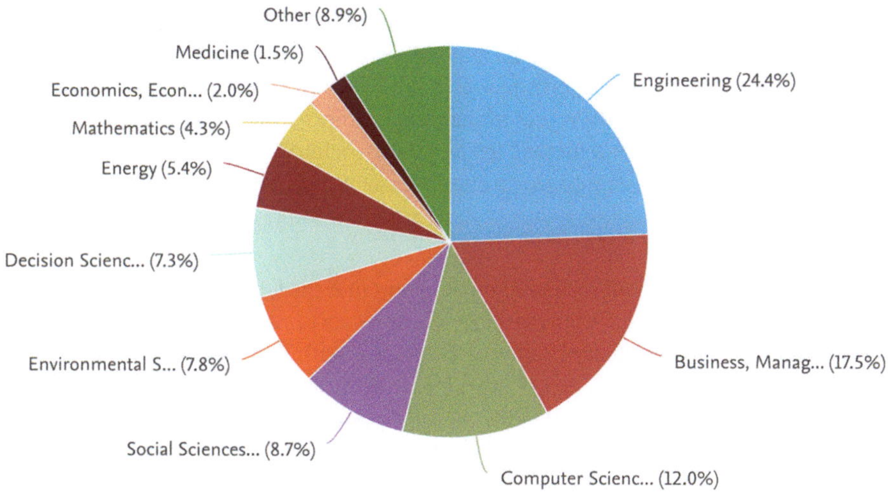

Fig. 2 The category of published documents based on subject areas

The following section provides a comprehensive overview of the analysis, including details of the methodology employed, the results obtained, key observations, and in-depth discussions of the findings. This section aims to shed light on the significance of the research and its implications for future studies in the field.

3 Results and Discussions

This section conducts a comprehensive bibliometric analysis, focusing on four key types of research.

The first type of analysis involves examining co-authorship patterns, specifically looking at the contributions and networks of authors from various countries worldwide. This analysis helps identify the collaborative relationships and trends within the global research community.

The second type of analysis evaluates the contributions of research centres and their international networks. By assessing these centres' research output and collaborations, we can gain insights into the extent and impact of their partnerships on a global scale.

The third type of analysis involves assessing authorship contributions based on the total number of documents and citations. This analysis helps identify prolific authors and their influence within the field of study, indicating their research output and impact.

The fourth type of analysis focuses on bibliographic coupling, aiming to provide an overview of related research articles based on recurring patterns. By identifying common themes and connections among the literature, this analysis enables us to understand the existing knowledge base and research trends in the field.

Finally, an assessment determines the co-occurrence of the most influential keywords in Operations and Project management improvement. This analysis helps identify the key concepts and themes frequently associated with these research areas, providing valuable insights into the prevailing trends and topics of interest.

- *Evaluating the global contributions of countries*

Based on the data presented in Fig. 3, it is evident that the China, USA, and UK have significantly contributed to improving Operations and Project management. These three countries have published many documents, with the China leading at 341, followed closely by the USA at 171, and UK at 88. Furthermore, when considering the SCP link, the China, USA, and UK exhibit respective values of 235, 132, and 57, aligning precisely with their published document numbers.

- *Evaluating the keywords analysis*

Keyword analysis is a crucial component of bibliometric analysis, enabling researchers to identify and comprehend the principal themes, topics, and trends within a specific body of literature. Figure 4 presents the most frequently occurring keywords, with "decision-making," "project management," and "construction industry" having the highest frequencies of 1499, 1098, and 188, respectively. Likewise, Fig. 5 showcases word clouds where "decision-making" and "project management" are positioned at the centre, surrounded by words such as "construction industry," "risk assessment," "sustainable development," and others. Furthermore, Fig. 6 demonstrates trend analysis, indicating that in 2015, the general research topic in operations and project management was "program development," which evolved into "multi-criteria decision-making" by 2023.

- *Evaluating the impact authors*

The initial analysis focuses on identifying the most relevant authors in the field under study. According to Fig. 7, Li H, Wang Y, and Li Y stand out as highly productive authors with a total number of publications of 20, 14, and 13, respectively. These authors have made significant contributions to the field.

Figure 8 illustrates the variation in author productivity over time. The initially identified prolific authors have consistently contributed to operations and project management. Their productivity has remained stable or shown a positive trend over time.

In this analysis, Lotka's Law, also known as Lotka's power law or Lotka's frequency distribution (Fig. 9), describes the frequency distribution of a specific type of published record. The findings indicate that approximately 75% of authors in this field have published two papers. Lotka's Law provides valuable insights into the productivity patterns and distribution of authors in the area.

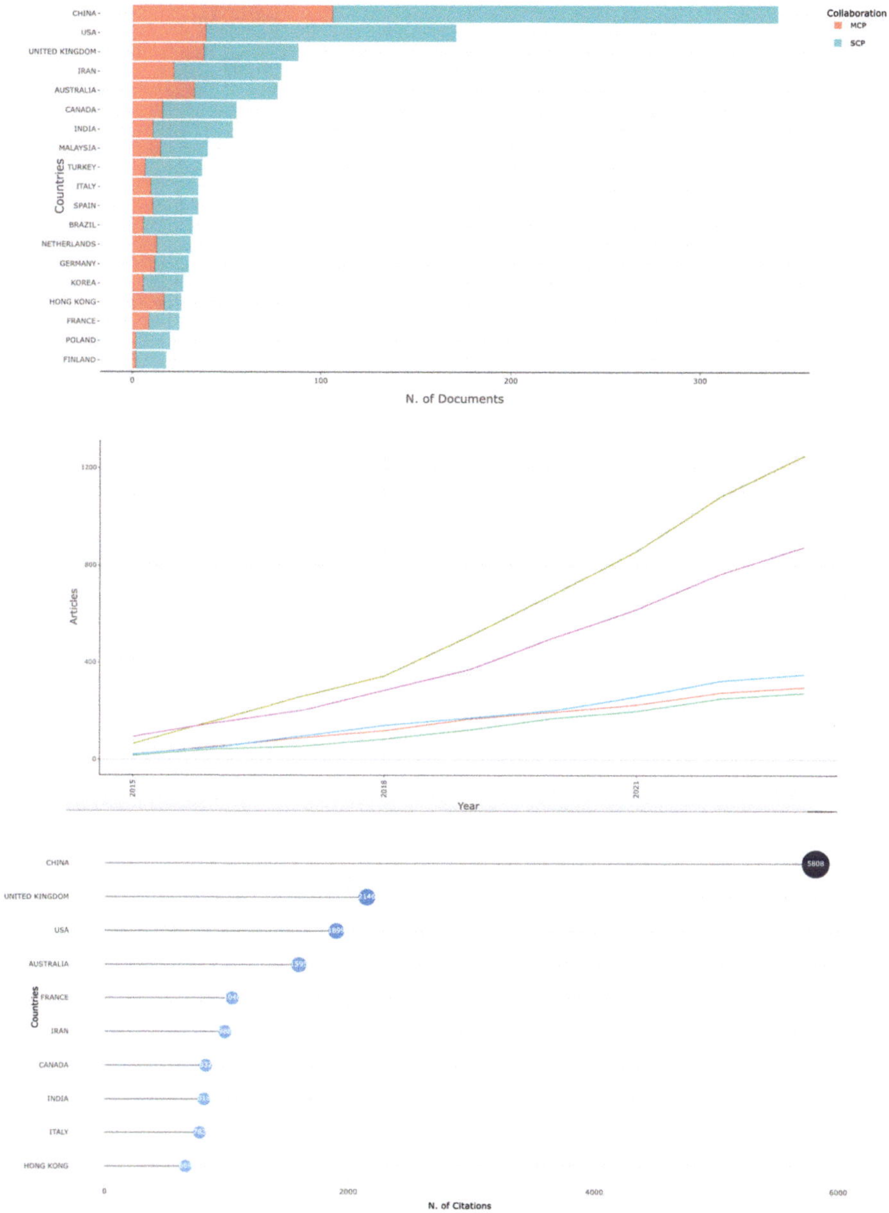

Fig. 3 The worldwide countries' contributions

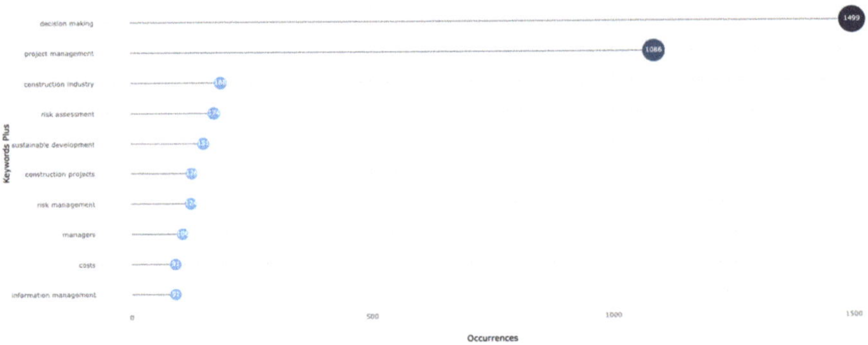

Fig. 4 The most frequent words

Fig. 5 The most frequent words (word clouds analysis)

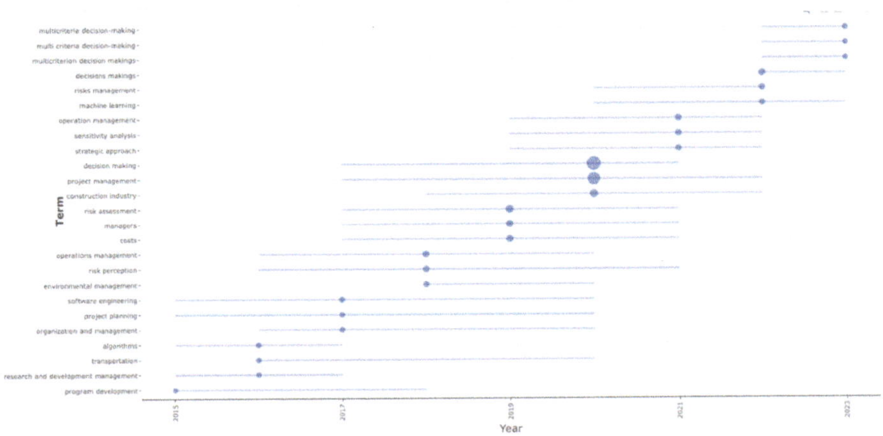

Fig. 6 The keywords trend analysis

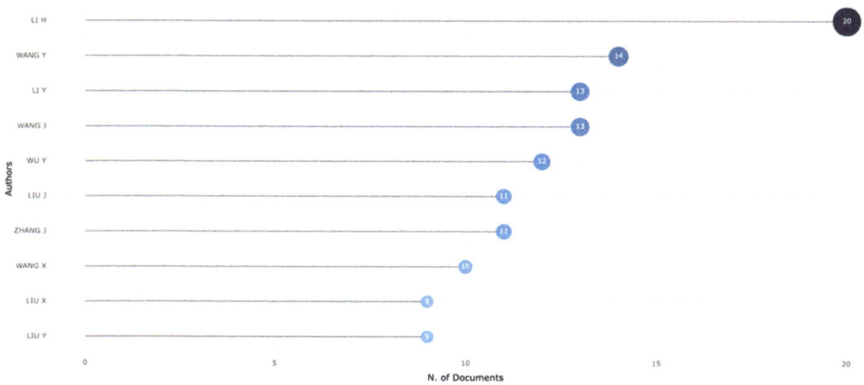

Fig. 7 The most relevant authors in the area operations and project management

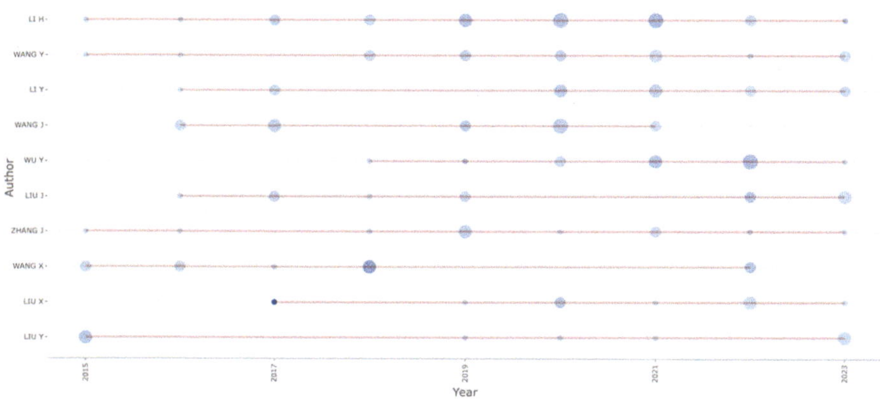

Fig. 8 The authors' productivity over time in the area operations and project management

- *Recourse analysis*

The analysis identifies three journals as the most relevant resources for operations and project management: Sustainability (Switzerland), International Journal of Project Management, and Engineering, Construction and Architectural Management. These journals have published 108, 50, and 47 articles, respectively. The results of this analysis are depicted in Fig. 10.

In addition to the analysis mentioned above, Bradford's Law, also known as Bradford's distribution or Bradford's scatter, was applied. Bradford's Law is an empirical pattern observed in the library and information science field, which demonstrates that a small core of journals, known as "core journals," contain a significant portion of the articles published in a particular area. On the other hand, a more substantial number of journals, referred to as "scatter journals," contribute to a smaller proportion of articles.

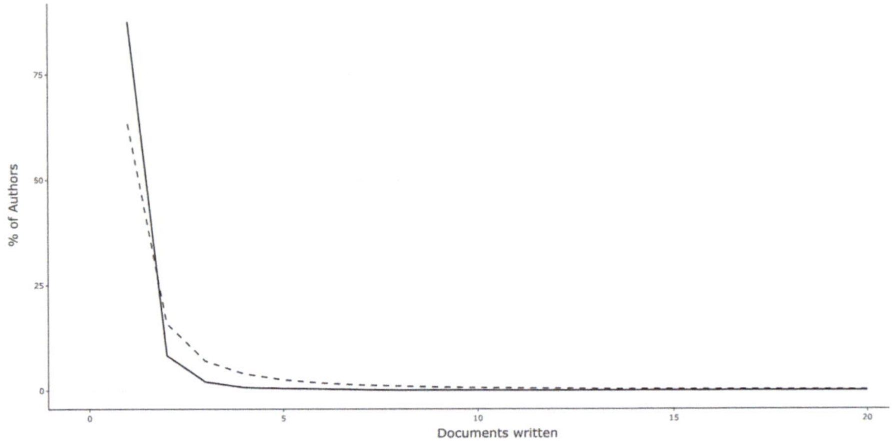

Fig. 9 Author productivity through Lotka's Law

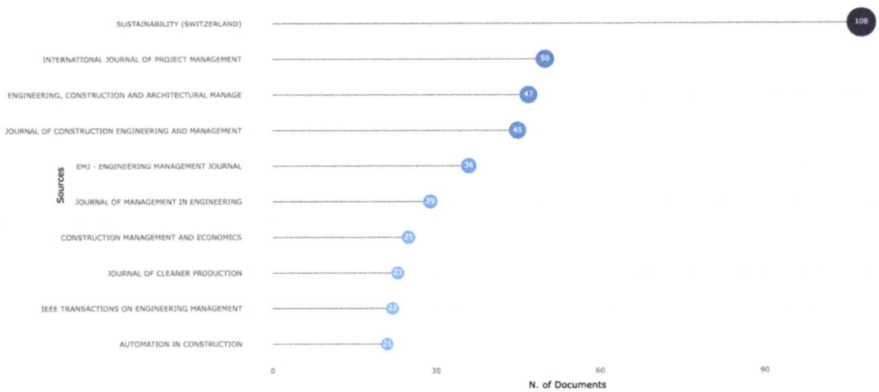

Fig. 10 The most relevant recourses

Figure 11 displays the application of Bradford's Law to the analysis. It shows that the three journals identified earlier, Sustainability (Switzerland), International Journal of Project Management, and Engineering, Construction and Architectural Management, continue to be core sources based on Bradford's Law. This indicates that these journals play a vital role in operations and project management by publishing many articles.

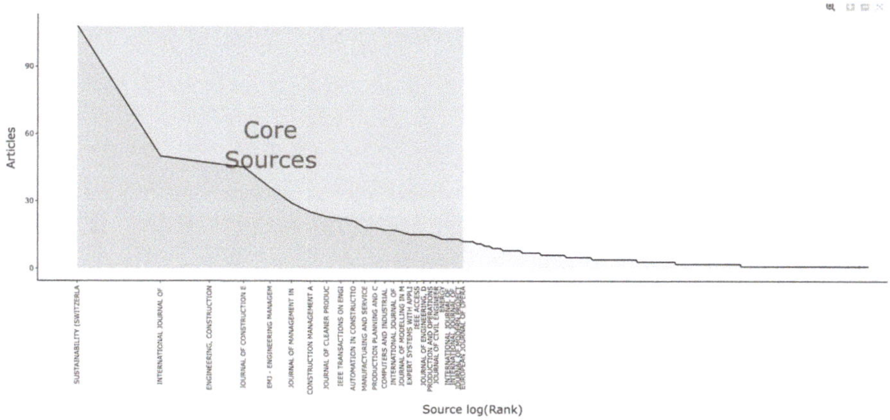

Fig. 11 Core sources by Bradford's Law

Here, the subsequent analysis aims to assess the impact of resources. Figure 12 illustrates the impact measure of three journals: the Journal of Construction Engineering and Management, the Journal of Cleaner Production, and the Journal of Construction Engineering and Management, with impact scores of 29, 19, and 19, respectively. These journals have demonstrated the highest influence within the scientific community.

Furthermore, Fig. 13 provides insights into the cumulative production of resources over time. Notably, the Journal of Sustainability has exhibited the most substantial growth over the years, followed by the International Journal of Project Management.

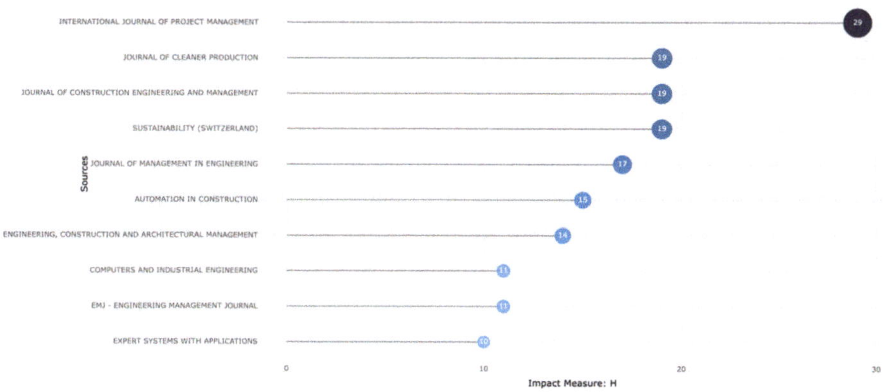

Fig. 12 The impact of recourses

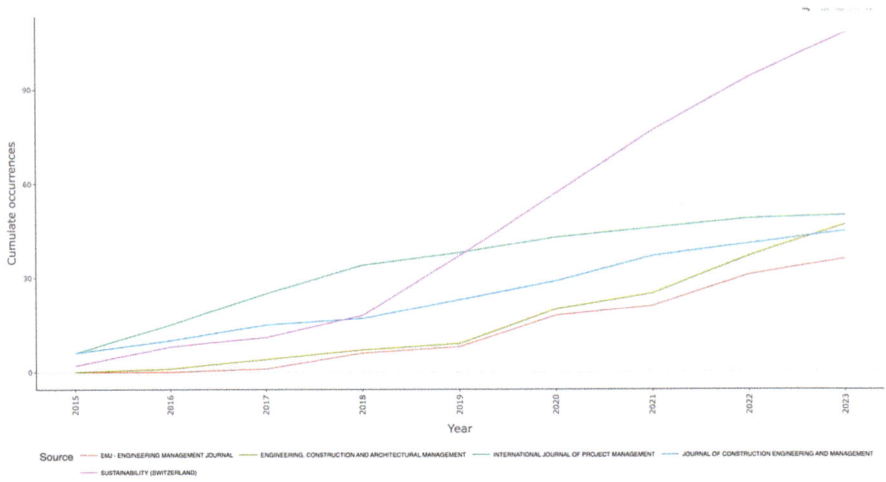

Fig. 13 The Sources' production over Time

This indicates that these two journals have experienced a significant increase in their contributions and influence within the field.

- *Countries collaborations*

A collaborative world map of countries is presented in Fig. 14 in the subsequent analysis, considering a minimum edge threshold of 10. The map highlights the extent of collaborations among countries in operations and project management. The analysis reveals that the United States, China, and Australia emerge as the countries with the most significant collaborative efforts in this domain. These nations have demonstrated substantial partnerships and joint initiatives, indicating their prominent role in fostering cooperation and advancing the field of operations and project management.

- *Factorial analysis*

Factorial analysis, or factor analysis, is a statistical technique to identify underlying factors or dimensions that explain the interrelationships among a set of observed variables. It can be observed that among all topic dendrogram, the "risk management" has the highest relationships with others (Fig. 15).

4 Conclusions

Based on the comprehensive review of 1994 collected and identified documents, a bibliometric analysis was conducted to examine the improvement areas related to Operations and Project Management. The key findings and significant conclusions derived from this analysis are summarized as follows.

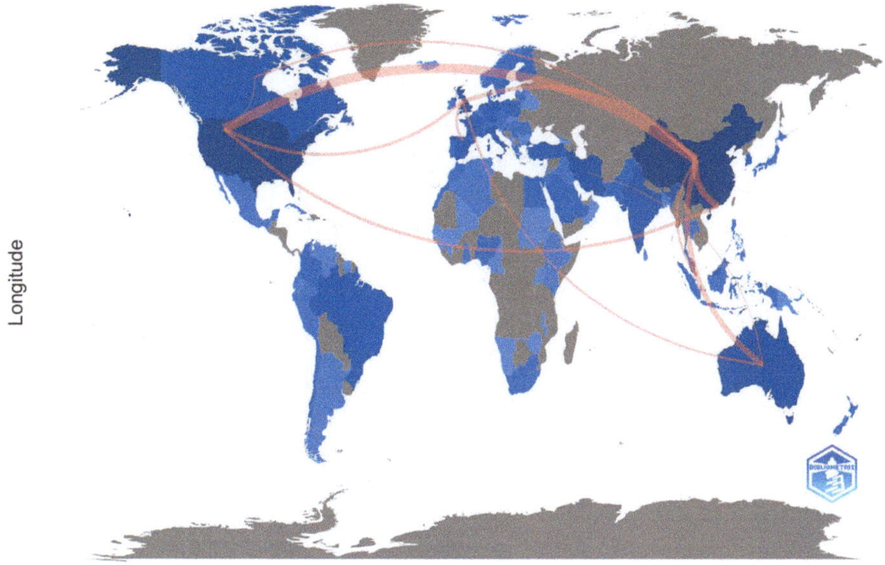

Fig. 14 Countries' collaboration world map

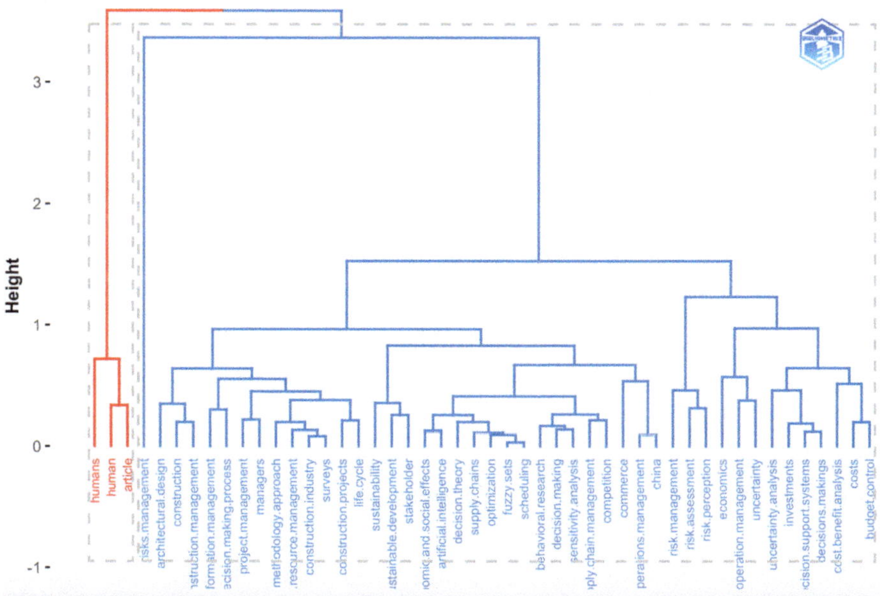

Fig. 15 Factorial analysis

Key findings from the analysis of co-authorship patterns, research centres, author contributions, bibliographic coupling, and keyword co-occurrence in the field of Operations and Project Management are as follows:

1. Global Contributions of Countries: China, the USA, and the UK are the leading contributors to improving Operations and Project Management, with significant numbers of published documents and solid collaborative networks.
2. Keyword Analysis: "Decision-making," "project management," and "construction industry" are the most frequently occurring keywords, indicating their prominence and relevance in the field.
3. Trend Analysis: The research topic in Operations and Project Management shifted from "program development" in 2015 to "multi-criteria decision-making" in 2023, indicating evolving research interests.
4. Prolific Authors: Li H, Wang Y, and Li Y are identified as highly productive authors with substantial contributions to the field.
5. Author Productivity: The productivity of identified prolific authors has remained stable or shown a positive trend over time.
6. Lotka's Law: Around 75% of authors in the field have published two papers illustrating the productivity patterns and distribution of authors.
7. Core Journals: "Sustainability" (Switzerland), International Journal of Project Management, Engineering, Construction, and Architectural Management are core journals with many published articles.
8. Bradford's Law: The previously mentioned core journals remain primary sources in Operations and Project Management, indicating their importance in the field.
9. Journal Impact: The Journal of Construction Engineering and Management, the Journal of Cleaner Production, and the Journal of Sustainability have the highest impact scores, demonstrating their influence in the scientific community.
10. Cumulative Production: The Journal of Sustainability and the International Journal of Project Management have shown substantial growth in contributions over time.
11. Collaborative World Map: The United States, China, and Australia are significant countries in fostering collaborations and joint initiatives in Operations and Project Management.

However, the study encountered several challenges that must be appropriately addressed in future research. These challenges are outlined as follows:

1. Limited dynamic visualization capabilities of biblioshiny: The biblioshiny software used in this study has limited capacity to display the dynamic behaviour of published documents. To overcome this limitation, it is recommended for future studies consider alternative tools such as CiteSpace, which provide more advanced features for visualizing temporal changes in scholarly literature. Researchers may develop in-house codes or customized software solutions to meet their specific visualization requirements.

2. Influence of decision-makers subjectivity: In this review work, decision-makers subjectivity played a significant role in analyzing the relevant published documents. It is advised to employ a comprehensive reference analysis structure to enhance the objectivity and reliability of future studies. This structure can minimize subjective biases by providing a systematic and standardized document selection, analysis, and interpretation approach. This may involve establishing clear criteria for document inclusion, employing multiple reviewers to ensure inter-rater reliability, and utilizing objective metrics or methodologies for data extraction and synthesis.

References

1. Kazi, M.K., Eljack, F., Al-Sobhi, S.A., Kazantzis, N., Kazantzi, V.: Application of i-SDT for safer flare management operation. Process Saf. Environ. Prot. **132**, 249–264 (2019). https://doi.org/10.1016/j.psep.2019.10.023
2. Little, J.D.C.: Models and managers: the concept of a decision calculus. Manage. Sci. **50**, 1841–1853 (2004). https://doi.org/10.1287/mnsc.1040.0267
3. Chen, K., Wei, X., Li, H., Lin, H., Khan, F.: Operational risk analysis of blowout scenario in offshore drilling operation. Process Saf. Environ. Prot. **149**, 422–431 (2021). https://doi.org/10.1016/j.psep.2020.11.010
4. Li, H., Yazdi, M.: An advanced TOPSIS-PFS method to improve human system reliability BT. In: Li, H., Yazdi, M. (eds.) Advanced Decision-Making Methods and Applications in System Safety and Reliability Problems: Approaches, Case Studies, Multi-criteria Decision-Making, Multi-objective Decision-Making, pp. 109–125. Springer International Publishing, Cham (2022). https://doi.org/10.1007/978-3-031-07430-1_7
5. Yazdi, M., Khan, F., Abbassi, R., Rusli, R.: Improved DEMATEL methodology for effective safety management decision-making. Saf. Sci. **127**, 104705 (2020). https://doi.org/10.1016/j.ssci.2020.104705
6. Li, H., Yazdi, M.: Advanced decision-making neutrosophic fuzzy evidence-based best–worst method BT. In: Li, H., Yazdi, M. (eds.) Advanced Decision-Making Methods and Applications in System Safety and Reliability Problems: Approaches, Case Studies, Multi-criteria Decision-Making, Multi-objective Decision-Making, Fuzzy Risk-Based Models, pp. 153–184. Springer International Publishing, Cham (2022). https://doi.org/10.1007/978-3-031-07430-1_9
7. Adesina, K.A., Yazdi, M., Omidvar, M.: Emergency decision making fuzzy-expert aided disaster management system BT. In: Yazdi, M. (ed.) Linguistic Methods Under Fuzzy Information in System Safety and Reliability Analysis, pp. 139–150. Springer International Publishing, Cham (2022). https://doi.org/10.1007/978-3-030-93352-4_6
8. Li, H., Yazdi, M.: Stochastic game theory approach to solve system safety and reliability decision-making problem under uncertainty BT. In: Li, H., Yazdi, M. (eds.) Advanced Decision-Making Methods and Applications in System Safety and Reliability Problems: Approaches, Case Studies, Multi-criteria Decision-Making, Multi-objective Decision-Making, Fuzzy Risk-Based Models, pp. 127–151. Springer International Publishing, Cham (2022). https://doi.org/10.1007/978-3-031-07430-1_8
9. Yazdi, M., Ding, Y., Adumene, S., Shafie, P.: Progressive system safety and reliability analysis: a sustainable game theory approach, Qual. Reliab. Eng. Int. n/a (2022). https://doi.org/10.1002/qre.3247
10. Li, H., Yazdi, M.: What are the critical well-drilling blowouts barriers? A progressive DEMATEL-game theory BT. In: Li, H., Yazdi, M. (eds.) Advanced Decision-Making Methods and Applications in System Safety and Reliability Problems: Approaches, Case Studies,

Multi-criteria Decision-Making, Multi-objective Decision-Making, Fuzzy Risk-Based Models, pp. 29–46. Springer International Publishing, Cham (2022). https://doi.org/10.1007/978-3-031-07430-1_3

11. Li, H., Peng, W., Adumene, S., Yazdi, M.: Operations management of critical energy infrastructure: a sustainable approach BT. In: Li, H., Peng, W., Adumene, S., Yazdi, M. (eds.) Intelligent Reliability and Maintainability of Energy Infrastructure Assets, pp. 39–52. Springer Nature Switzerland, Cham (2023). https://doi.org/10.1007/978-3-031-29962-9_3
12. Li, H., Yazdi, M., Huang, C.-G., Peng, W.: A reliable probabilistic risk-based decision-making method: Bayesian technique for order of preference by similarity to ideal solution (B-TOPSIS). Soft Comput. **26**, 12137–12153 (2022). https://doi.org/10.1007/s00500-022-07462-5
13. Liu, P., Wang, P., Liu, J.: Normal neutrosophic frank aggregation operators and their application in multi-attribute group decision making. Int. J. Mach. Learn. Cybern. **10**, 833–852 (2019). https://doi.org/10.1007/s13042-017-0763-8
14. Roldán López de Hierro, A.F., Sánchez, M., Roldán, C.: Multi-criteria decision making involving uncertain information via fuzzy ranking and fuzzy aggregation functions. J. Comput. Appl. Math. **404**, 113138 (2022). https://doi.org/10.1016/j.cam.2020.113138
15. Yazdi, M., Golilarz, N.A., Nedjati, A., Adesina, K.A.: Intelligent fuzzy Pythagorean Bayesian decision making of maintenance strategy selection in offshore sectors BT. In: Kahraman, C., Cebi, S., Cevik Onar, S., Oztaysi, B., Tolga, A.C., Sari, I.U. (eds.) Intelligent and Fuzzy Techniques for Emerging Conditions and Digital Transformation, pp. 598–604. Springer International Publishing, Cham (2022)
16. Hayes, A.: Operations Management: Understanding and Using It. https://www.investopedia.com/terms/o/operations-management.asp (n.d.)
17. Li, H., Yazdi, M.: Advanced Decision-Making Methods and Applications in System Safety and Reliability Problems. Springer Cham. https://link.springer.com/book/9783031074295 (2022)
18. Frisanco, T., Anglberger, N.: Operations management vs. project management—the operations services universe and its new project manager. In: 2008 4th IEEE International Conference on Management of Innovation and Technology, pp. 323–326 (2008). https://doi.org/10.1109/ICMIT.2008.4654384
19. Govindan, K., Jafarian, A., Khodaverdi, R., Devika, K.: Two-echelon multiple-vehicle location–routing problem with time windows for optimization of sustainable supply chain network of perishable food. Int. J. Prod. Econ. **152**, 9–28 (2014). https://doi.org/10.1016/j.ijpe.2013.12.028
20. Mardani, A., Weber, G.W., Khan, S.A.R.: Guest editorial: editorial for special issue on the role of big data on the transition to circular economy and sustainable operations management. J. Enterp. Inf. Manag. **35**, 949–954 (2022). https://doi.org/10.1108/JEIM-07-2022-559
21. Song, Y., Ou, J., Wu, J., Wu, Y., Xing, L., Chen, Y.: A cluster-based genetic optimization method for satellite range scheduling system. Swarm Evol. Comput. **79**, 101316 (2023). https://doi.org/10.1016/j.swevo.2023.101316
22. Karimi, H., Jadid, S.: Multi-layer energy management of smart integrated-energy microgrid systems considering generation and demand-side flexibility. Appl. Energy **339**, 120984 (2023). https://doi.org/10.1016/j.apenergy.2023.120984
23. Wu, Z.Y., Chew, A., Meng, X., Cai, J., Pok, J., Kalfarisi, R., Lai, K.C., Hew, S.F., Wong, J.J.: High fidelity digital twin-based anomaly detection and localization for smart water grid operation management. Sustain. Cities Soc. **91**, 104446 (2023). https://doi.org/10.1016/j.scs.2023.104446
24. Cravero, C., De Domenico, D., Marsano, D.: The use of uncertainty quantification and numerical optimization to support the design and operation management of air-staging gas recirculation strategies in glass furnaces. Fluids **8** (2023). https://doi.org/10.3390/fluids8020076
25. Zeng, G., Wang, J., Zhang, L., Xie, X., Wang, X., Chen, G.: Multi-domain modeling and analysis of marine steam power system based on digital twin. J. Mar. Sci. Eng. **11** (2023). https://doi.org/10.3390/jmse11020429

26. Peng, Z., Zhang, Y., Tan, F., Lv, J., Li, L.: Variable-weight suitability evaluation of underground space development considering socioeconomic factors. Sustainability **15** (2023). https://doi.org/10.3390/su15043574
27. Huang, J., You, J.X., Liu, H.C., Song, M.S.: Failure mode and effect analysis improvement: a systematic literature review and future research agenda. Reliab. Eng. Syst. Saf. **199**, 106885 (2020). https://doi.org/10.1016/j.ress.2020.106885
28. Huang, J., Mao, L.X., Liu, H.C., Shun Song, M.: Quality function deployment improvement: a bibliometric analysis and literature review. Qual. Quant. **56**, 1347–1366 (2022). https://doi.org/10.1007/s11135-021-01179-7
29. Li, H., Peng, W., Adumene, S., Yazdi, M.: Cutting edge research topics on system safety, reliability, maintainability, and resilience of energy-critical infrastructures BT. In: Li, H., Peng, W., Adumene, S., Yazdi, M. (eds.) Intelligent Reliability and Maintainability of Energy Infrastructure Assets, pp. 25–38. Springer Nature Switzerland, Cham (2023). https://doi.org/10.1007/978-3-031-29962-9_2
30. Yazdi, M., Mohammadpour, J., Li, H., Huang, H.-Z., Zarei, E., Pirbalouti, R.G., Adumene, S.: Fault tree analysis improvements: a bibliometric analysis and literature review. Qual. Reliab. Eng. Int. n/a (2023). https://doi.org/10.1002/qre.3271
31. Li, H., Peng, W., Adumene, S., Yazdi, M.: Advances in intelligent reliability and maintainability of energy infrastructure assets BT. In: Li, H., Peng, W., Adumene, S., Yazdi, M. (eds.) Intelligent Reliability and Maintainability of Energy Infrastructure Assets, pp. 1–23. Springer Nature Switzerland, Cham (2023). https://doi.org/10.1007/978-3-031-29962-9_1
32. Garza-Reyes, J.A.: Lean and green—a systematic review of the state of the art literature. J. Clean. Prod. **102**, 18–29 (2015). https://doi.org/10.1016/j.jclepro.2015.04.064
33. Yazdi, M., Khan, F., Abbassi, R., Quddus, N., Castaneda-Lopez, H.: A review of risk-based decision-making models for microbiologically influenced corrosion (MIC) in offshore pipelines. Reliab. Eng. Syst. Saf. 108474 (2022). https://doi.org/10.1016/j.ress.2022.108474
34. Ganbat, T., Chong, H.-Y., Liao, P.-C., Wu, Y.-D.: A bibliometric review on risk management and building information modeling for international construction. Adv. Civ. Eng. **2018**, 8351679 (2018). https://doi.org/10.1155/2018/8351679
35. Aria, M., Cuccurullo, C.: bibliometrix: An R-tool for comprehensive science mapping analysis. J. Informetr. **11**, 959–975 (2017). https://doi.org/10.1016/j.joi.2017.08.007

Decision-Making in Project and Operations Management

Reza Esmaeili, Mahnaz Shakerian, and Mohammad Yazdi

Abstract Operations management refers to a series of activities that add value by converting inputs into outputs in the form of goods and services. The purview of Operations Management encompasses a multitude of tasks, including but not limited to Planning, Scheduling, Organizing, Controlling, Quality control, and Inventory control. Project management is a term that is used in conjunction with operations management. A project is an exclusive and exceptional arrangement of tasks planned to accomplish a particular objective within a specific timeframe. The distinctiveness of a project lies in its unique characteristics, which differentiate it from other routine operations or activities. The domains of decision making within the realm of operations and project management present a multitude of challenges that require a meticulous evaluation. These challenges commonly possess established and comprehended solutions that have been developed through empirical experimentation. The notion of decision-making pertains to the sequence of actions taken to arrive at a specific resolution for a given predicament. The efficacy of an enterprise is significantly dependent on the caliber of decisions made by its managerial personnel. In the realm of Operations Management, the act of Decision Making is regarded as a purposeful process that involves selecting the most appropriate course of action from a vast array of possibilities. The importance of arriving at the 'correct' decision cannot be overstated, as it holds the potential to significantly influence the triumph or downfall of an organization's operations. Hence, it is essential for managers to possess the essential tools and strategies to enable them to make well-informed decisions.

R. Esmaeili
Student Research Committee, Department of Occupational Health and Safety Engineering, School of Health, Isfahan University of Medical Sciences, Isfahan, Iran

M. Shakerian
Department of Occupational Health and Safety Engineering, School of Health, Isfahan University of Medical Sciences, Isfahan, Iran

M. Yazdi (✉)
School of Computing, Engineering and Physical Sciences, University of the West of Scotland, London, UK
e-mail: mohammad.yazdi@mq.edu.au

Faculty of Science and Engineering, Macquarie University, Sydney, Australia

M. Yazdi (ed.), *Progressive Decision-Making Tools and Applications in Project and Operation Management*, Studies in Systems, Decision and Control 518,
https://doi.org/10.1007/978-3-031-51719-8_2

Keywords Operations management · Project management · Production · Decision making

1 Operations Management Defined

Operations management (OM) involves overseeing business practices to achieve optimal efficiency within an organization. Its primary focus is on efficiently transforming materials and labor into goods and services, ultimately aiming to maximize the organization's profitability [1]. Hence, operations management refers to a series of activities that add value by converting inputs into outputs in the form of goods and services. These activities are present in all types of organizations [2]. In manufacturing companies, the production activities involved in creating goods are typically more apparent. In an organization where a tangible good or product is not created, the production function may not be as conspicuous. Such activities are typically referred to as services. These services may be inconspicuous to the public and clientele alike. The product itself may take the form of various activities, such as the transfer of funds between bank accounts, liver transplants, filling an empty seat on an airplane, or educating a student. Regardless of whether the end product is a good or service, the production activities performed within the organization are often referred to as operations, or operations management [3, 4]. All organizations engage in three fundamental functions, in order to produce goods and services. These functions are indispensable not only for production, but also for the survival of the organization [5]. The three functions are as follows: **firstly**, marketing, which generates demand and takes orders for products and services, thereby constituting a prerequisite for any sale to occur; **secondly**, production/operations, which is responsible for creating, producing, and delivering the product; and **lastly**, finance/accounting, which monitors the organization's performance, settles bills, and manages finances [6]. Figure 1 illustrates the role of operations management in the services/good production strategy of a company.

The key roles of operations management are exemplified in Fig. 2 [8]. Accordingly, **Planning** constitutes a crucial process that encompasses selecting a site for the enterprise and coordinating production. The location of a business has a direct association with its level of achievement. This holds true for both a fledgling enterprise initiating its first factory or store and an established business seeking to venture into a new region. Several factors warrant consideration, including proximity to markets, accessibility to raw materials, labor force availability, and transportation infrastructure, when deciding on the location of a business. Each of these elements contributes to the success or failure of a company, and as such, they should be evaluated exhaustively before making a decision. The optimal location for a business is one that balances these factors and aligns with the company's overall objectives.

Scheduling is an essential aspect of Operations that entails the establishment of start and finish times for each phase of the production process. This process involves planning and monitoring the deployment of labor, machinery, and resources

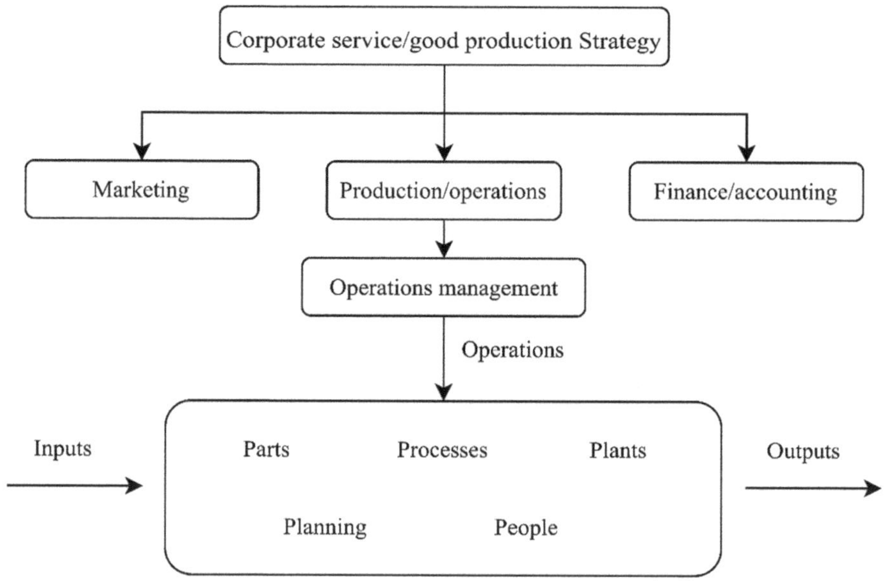

Fig. 1 Corporate service/good production strategy [7]

to ensure seamless production. The scheduling process guarantees timely completion of work, whether it involves the production of vehicles, books, or even dry cleaning a blouse or shirt. Therefore, efficient scheduling plays a crucial role in ensuring optimal productivity and customer satisfaction.

Organizing refers to the execution of activities that serve to establish a framework of tasks and authority within an organization. In the context of operations management, this entails the creation of a structure of roles and the establishment of a flow of information within the operations subsystem. The responsibilities of operation managers in this regard include identifying the specific activities that are necessary for realizing the organization's goals, as well as assigning the requisite authority and responsibility to individuals for carrying out these activities. Through effective organization, operations managers can ensure that their subordinates are able to work together efficiently and cohesively towards the achievement of organizational objectives.

In the course of conducting business, an organization requires various resources to produce goods or provide services. These resources include raw materials, machinery, office supplies, and other necessities. The process of acquiring these resources is commonly known as **purchasing**. This purchasing function is a vital component of the production process and is specifically focused on securing the most advantageous deals for the company. The individuals responsible for buying goods must make informed decisions regarding what to purchase, from whom to purchase, and at what price point. Ultimately, the purchasing function is an essential aspect of any

Fig. 2 Roles and responsibilities of operations management [9]

successful business operation as it ensures that the necessary resources are secured in an efficient and cost-effective manner.

Controlling refers to the range of activities that are undertaken to ensure that actual performance aligns with planned performance. In order to achieve the objectives for the operations subsystems, it is imperative that the operations manager exercise control by measuring actual outputs and comparing them to planned operations management. The three critical functions of cost control, quality control, and schedule control must be carefully monitored to ensure that the desired outcomes are achieved. The implementation of effective control mechanisms is an essential element of successful operations management.

Quality control is the process of evaluating the quality of goods that have been manufactured. This includes monitoring various factors such as the freshness and grade of the goods, their strength and workability, their workmanship and design, as well as their compliance with federal or industry standards, among other things. The implementation of quality control systems may range from testing a random sample of the items produced to testing every single product as it is completed. Quality control is a vital component in ensuring that the goods produced meet the required standards of quality and adhere to the relevant regulations.

Inventory control is a crucial aspect for a majority of manufacturers and service businesses, such as dry cleaners, who require inventories or stockpiles of materials for their products or services. Additionally, finished goods are also kept in stock for sale purposes. However, maintaining inventories can be expensive, and can restrict a business's capital for other activities. Decisions regarding inventory control must take into account other costs, such as potential price increases for raw materials, as well as discounts offered by suppliers for large orders. In some cases, businesses may determine that the benefits of discounts outweigh the expenses associated with maintaining a large inventory.

2 History of Operations Management

The epic tale of the emergence and evolution of the realm of operations management science can be described as follow [10].

Whitney is renowned for popularizing interchangeable parts in the early 1800s. This feat was achieved through a combination of standardization and quality control techniques. By signing a contract with the U.S. government to produce 10,000 muskets, Whitney was able to charge a premium price due to the use of interchangeable parts.

Frederick W. Taylor (1881), widely recognized as the originator of scientific management, made a significant contribution to personnel selection, planning and scheduling, motion study, and the presently renowned field of ergonomics. His noteworthy contribution lies in his conviction that management should be more resourceful and aggressive in the advancement of work methods. Taylor, along with his colleagues, Henry L. Gantt and Frank and Lillian Gilbreth, were among the pioneers who methodically sought out the most efficient means of production.

In 1913, the utilization of standardized parts and the quasi-assembly lines of the meatpacking and mail-order industries were combined by Henry Ford and Charles Sorensen, who then introduced the revolutionary concept of the assembly line, where workers remained stationary while materials were moved.

Quality control has also been a significant contribution to the field of operations management throughout history. In 1924, Walter Shewhart combined his statistical knowledge with the need for quality control, laying the foundations for statistical sampling in quality control. W. Edwards Deming, on the other hand, shared the belief with Frederick Taylor that management has a responsibility to enhance the work environment and processes in order to improve quality. According to Deming, this is an essential aspect of quality control.

Innovations emerging from the physical sciences, namely biology, anatomy, chemistry and physics, have been pivotal in the progress of operations management

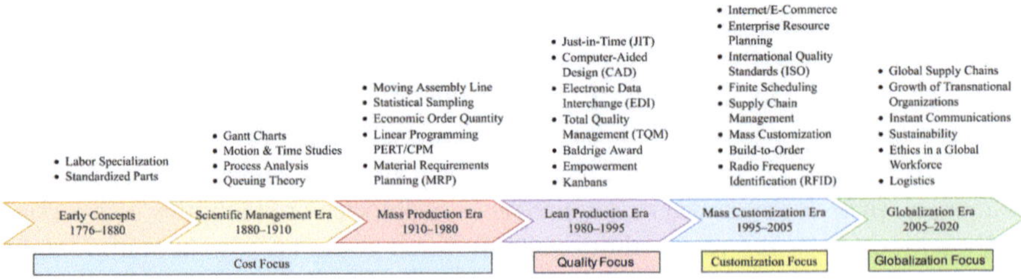

Fig. 3 Significant historical events in operations management [10]

(OM), leading to new adhesives, faster integrated circuits, gamma rays to sanitize food products and specialized glass for iPhones and plasma TVs. OM has significantly depended on the advancements made in physical sciences, thereby resulting in innovative products and processes. One of the major contributors to OM is information technology, which can be defined as the systematic processing of data to yield information. With the advent of wireless links, the internet, and other technological developments, information technology has become a paramount component in the development of OM. The implementation of information technology has revolutionized the way in which OM operates, leading to increased efficiency and productivity. Additionally, the integration of information technology has resulted in significant progress in supply chain management, such as real-time tracking and efficient inventory management. As a result, the advancements made in physical sciences and information technology have been vital in the development of OM.

Figure 3 portrays a visual representation of significant occurrences in the realm of Operations Management throughout history.

Furthermore, discipline of operation management boasts a rich and extensive chronicle, traceable back to the XVIII century. Table 1 elucidates the fundamental historical stages of operation management.

3 Project Management Defined

A project is an exclusive and exceptional arrangement of tasks planned to accomplish a particular objective within a specific timeframe. The project's length varies, ranging from a few weeks to several months or even years, and it involves the utilization of various resources, including human labor, capital, materials, and equipment [12]. The distinctiveness of a project lies in its unique characteristics, which differentiate it from other routine operations or activities. Projects are characterized by a diverse range of objectives, yet they share a fundamental life cycle consisting of a sequence of activities [13]. The life cycle commences with the inception of an initial concept, which is then pursued by an assessment of feasibility, the formulation of a project plan, the execution of the plan, and ultimately, the project's termination (Fig. 4). This

Table 1 Fundamental historical stages of operation management [11]

Year	Description	Author
1776	Labor specialization in manufacturing has become a focus of attention	Adam Smith
1799	Overall, the implementation of interchangeable parts and cost accounting has led to significant advancements in the manufacturing industry, improving efficiency and reducing costs while maintaining product quality	Eli Whitney et al.
1832	The division of labor based on skill and the allocation of tasks according to skill level are fundamental concepts within the realm of labor management. Furthermore, a thorough understanding of the basics of time study is essential in this field	Charles Babbage
1900	The principles of scientific management, which encompass time study and work study, were conceived by Frederick W. Taylor. As a means of dividing the planning and execution of work, Taylor's approach revolutionized the management of labor	Frederick W. Taylor
1900	Examination of jobs motion	Frank B. Gilbreth
1901	In the realm of manufacturing, the scheduling of employees, machines, and jobs is a crucial aspect of operational efficiency	Henry L. Gantt
1915	Optimizing inventory management practices and achieving operational efficiency is a priority for businesses, and determining the economic lot size plays a vital role in achieving this objective	F.W. Harris
1927	The topic of human relations has been the subject of extensive research, including the well-known Hawthorne studies. These studies, conducted in the early twentieth century, aimed to investigate the effects of various factors on worker productivity and morale	Elton Mayo
1931	The implementation of statistical inference in the context of product quality is exemplified by the use of quality control charts, which serve as a valuable tool for monitoring and evaluating the production process	W.A. Shewart
1935	The utilization of Statistical Sampling in the context of quality control is exemplified through the employment of inspection sampling plans	H.F. Dodge and H.G. Roming

(continued)

Table 1 (continued)

Year	Description	Author
1940	The utilization of operations research in the context of World War II represents a significant application of this discipline in a historical and practical context	P.M. Blacker et al.
1946	Digital computer technology is increasingly relevant in the field of operations management, as it provides the means to analyze complex data and optimize processes with greater efficiency	John Mauchlly and J.P. Eckert
1947	The intersection of Linear Programming and Operations management has been a topic of great interest in the field of optimization and decision-making, and has been widely studied by scholars and practitioners alike	G.B. Dantzig, Williams et al.
1950	The study of mathematical programming as it pertains to both linear and stochastic processes within the realm of Operations management is a complex and multifaceted field that requires a deep understanding of mathematical principles and their application to real-world scenarios	A. Charnes, W.W. Cooper et al.
1951	Commercial digital computers enable large-scale computations available in operations management	Sperry Univac
1960	The study of individuals in a work setting within the field of Operations management, commonly known as Organizational behavior, warrants further investigation	L. Cummings, L. Porter
1970	The incorporation of operations within a comprehensive strategy and policy, including computer applications for manufacturing, scheduling, and control, as well as Material Requirement Planning (MRP), is essential for effective management in modern organizations	W.S. Kinner J. Orlicky and G. Wright
1980	From Japan, there are a variety of quality and productivity applications available, such as robotics and CAD-CAM, which can be utilized in the management of operations	W.E. Deming and J. Juran
1990	The concept of lean management in operation management has emerged after the mass production philosophy, in the context of the Internet	Wormack, Jones

(continued)

Table 1 (continued)

Year	Description	Author
2000	The management of operations on mobile devices can be defined as the planning, organizing, directing and controlling of key business processes involving the use of mobile devices such as smartphones and tablets	–
2010	The implementation of Internet of Things (IoT) for operational management has facilitated global communication	–

sequential progression of activities forms the foundation of project management and serves as a framework for project managers to follow [14].

The implementation of project management strategies proves beneficial in instances where a project encompasses multiple tasks, some of which occur concurrently while others follow a sequential order. Project management techniques offer a framework that enables the project manager to monitor the progress of various activities that are required to complete a project [15]. Emphasis is placed on critical path activities, which represent the longest connected path through the project network. Any delay in a critical path activity can significantly impact the project completion time. These techniques also provide information on the expected completion time and cost of a project [16, 17]. The project manager critically reviews this information to ensure that adequate resources are available, and that the expected completion time is reasonable. If the expected completion date has a high probability of

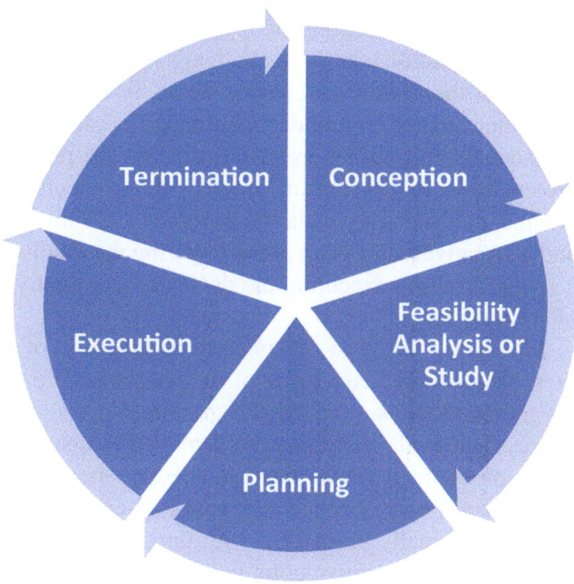

Fig. 4 Project life cycle components

exceeding a contractual due date and incurring penalty costs, the project manager considers different methods for crashing the project, which involves reducing the project completion time. This evaluation is conducted to economically justify the application of additional resources or to acknowledge a high probability of facing penalty costs [17].

It is imperative that project managers are well-versed in project management techniques, as they provide an essential foundation for project planning and execution. The use of these techniques is vital to ensure that projects are completed within the given time frame and budget [18]. By monitoring critical path activities, the project manager can identify potential bottlenecks and take corrective action to mitigate any delays. This approach ensures that the project is completed on time and within budget. Additionally, the project manager can use this information to evaluate the feasibility of the project and make informed decisions about resource allocation [16]. Ultimately, the use of project management techniques provides a structured approach to project management, which improves efficiency and reduces the risk of project failure. In addition to other responsibilities, the project manager endeavors to examine the resource load profile with the aim of equalizing the resource requisites for the project, thereby optimizing project resource allocation [19].

Table 2 presents a comprehensive comparison between operations management and project management, elucidating the various components of each management approach.

4 Decision-Making in Project and Operations Management

The concept of decision-making refers to the series of steps taken to arrive at a particular solution to a given challenge. The effectiveness of a corporation is heavily reliant on the quality of choices made by its managerial staff. It follows that the implementation of sound decision-making practices is crucial to the overall success and sustainability of an organization [20]. The process of making a decision can be broken down into seven fundamental steps, which include recognizing the problem at hand, determining available resources and limitations, creating various solutions and alternatives, assessing and comparing these solutions and alternatives, choosing a single solution, executing the decision, and finally, monitoring the chosen course of action [21].

Managerial decision-making is a crucial aspect of modern organizational management. Two fundamental types of such decisions are programmed decisions and non-programmed decisions [22]. Programmed decisions are primarily concerned with addressing mundane, routine problems that frequently arise in organizational settings. These problems typically have well-established and understood solutions that have been devised over time through trial-and-error. In contrast, non-programmed decisions are typically associated with unusual or exceptional problems that require a higher level of cognitive complexity and creativity on the part of decision-makers. These decisions often lack well-established and understood solutions and require a

Table 2 Operations management versus project management

Operations management	Project management
Operations management involves the administration, management, and modification of business practices with the aim of enhancing productivity and optimizing profit	Project management involves the systematic planning, execution, monitoring, control, and delivery of a specific objective within a predetermined timeframe, scope, and budget
Components	
Task management In the realm of task management, one possesses the power to craft intricate plans and subplans, accompanied by relevant documents. Furthermore, assignees are graciously bestowed upon these endeavors with their corresponding deadlines firmly established	**Workflow management** Devise and mechanize the operational systems of commerce. Integrate assignments within these systematic arrangements
Real-time communication The nearly instantaneous transmission of information across various telecommunications services, from the sender to the recipient, within a connection characterized by extremely low latency	**Quality management** Effortlessly incorporate compliance forms within the fluidity of workflows and engender a seamless integration. Streamline quality management processes through the marvels of automation, ensuring meticulous follow-ups and gentle reminders along the way
Collaboration Engage in dynamic collaboration on an array of textual materials, conversational exchanges, and exclusive domains for projects with instantaneous responsiveness	**Inventory management** Incorporate new product data and oversee its management with utmost efficiency. Keep an intricate record of inventory expenses, product abundance, duplicate occurrences, as well as order details
Reporting Receive bespoke reports on the performance of team, the progress of projects, as well as detailed insights into budgets and expenses	**Asset tracking** Monitor the movements of assets through employment of either barcode or RFID tags. Assure that said assets retain legally-binding service agreements, licenses, and contracts
Project planning By employing the aid of Gantt diagrams, one possesses the ability to skillfully conceive, modify, and disseminate visual project plans. This comprehensive tool facilitates the clear delineation of both project scope and essential resource needs	**Business performance management** Track and diligently oversee the all-encompassing performance of the enterprise, while vigilantly scrutinizing and modifying the business strategy with utmost attention to bolster financial gain
Business functions	
Incorporate due diligence in overseeing project scope, time management, and budgetary matters. Endeavor to enhance the overall efficiency of the team	It is imperative to guarantee the seamless functionality of business operations while upholding resource efficiency

(continued)

Table 2 (continued)

Operations management	Project management
Summary	
Investment in operations management software facilitates businesses in the planning, regulation, and monitoring of resources to ensure the smooth functioning of all operational processes	Investment in project management software greatly benefits businesses, as it facilitates the precise organization and surveillance of projects. Such software guarantees the greatest utilization and efficiency of resources while fostering an atmosphere of optimum productivity

more in-depth analysis of the underlying causes and potential consequences of the problem [23, 24].

When it comes to decision-making, there are three distinct categories that are determined by the level of certainty involved. These categories are certainty, risk, and uncertainty. In situations that fall under the certainty category, the decision-maker is fully aware of the available alternatives, the conditions associated with each alternative, and the potential outcome of each alternative. When dealing with risk, the decision-maker has incomplete information about the available alternatives, but has a fair understanding of the probability of each alternative's potential outcomes. Uncertainty arises when the future environment is unpredictable and the state of affairs is in flux. It is in these instances that decision-makers will have the most difficulty in reaching a conclusion, as the potential outcomes of any given alternative may be difficult to predict. Despite the inherent complexities of each category, understanding them can greatly assist decision-makers in making the most informed decision possible [25, 26].

In real-world scenarios, a decision-maker's goals are frequently only attainable at the expense of other goals, and these goals may be in conflict with one another. As a result, a hierarchy of importance must be established among these opposing goals, prioritizing the most significant objectives to be met or the point at which no further enhancements are feasible [27]. So, decision-making is a crucial aspect of organizational management that involves the selection of an optimal course of action from a range of alternatives. It is a process that necessitates the consideration of various factors such as time, situation, organizational process, and suitability. A problem can be approached in several ways, but the decision maker must evaluate all the available alternatives to determine the best course of action. The selection of the most appropriate course of action is critical for the smooth operation of an organization and the attainment of its objectives [28]. Thus, decision-making is a vital component of organizational success that requires careful consideration and evaluation of all possible alternatives. Ultimately, the decision-making process must prioritize organizational goals and objectives above all else to ensure the best possible outcome [29].

The domain of operations management extends throughout the organization, with operations management personnel participating in product and service design, process selection, technology selection and management, work systems design, location planning, facilities planning, and quality improvement of their organization's products or services. The operations function involves several interconnected activities, including forecasting, capacity planning, scheduling, inventory management, quality assurance, employee motivation, facility location decisions, and more. These activities are critical to the organization's success and must be carried out efficiently and effectively to achieve the desired outcomes. Operations management is a complex and multifaceted field that requires a deep understanding of the organization's goals, processes, and systems [30].

Under the purview of Operation Management, Decision making is deemed to be the process of judiciously choosing the most optimal course of action from a plethora of alternatives, including but not limited to Selection of Best Product/Service, Plant Location, Selection of vendor/supplier etc. This plays a pivotal role in supporting various managerial functions such as planning, organizing, directing and controlling through astute decision-making. The significance of making the 'right' decision cannot be overstated, as it can directly impact the success or failure of an organization's operations. Therefore, it is imperative for managers to be equipped with the necessary tools and strategies to make informed decisions [31, 32].

4.1 Distinguishing Between Strategic and Tactical Operations Management Decisions

In operations management, there are two distinct types of decision-making: strategic and tactical. Each serves a different purpose and timeframe. Strategic decisions are focused on achieving long-term success, while tactical decisions are geared towards short-term adjustments. For optimal outcomes, it is advisable to combine these two approaches [33].

Strategic operations decisions involve making estimates, predictions, setting goals, and conducting comparative analyses. This approach emphasizes future-oriented planning to meet long-term objectives and ensure sustainable success. A crucial element of this approach is the utilization of operational data to analyze outcomes and make necessary adjustments to long-term strategic plans [34].

On the other hand, tactical operations decisions revolve around responding to changing operational conditions. For instance, a company may need to adapt its production schedules or workforce planning in response to fluctuations in the marketplace. This type of decision-making is focused on the immediate requirements and challenges faced by the organization. Hence, the purview of decision-making in operations management spans operational management to strategic management, as shown in Fig. 5, facilitating a comprehensive approach to the field [35].

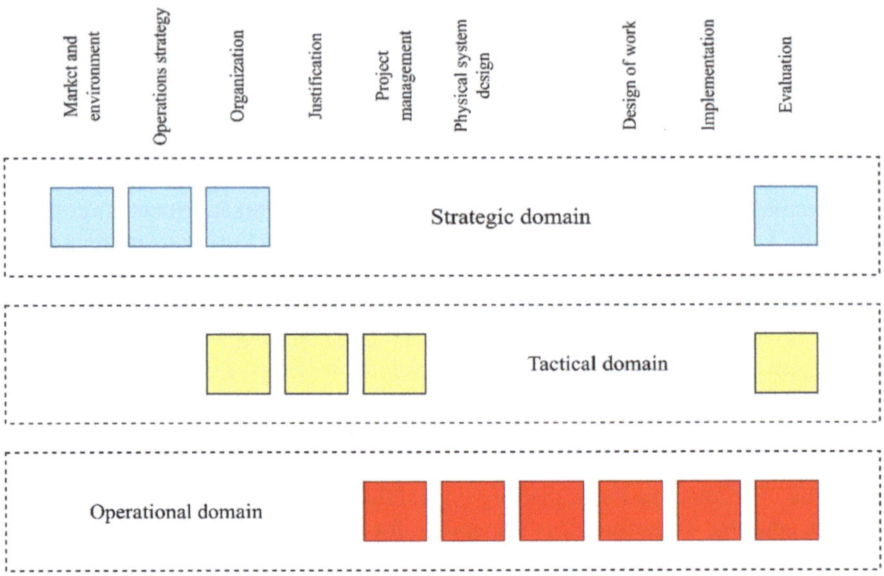

Fig. 5 The decision-making process levels in operations management [36]

4.2 *Decision-Making Tools and Applications in Project and Operations Management*

The planner/decision maker role of the operations manager is of utmost importance as it significantly affects the achievement of an organization's goals and objectives. Numerous alternatives are often available when decisions have to be made, and the resulting impact on profits or costs can vary widely. Therefore, making well-informed decisions is vital [37]. To achieve this, several approaches to decision-making can be employed, such as quantitative methods, models, trade-off analysis, establishing priorities, the systems approach, and ethics. The use of models involves developing mathematical representations of the problem at hand, enabling the identification of the best course of action. Quantitative methods involve the application of statistical and mathematical techniques to analyze data and make decisions [38–40]. Trade-off analysis involves considering the advantages and disadvantages of each option, while establishing priorities involves deciding which goals are more important than others [41]. The systems approach involves taking a holistic view of the organization, considering the interrelationships between different parts [40–45].

5 Conclusion

Operations management is a discipline of management that pertains to the conception and administration of commodities, procedures, amenities and the supply chain. Its aim is to acquire, cultivate and employ assets and basic materials to attain the wares or amenities required by a business's clientele. As previously noted, the act of decision-making within the context of operations management holds great significance, as a judicious choice can result in the amelioration of the organization's standing and advancement, while a misguided choice can potentially culminate in the financial insolvency of the organization. As a result, it can be maintained that the employment of appropriate methodologies and tools that assist executives in reaching a well-reasoned and accurate decision within a timely manner is extremely important. For instance, an effective tool that greatly assists managers in decision-making is the use of multi-criteria decision-making systems. These systems aid researchers and managers in selecting the most desirable alternative from the available options, thus leading to an optimal choice. Therefore, it is suggested that researchers develop and introduce methodologies that can improve the accuracy of decision-making. Thus, it is recommended that scholars in the forthcoming times devise and introduce methodologies that enhance the precision of decision-making.

References

1. McIntosh, L.D., Zabarovskaya, C., Uhlmansiek, M.: It's just (academic) business: a use case in improving informatics operations with business intelligence. In: AMIA Summits on Translational Science Proceedings, pp. 117 (2015)
2. Haddadpoor, A., Taheri, B., Nasri, M., Heydari, K., Bahrami, G.: Process documentation: a model for knowledge management in organizations. Materia Socio-Medica **27**(5), 347 (2015)
3. Sutherland, V., Buffo, M.J., Whiteside, T.L.: Impact of contracted manufacturing organization protocols on operations in an academically based current good manufacturing practice facility. Cytotherapy **24**(1), 32–36 (2022)
4. Boland, G.W., Glenn, L., Goldberg-Stein, S., Jha, S., Mangano, M., Patel, S., et al.: Report of the ACR's economics committee on value-based payment models. J. Am. Coll. Radiol. **14**(1), 6–14 (2017)
5. Sartori, R., Costantini, A., Ceschi, A., Tommasi, F.: How do you manage change in organizations? Training, development, innovation, and their relationships. Front. Psychol. **9**, 313 (2018)
6. Lee, C.K.H.: A review of applications of genetic algorithms in operations management. Eng. Appl. Artif. Intell. **76**, 1–12 (2018)
7. Knod, E.M., Schonberger, R.: Operations management: meeting customers' demands. (No Title) (2001)
8. Wolniak, R.: Main functions of operation management. Prod. Eng. Arch. **26**(1), 11–14 (2020)
9. Kumar, S.A., Suresh, N.: Production and Operations Management. New Age International (2006)
10. Render, B., Heizer, J.: Principles of Operations Management: Sustainability and Supply Chain Management. Pearson (2014)
11. Kumar, S., Suresh, N.: Operations Management. New Age International (P) Ltd. Publishers, New Delhi (2009)

12. Abello, M.B., Michalewicz, Z.: Multiobjective resource-constrained project scheduling with a time-varying number of tasks. Sci. World J. (2014)
13. Thiele, B., Ryan, M., Abbasi, A.: Developing a dataset of real projects for portfolio, program and project control management research. Data Brief **34**, 106659 (2021)
14. Westland, J.: The Project Management Life Cycle: A Complete Step-by-Step Methodology for Initiating Planning Executing and Closing the Project. Kogan Page Publishers (2007)
15. Li, F., Xu, Z.: A multi-agent system for distributed multi-project scheduling with two-stage decomposition. PLoS One **13**(10), e0205445 (2018)
16. Vazquez, A., Pozzana, I., Kalogridis, G., Ellinas, C.: Activity networks determine project performance. Sci. Rep. **13**(1), 509 (2023)
17. Cohen Kashi, S., Rozenes, S., Ben-Gal, I.: Project management monitoring based on expected duration entropy. Entropy **22**(8), 905 (2020)
18. Yazdi, M., Moradi, R., Pirbalouti, R.G., Zarei, E., Li, H.: Enabling safe and sustainable hydrogen mobility: circular economy-driven management of hydrogen vehicle safety. Processes **11** (2023). https://doi.org/10.3390/pr11092730
19. Fernandes, G., O'sullivan, D.: Project management practices in major university-industry R&D collaboration programs—a case study. J. Technol. Transf. **48**(1), 361–391 (2023)
20. Boutkhoum, O., Hanine, M., Boukhriss, H., Agouti, T., Tikniouine, A.: Multi-criteria decision support framework for sustainable implementation of effective green supply chain management practices. SpringerPlus **5**, 1–24 (2016)
21. Yazdi, M., Golilarz, N.A., Nedjati, A., Adesina, K.A.: Intelligent fuzzy Pythagorean Bayesian decision making of maintenance strategy selection in offshore sectors BT. In: Kahraman, C., Cebi, S., Cevik Onar, S., Oztaysi, B., Tolga, A.C., Sari, I.U. (eds.) Intelligent and Fuzzy Techniques for Emerging Conditions and Digital Transformation, pp. 598–604. Springer International Publishing, Cham (2022)
22. Cristofaro, M., Giardino, P.L., Malizia, A.P., Mastrogiorgio, A.: Affect and cognition in managerial decision making: a systematic literature review of neuroscience evidence. Front. Psychol. **13**, 762993 (2022)
23. Foxall, G.R.: Cognitive requirements of competing neuro-behavioral decision systems: some implications of temporal horizon for managerial behavior in organizations. Front. Hum. Neurosci. **8**, 184 (2014)
24. Yazdi, M.: Linguistic Methods Under Fuzzy Information in System Safety and Reliability Analysis. Springer Cham (2022). https://doi.org/10.1007/978-3-030-93352-4%0A%0A
25. Lee, D.G., Hare, T.A.: Value certainty and choice confidence are multidimensional constructs that guide decision-making. Cogn. Affect. Behav. Neurosci. 1–19 (2023)
26. Fu, M., Wang, L., Cao, X., Zheng, B., Zhou, X., Yin, S.: A closed-loop method for multiperiod intelligent information processing with cost constraints under the fuzzy environment. Comput. Intell. Neurosci. (2022)
27. Badeel, R., Subramaniam, S.K., Muhammed, A., Hanapi, Z.M.: A multicriteria decision-making framework for access point selection in hybrid LiFi/WiFi networks using integrated AHP–VIKOR technique. Sensors **23**(3), 1312 (2023)
28. Dimara, E., Zhang, H., Tory, M., Franconeri, S.: The unmet data visualization needs of decision makers within organizations. IEEE Trans. Vis. Comput. Graph. **28**(12), 4101–4112 (2021)
29. Greve, H.R.: Structuring the situation: organizational goals trigger and direct decision-making. Front. Psychol. **14**, 1140408 (2023)
30. Reid, R.D., Sanders, N.R.: Operations Management: An Integrated Approach. Wiley (2019)
31. Altubaishe, B., Desai, S.: Multicriteria decision making in supply chain management using FMEA and hybrid AHP-PROMETHEE algorithms. Sensors **23**(8), 4041 (2023)
32. Yazdi, M.: Acquiring and sharing tacit knowledge in failure diagnosis analysis using intuitionistic and Pythagorean assessments. J. Fail. Anal. Prev. **19**, 369–386 (2019). https://doi.org/10.1007/s11668-019-00599-w
33. Gracey, A.: Building an organisational resilience maturity framework. J. Bus. Contin. Emerg. Plan. **13**(4), 313–327 (2020)

34. Yazdi, M., Zarei, E., Pirbalouti, R.G., Li, H.: A comprehensive resilience assessment framework for hydrogen energy infrastructure development. Int. J. Hydrog. Energy (2023). https://doi.org/10.1016/j.ijhydene.2023.06.271
35. Teimoury, E., Jabbarzadeh, A., Babaei, M.: Integrating strategic and tactical decisions in livestock supply chain using bi-level programming, case study: Iran poultry supply chain. PLoS One **12**(10), e0185743 (2017)
36. Bennett, D.J., Forrester, P.L.: Market-Focused Production Systems: Design and Implementation. Prentice Hall Hertfordshire (1993)
37. Hofstra, N., Dullaert, W., De Leeuw, S., Spiliotopoulou, E.: Individual goals and social preferences in operational decisions: behavioural insights from transport planning. Int. J. Oper. Prod. Manag. **39**(1), 116–137 (2019)
38. Thompson, B.P., Bank, L.C.: Use of system dynamics as a decision-making tool in building design and operation. Build. Environ. **45**(4), 1006–1015 (2010)
39. Martínez, L., Ruan, D., Herrera, F., Herrera-Viedma, E., Wang, P.P.: Linguistic Decision Making: Tools and Applications, pp. 2297–2298. Elsevier (2009)
40. Helleno, A., Pimentel, C., Ferro, R., Santos, P., Oliveira, M., Simon, A.: Integrating value stream mapping and discrete events simulation as decision making tools in operation management. Int. J. Adv. Manuf. Technol. **80**, 1059–1066 (2015)
41. Inostroza, L., König, H.J., Pickard, B., Zhen, L.: Putting Ecosystem Services into Practice: Trade-Off Assessment Tools, Indicators and Decision Support Systems, pp. 303–305. Elsevier (2017)
42. Li, H., Yazdi, M.: Advanced Decision-Making Methods and Applications in System Safety and Reliability Problems. Springer Cham (2022)
43. Yazdi, M., Khan, F., Abbassi, R., Rusli, R.: Improved DEMATEL methodology for effective safety management decision-making. Saf. Sci. **127**, 104705 (2020). https://doi.org/10.1016/j.ssci.2020.104705
44. Li, H., Yazdi, M.: What are the critical well-drilling blowouts barriers? A progressive DEMATEL-game theory BT. In: Li, H., Yazdi, M. (eds.) Advanced Decision-Making Methods and Applications in System Safety and Reliability Problems: Approaches, Case Studies, Multi-criteria Decision-Making, Multi-objective Decision-Making, Fuzzy Risk-Based Models, pp. 29–46. Springer International Publishing, Cham (2022). https://doi.org/10.1007/978-3-031-07430-1_3
45. Yazdi, M., Ding, Y., Adumene, S., Shafie, P.: Progressive system safety and reliability analysis: a sustainable game theory approach. Qual. Reliab. Eng. Int. n/a (2022). https://doi.org/10.1002/qre.3247

Nash Equilibrium-Based FMEA for Risk Prioritization in Hydrogen Refueling Station Design

He Li, Rosita Moradi, Mohammad Yazdi, and Esmaeil Zarei

Abstract This chapter focuses on developing a modified Failure Mode and Effect Analysis (FMEA) method for prioritising potential failure modes in the design stage of hydrogen refuelling stations, explicitly targeting the liquid hydrogen storage (Dewar) system. The conventional FMEA method is widely used but has limitations, such as equal importance weights for risk factors, limited consideration of risk factors, and static risk assessment. To address these shortcomings, a new approach based on the "Nash equilibrium" concept is proposed. The proposed method incorporates the perspectives of multiple decision-makers (DMs) who assess the risk factors associated with each failure mode using Pythagorean fuzzy uncertain linguistic variables. Payoff matrices are constructed based on the assessments provided by the DMs, representing a zero-sum game between failure and success. The FMEA process is then formulated as a combination of game theory and risk assessment, aiming to prioritise failure modes for intervention actions. To demonstrate the application of the proposed method, a case study of the liquid hydrogen storage (Dewar) system in a hydrogen refuelling station is presented. Failure modes are identified, and their severity, occurrence, and detection are assessed. The risk priority number (RPN) is calculated for each failure mode based on the proposed method, enabling the

H. Li
Centre for Marine Technology and Ocean Engineering (CENTEC), Instituto Superior Técnico, Universidade de Lisboa, Lisbon, Portugal

R. Moradi
University of Wollongong, Northfields Avenue, Wollongong, NSW 2522, Australia

M. Yazdi (✉)
School of Computing, Engineering and Physical Sciences, University of the West of Scotland (UWS), London, UK
e-mail: mohammad.yazdi@mq.edu.au

Faculty of Science and Engineering, Macquarie University, Sydney, Australia

E. Zarei
Department of Safety Science, College of Aviation, Embry-Riddle Aeronautical University, Prescott, AZ, USA

Robertson Safety Institute (RSI), Embry-Riddle Aeronautical University, Prescott, AZ, USA

M. Yazdi (ed.), *Progressive Decision-Making Tools and Applications in Project and Operation Management*, Studies in Systems, Decision and Control 518,
https://doi.org/10.1007/978-3-031-51719-8_3

prioritisation of failure modes for appropriate intervention actions. The proposed FMEA method offers several advantages over the conventional approach. It allows for unequal importance weights of risk factors, incorporates a comprehensive set of risk factors, avoids the issue of different combinations producing the same RPN, and provides a systematic framework for considering the dynamic nature of risk assessment over time. By integrating game theory principles and FMEA, this study offers a novel perspective on risk prioritisation in the design stage of hydrogen refuelling stations. The proposed method provides a more comprehensive and flexible approach to assessing and prioritising failure modes, contributing to developing safer and more reliable hydrogen infrastructure.

Keywords Risk management · FMEA · Nash equilibrium · Hydrogen refueling station · Liquid hydrogen storage · Decision-makers

1 Introduction

Different methods concerning the given type of system can be considered to obtain the risk priorities of identified hazards in a system. Some critical methods include what-if analysis, checklists, hazard, and operability study (HAZOP) [1–3], failure mode and effect analysis (FMEA) [4–9], fault tree analysis [10–15], and risk matrix [16–19]. Each method can be utilized to derive the priority of identified hazards in a system. In all risk management methods, the initial step is identifying the potential hazards and threats related to the study object, such as accidents, near misses, incidents, job diseases, and environmental results. Typically, the likelihood, as well as the severity of identified hazards, are determined qualitatively or quantitively [16, 20–22]. Likelihood shows the chance of a hazard occurrence, and the severity represents the seriousness of the consequences. Next, the risk profile is obtained based on the qualitative or quantitative scores derived in the previous step. According to the risk profile for identified hazards, intervention actions may be needed to reduce the risk to an acceptable level [18, 23–28].

The conventional FMEA is a widely accepted proactive risk management method, which is commonly used to identify the failure modes of a system [29–36]. In conventional FMEA techniques, three risk factors for each failure mode (known as Severity, Occurrence, and Detection) are scored on a scale of 1–10. The score of 1 and 10 correspond to the lowest and highest importance in terms of risk factors, respectively. By multiplying the risk factor score, the risk priority number for each failure mode is obtained, and the failure modes are prioritized in descending order to receive intervention actions [37–43].

However, the conventional FMEA method suffers from several limitations. These shortcomings, as well as proposed improvements to this method, are outlined by scholars as follows: (A) The importance weights of Severity, Occurrence, and Detection are assumed to be equal [44, 45]; (B) risk priority numbers only take into account Severity, Occurrence, and Detection [46, 47]; (C) different combinations of Severity,

Occurrence, and Detection produce the same risk priority number [48, 49]; (D) there is not enough information to judge risk priority number computation [50–52]; (E) in some systems, failure modes can receive some intervention actions on which they are entirely dependent; and (F) conventional FMEA (like many risk management methods) generally assesses the risk of a system statically and does not consider the effect of time in the system profile [19, 53–57]. A comprehensive review of the limitations of conventional FMEA is presented in [35].

In this study, we developed FMEA method underlying the umbrella of "*Nash equilibrium*".

A "*Nash equilibrium*" [21, 58, 59], in general, is a law that none of the players would like to break the rules even in the absence of law enforcement. As an example, two players Matt and Rose, select two strategies Δ and ∇. . Therefore, (Δ, ∇) is known as a "*Nash equilibrium*" if Matt is without any strategies, should be better than Δ maximizing his payoff responding to Rose to select ∇. Similarly, Matt is without any strategies should be better than ∇ at maximizing his payoff responding to Rose selection. In another example, in case that Shahrazad and Donyazad are added as two new players to the game played by Matt and Rose, $(\Delta, \nabla, \Lambda\Lambda, \Psi)$ is a "*Nash equilibrium*" if Δ is Matt's best response to $(\nabla, \Lambda\Lambda, \Psi)$, and ∇ is Arman's best response to $(\Delta, \Lambda\Lambda, \Psi)$.

2 Study Example

Let us, for an FMEA worksheet, take a set of failure modes a $FM_i = \{FM_1, FM_2, FM_3, \ldots, FM_i\}$, which is known in FMEA procedure as the "*strategy of failure*" for l number of decision-makers. Also, let us take $DM = \{DM_1, DM_2, DM_3, \ldots, DM_l\}$, as a set of risk factors, $RF_i = \{S_{FMi}, O_{FMi}, D_{FMi}\}$ aa s "*strategy of success*", and $\omega_j = \{\omega_1, \omega_2, \omega_3, \ldots, \omega_n\}$ as the importance weight of decision-makers based on their quality profile in which $0 \le \omega_j \le 1$ and $\sum_{j=1}^{n} \omega_j = 1$.

Thus, all employed decision-makers share their individual payoff judgments of FM_i regarding the risk factors (RF_j) using Pythagorean fuzzy uncertain linguistic variables. In addition, the l payoff matrices can be derived from $\mathbb{P}_k = \left[\shortmid_{ij}^{k}\right]_{m \times n}$ $(k = 1, 2, \ldots, l)$, where, $\shortmid_{ij}^{k} = \langle\left[\mathbb{S}_{\theta_{ij}^k}, \mathbb{S}_{\tau_{ij}^k}\right], (\mu_{ij}^k, \upsilon_{ij}^k)\rangle$, and $\left[\mathbb{S}_{\theta_{ij}^k}, \mathbb{S}_{\tau_{ij}^k}\right]$ is the uncertain linguistic assessment. Uncertain linguistic assessment obtained from the decision-makers over FM_i with respect to the risk factors RF_j collected from decision-maker l according to the linguistic term set $\mathbb{P} = \left\{\mathbb{S}_0, \mathbb{S}_1, \ldots, \mathbb{S}_g\right\}$. According to the obtained outcomes, a zero-sum game between failure and success can be demonstrated as $B = \{\text{Failure, Success, FM, RF}, \mathbb{P}\}$. Thus, the comprehensive processes of the proposed FMEA method are provided in the following descriptions. It is discussed that the FMEA can be seen as a zero-sum game; however, as a practical issue, it can be studied how the combination of game theory and FMEA works. Ding and Liu (2019) [60] used "*Nash equilibrium*" for ranking the set of alternatives as emergency decision-making problems. Thus, in this chapter, we assume that the FMEA is a zero-sum game problem and then develop the methodology.

The example is provided to study the potential failure modes in the hydrogen refueling station design stage. It is vital for hydrogen energy planners (decision-makers) to ensure that establishing a hydrogen refueling station is safe for the public and the surrounding environment. Therefore, the hydrogen refueling station, including storage and process devices must be constantly protected. The first meaningful way to reach a safe design is using guidelines, codes, and standards where required, such as NFPA (50A "Gaseous H_2 at consumer sites", 50B "Liquid H_2 at consumer sites", 54 "Fuel gas piping installation (Fuel gas code)"), ISO (13984 "Liquid H_2, land vehicle fuelling system interface", 14687 "Quality characteristics of H_2 fuel"), ISO/AWI TS 20012 "Gaseous H_2 service stations", ISO/DIS (11119 "Composite tanks for storing H_2 at pressures up to 600 bar", and 13985 "Liquid H_2, land vehicle fuel tanks") [61]. This system also can use FMEA technique for system safety and reliability analysis. FMEA can be utilized in hydrogen refueling station design to identify the system's weakness and provide different system justification to address the critical safety issues during the risk assessment procedure.

In fact, most of the guidelines, codes, and standards are not freely available or not available at all since they may currently be under development. Therefore, the decisions can be conducted based on system safety and reliability analysis rather than relying on existing guidelines, codes, and standards [62–65]. This motivate us to use the proposed innovative approach in the safe design of the hydrogen refueling station, which is restricted by the existing document.

The fuelling Process is explicitly explained in [61]: "The station fuelling process begins with the bulk delivery of liquid hydrogen. There is a dedicated lane around the station for liquid hydrogen tanker operation. The tanker will pull into the station, park under the storage and distribution system, and connect its hoses to the liquid hydrogen double-walled vacuum storage tank, or Dewar. When hydrogen is needed, it is drawn from the Dewar by a cryogenic pump. The pump increases the liquid hydrogen pressure up to approximately 9500 psi. The hydrogen then passes through a vaporizer, where its temperature is increased. The cold, compressed hydrogen is stored in buffer storage tanks at approximately 9500 psi. This higher pressure allows for a largely passive fuelling process to occur from that point on. The low temperature of the compressed hydrogen gas serves several important functions. First, buffer storage volume decreases due to the higher H_2 density at these conditions. More important, however, is that the filling time can be much shorter since the tank temperature will not rise as rapidly. The buffer storage tanks, thermally insulated to passively maintain the low hydrogen temperature, are connected to the dispenser system through an insulated piping line". A schematic illustration of the fuelling process, including all the major components is available in the existing literature [61].

Many safety measures are taken into account at the storage design stage of the hydrogen refuelling station, as "*(i) vent stack, (ii) sprinkler, (iii) split roof style allows for ventilation, (iv) multiple temperature and flame detectors located in storage area,*

(v) sealed containment brim (2 × Dewar volume capacity), and (vi) cross ventilation" [61]. The station is designed to minimize public exposure to the hydrogen process components. The design contains an elevated and isolated liquid hydrogen storage area within all process equipment. The design would keep the spaces between the hydrogen process components and the public. Thus, the dispenser would be the only component accessible to the public. All liquid hydrogen is stored in the confined space, and the hydrogen release is entirely controlled. The Dewar is 40 feet far away from buildings. In addition, a potential leakage would have less chance of leading to fire or explosion due to the open-air design of the hydrogen storage area. This simply means that the hydrogen would quickly disperse into the atmosphere, safely. In case of fire or explosion, the hydrogen storage area is adequately isolated and located with a higher elevation than the public areas, resulting in less severe consequences.

As mentioned earlier, FMEA can formulate all potential failure modes and their effect on the system and prioritize the critical failure modes to receive intervention actions. For example, FMEA can be used to develop liquid hydrogen storage (Dewar) presented in Table 1.

The four assumed decision-makers $\#DM = \{\#DM_1, \#DM_2, \#DM_3, \#DM_4\}$ evaluate the S, O, and D for 24 failure modes of hydrogen refuelling station. Thus, for this case, the importance weights of four decision-makers can be calculated as 0.260, 0.240, 0.431, and 0.069, respectively. Considering the importance weights of decision-makers, and utilizing the BWM method, the importance weights of risk factors S, O, and D are derived as(0.432, 0.385, 0.183), respectively.

The crisp single weighted payoff matrix $\#\overline{R} = \left[\overline{r}_{ij}\right]_{25\times 3}$ is computed and listed in Table 2.

Subsequently, by eliminating $\mathcal{G}_i \neq 0$, the procedure is continued, and the final ranking of failure modes is $\#FM7 \succ \#FM8 \succ \#FM24 \succ \#FM10 \succ \#FM5 \succ \#FM14 \succ \#FM17 \succ \#FM18 \succ \#FM21 \succ \#FM23 \succ \#FM22 \succ \#FM25 \succ \#FM2 \succ \#FM3 \succ \#FM1 \succ \#FM4 \succ \#FM9 \succ \#FM12 \succ \#FM11 \succ \#FM16 \succ \#FM15 \succ \#FM13 \succ \#FM6 \succ \#FM19 \succ \#FM20$, and illustrated in Fig. 1. Therefore, the failure modes #FM7 "Mechanical damage (structural)", #FM24 "Electrical Damage (shock)", #FM8 "Operator Error" are the critical failure modes for the decision-maker to be selected for further intervention actions. As shown in Fig. 1, most of the failure modes have different rankings due to the inherent capabilities of the proposed approach to deal with the uncertainty and Game-theoretic-based context. To improve the safety performance of the system, considering the supports from exiting state of art models [66–68] and to the best of the authors understanding, the following actions can be performed to either reduce the severity of any consequences or probability of a fire or explosion:

- Different barriers as a wall can be constructed to distance between the storage and public areas,
- Several sensors that are sensitive to the temperature, flame, and leakages should be placed in the storage area,
- A sprinkler system such as a helium purging system should be established to use whenever required,

Table 1 Developed FMEA for liquid hydrogen storage (Dewar)

Failure modes tag	Failure mode description	Causes of failure modes	Effect of failure modes	Current control activities
#FM1	Leakage/rupture	H_2 embrittlement	Spill to storage area and dissipation of H_2	Specification: supplier
#FM2		Corrosion	Spill to storage area and combustion of vapor	Specification: supplier
#FM3		Stress cycling/ fatigue	Spill to storage area and pooling of LH_2	Specification: supplier
#FM4		Mechanical damage (collision)	NA	Design: barrier Procedure: restricted access
#FM5		Mechanical damage (impact)	NA	NA
#FM6		Mechanical damage (attack)	NA	NA
#FM7		Mechanical damage (structural)	NA	NA
#FM8		Operator error	NA	Procedure: operator training
#FM9	Allow excess heat conduction	Vacuum jacket failure	Excess vapour pressure	Control: vent to stack, specification: supplier
#FM10		Mechanical damage (impact)	NA	Procedure/ design: restricted access Procedure: operator training
#FM11	PRD fails	H2 embrittlement	Excess pressure buildup	Specification: supplier design: rupture disk—vent to stack
#FM12		Corrosion	NA	Specification: supplier
#FM13		Mechanical damage (impact)	NA	Procedure/ design: restricted access Procedure: operator training

(continued)

Table 1 (continued)

Failure modes tag	Failure mode description	Causes of failure modes	Effect of failure modes	Current control activities
#FM14		Debris		NA
#FM15	Vent line blockage	H_2 embrittlement	Excess pressure buildup	Specification: supplier design: rupture disk—vent to stack
#FM16		Corrosion	NA	Specification: supplier
#FM17		Debris	NA	NA
#FM18	Pressure signal, no signal	Electrical damage (shock)	System fault	Control: shutdown
#FM19		Out of range	NA	Specification: supplier
#FM20		Power failure	NA	NA
#FM21		Calibration error	NA	Procedure: supplier calibration
#FM22	Signal bias high	Electrical damage (shock)	NA	NA
#FM23		Calibration error	NA	Procedure: supplier calibration
#FM24	Signal bias low	Electrical damage (shock)	Excess vapor pressure	Design: rupture disk—vent to stack
#FM25		Calibration error	NA	Procedure: supplier calibration

- The ventilation stack should be constructed appropriately to flush out if there is malfunction detection,
- Conducting a seismic assessment of the storage area would be helpful to warranty that the structure is resilient to a severe earthquake.

To perform a sensitivity analysis, the other risk factors ("Severity" and "Occurrence") are considered as the expected values as well. Finally, the ranking of the failure modes is obtained based on risk priority number by using Pythagorean fuzzy uncertain linguistic variables. As demonstrated in Fig. 2, the specified failure modes such as $F14$, $F11$, and $F7$ that are ranked using Severity-based and Occurrence-based, have almost similar patterns. However, there is practically a reverse pattern for ranking of failure modes with Linguistic based. Thus, we restated our words that the ranking conformity of the proposed "*Nash equilibrium*" in this paper compared to

Table 2 The crisp single weighted payoff matrix (case study: hydrogen refueling station)

Failure modes tag	Severity	Occurrence	Detection
#FM1	0.03080	0.00427	0.00196
#FM2	0.03362	0.00480	0.00072
#FM3	0.03346	0.00399	0.00169
#FM4	0.03349	0.00332	0.00141
#FM5	0.03336	0.00459	0.00124
#FM6	0.38109	0.00470	0.00170
#FM7	0.04879	0.00043	0.00189
#FM8	0.05475	0.00678	0.00237
#FM9	0.03274	0.00455	0.00171
#FM10	0.03490	0.00347	0.00185
#FM11	0.02856	0.00461	0.00158
#FM12	0.03010	0.00489	0.00257
#FM13	0.03625	0.00284	0.00182
#FM14	0.03667	0.00488	0.00184
#FM15	0.02973	0.00509	0.00214
#FM16	0.02973	0.00488	0.00198
#FM17	0.02973	0.00448	0.00173
#FM18	0.03457	0.00424	0.00158
#FM19	0.03420	0.00357	0.00176
#FM20	0.03423	0.00378	0.00163
#FM21	0.03330	0.00439	0.00169
#FM22	0.03337	0.00427	0.00168
#FM23	0.04268	0.00243	0.00167
#FM24	0.03201	0.00243	0.00167
#FM25	0.03080	0.00427	0.00196

other approaches such as Severity-based and Occurrence-based is higher. Comparing "*Nash equilibrium*" with the other three methods, the conformity failure modes ranking confirms that the proposed hybrid model works appropriately the same as the Severity-based and Occurrence-based methods. However, the huge differences prove the Linguistic-based method's excellence due to its constructed raw and unprocessed data.

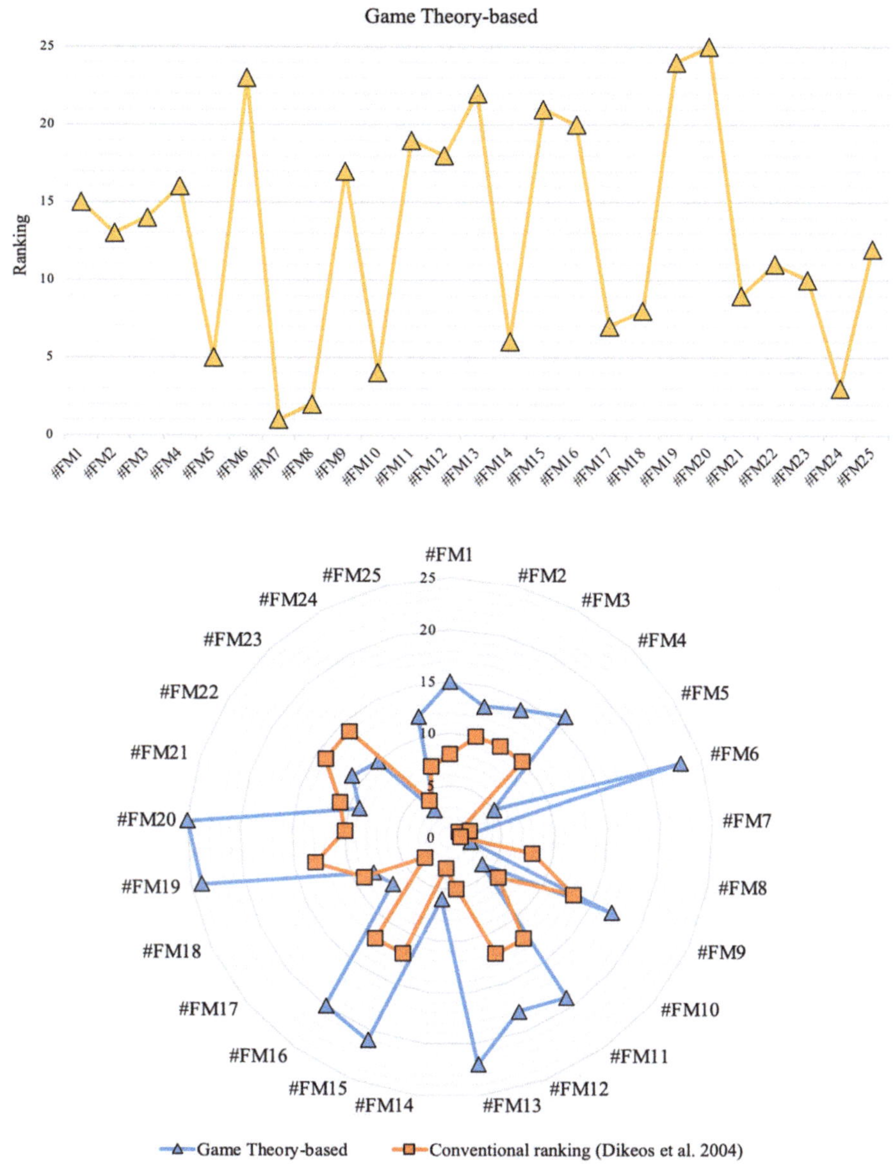

Fig. 1 The ranking failure modes (case study: hydrogen refueling station)

3 Conclusions

In conclusion, this study has presented a novel approach to FMEA by integrating the "Nash equilibrium" concept in the risk prioritization process for hydrogen refuelling station design, specifically focusing on the liquid hydrogen storage (Dewar) system.

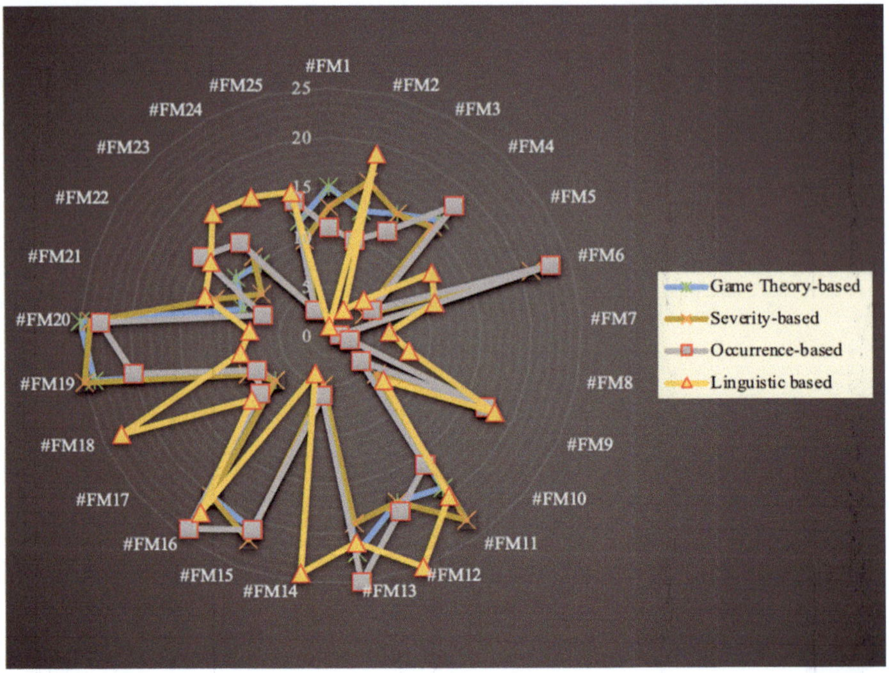

Fig. 2 Ranking conformity of failure modes based on different strategies (case study: hydrogen refuelling station)

The conventional FMEA method has limitations, which prompted the development of an improved approach. By considering multiple decision-makers and incorporating Pythagorean fuzzy uncertain linguistic variables, the proposed method allows for a more comprehensive assessment of risk factors associated with each failure mode. The formulation of FMEA as a zero-sum game between failure and success provides a systematic framework for prioritizing failure modes and determining intervention actions. Applying the proposed method to the liquid hydrogen storage case study in a hydrogen refuelling station demonstrates its effectiveness in identifying and prioritizing critical failure modes. Considering unequal importance weights for risk factors, comprehensive evaluation of risk factors, avoidance of repetitive risk priority numbers, and dynamic assessment over time contribute to a more robust risk management process. Integrating game theory principles and FMEA opens up new possibilities for enhancing risk assessment and decision-making in the design stage of hydrogen infrastructure. This approach facilitates the development of safer and more reliable hydrogen refuelling stations by addressing potential hazards and implementing appropriate intervention actions. Further research and validation of the proposed method in real-world applications are recommended to assess its practicality and effectiveness. Additionally, integrating other risk management techniques and considering additional factors could be explored to enhance further the accuracy and comprehensiveness of the risk prioritization process in hydrogen infrastructure design.

References

1. Kletz, T.: The history of process safety. J. Loss Prev. Process Ind. **25**, 763–765 (2012). https://doi.org/10.1016/j.jlp.2012.03.011
2. Ahn, J., Chang, D.: Fuzzy-based HAZOP study for process industry. J. Hazard. Mater. **317**, 303–311 (2016). https://doi.org/10.1016/j.jhazmat.2016.05.096
3. Li, H., Yazdi, M.: Reliability analysis of correlated failure modes by transforming fault tree model to Bayesian network: a case study of the MDS of a CNC machine tool BT. In: Li, H., Yazdi, M. (eds.) Advanced Decision-Making Methods and Applications in System Safety and Reliability Problems: Approaches, Case Studies, Multi-criteria Decision Making, Multi-objective Decision Making, Fuzzy Risk Based Model, pp. 15–28. Springer International Publishing, Cham (2022). https://doi.org/10.1007/978-3-031-07430-1_2
4. Jong, C.H., Tay, K.M., Lim, C.P.: Application of the fuzzy failure mode and effect analysis methodology to edible bird nest processing. Comput. Electron. Agric. **96**, 90–108 (2013). https://doi.org/10.1016/j.compag.2013.04.015
5. Lo, H.-W., Liou, J.J.H.: A novel multiple-criteria decision-making-based FMEA model for risk assessment. Appl. Soft Comput. **73**, 684–696 (2018). https://doi.org/10.1016/j.asoc.2018.09.020
6. Adesina, K.A., Yazdi, M., Zarei, E., Pouyakian, M.: Smart decision fuzzy-based data envelopment model for failure modes and effects analysis BT. In: Yazdi, M. (ed.) Linguistic Methods Under Fuzzy Information in System Safety and Reliability Analysis, pp. 151–170. Springer International Publishing, Cham (2022). https://doi.org/10.1007/978-3-030-93352-4_7
7. Yazdi, M., Daneshvar, S., Setareh H.: An extension to fuzzy developed failure mode and effects analysis (FDFMEA) application for aircraft landing system. Saf. Sci. **98** (2017). https://doi.org/10.1016/j.ssci.2017.06.009
8. Li H., Yazdi, M.: Developing failure modes and effect analysis on offshore wind turbines using two-stage optimization probabilistic linguistic preference relations BT. In: Li, H., Yazdi, M. (eds.) Advanced Decision-Making Methods and Applications in System Safety and Reliability Problems: Approaches, Case Studies, Multi-criteria Decision-Making, Multi-objective Decision-Making, Fuzzy Risk-Based Models, pp. 47–68. Springer International Publishing, Cham (2022). https://doi.org/10.1007/978-3-031-07430-1_4
9. Yazdi, M.: Improving failure mode and effect analysis (FMEA) with consideration of uncertainty handling as an interactive approach. Int. J. Interact. Des. Manuf. **13**, 441–458 (2019). https://doi.org/10.1007/s12008-018-0496-2
10. Cheng, C.Y., Li, S.F., Chu, S.J., Ych, C.Y., Simmons, R.J.: Application of fault tree analysis to assess inventory risk: a practical case from aerospace manufacturing. Int. J. Prod. Res. **51**, 6499–6514 (2013). https://doi.org/10.1080/00207543.2013.825744
11. Sarkar, A., Panja, S.C., Das, D.: Fault tree analysis of Rukhia gas turbine power plant. HKIE Trans. Hong Kong Inst. Eng. **22**, 32–56 (2015). https://doi.org/10.1080/1023697X.2015.1008394
12. Yazdi, M., Golilarz, N.A., Adesina, K.A., Nedjati A.: Probabilistic risk analysis of process systems considering epistemic and aleatory uncertainties: a comparison study. Int. J. Uncertain. Fuzziness Knowl.-Based Syst. **29**, 181–207 (2021). https://doi.org/10.1142/S0218488521500098
13. Yazdi, M., Darvishmotevali, M.: Fuzzy-based failure diagnostic analysis in a chemical process industry BT. In: Aliev, R.A., Kacprzyk, J., Pedrycz, W., Jamshidi, M., Sadikoglu, F.M. (eds.) 13th International Conference on Theory and Application of Fuzzy Systems and Soft Computing—ICAFS-2018, pp. 724–731. Springer International Publishing, Cham (2019)

14. Tyagi, S., Pandey, D., Tyagi, R.: Fuzzy set theoretic approach to fault tree analysis. Int. J. Eng. Sci. Technol. **2**, 276–283 (2010). https://doi.org/10.4314/ijest.v2i5.60165
15. Yazdi, M., Kabir, S., Kumar, M., Ghafir, I., Islam, F.: Reliability analysis of process systems using intuitionistic fuzzy set theory BT. In: Garg, H. (ed.) Advances in Reliability, Failure and Risk Analysis, pp. 215–250. Springer Nature Singapore, Singapore (2023). https://doi.org/10.1007/978-981-19-9909-3_10
16. Rausand, M., Hoyland, A.: System Reliability Theory: Models, Statistical Methods, and Applications, p. 664 (2004). https://doi.org/10.1109/WESCON.1996.554026
17. Markowski, A.S., Siuta, D.: Fuzzy logic approach for identifying representative accident scenarios. J. Loss Prev. Process Ind. **56**, 414–423 (2018). https://doi.org/10.1016/j.jlp.2018.10.003
18. Yazdi, M.: Risk assessment based on novel intuitionistic fuzzy-hybrid-modified TOPSIS approach. Saf. Sci. **110**, 438–448 (2018). https://doi.org/10.1016/j.ssci.2018.03.005
19. Yazdi, M., Khan, F., Abbassi, R., Rusli, R.: Improved DEMATEL methodology for effective safety management decision-making. Saf. Sci. **127**, 104705 (2020). https://doi.org/10.1016/j.ssci.2020.104705
20. Li, H., Yazdi, M.: Stochastic game theory approach to solve system safety and reliability decision-making problem under uncertainty BT. In: Li, H., Yazdi, M. (eds.) Advanced Decision-Making Methods and Applications in System Safety and Reliability Problems: Approaches, Case Studies, Multi-criteria Decision-Making, Multi-objective Decision-Making, Fuzzy Risk-Based Models, pp. 127–151. Springer International Publishing, Cham (2022). https://doi.org/10.1007/978-3-031-07430-1_8
21. Yazdi, M., Ding, Y., Adumene, S., Shafie, P.: Progressive system safety and reliability analysis: a sustainable game theory approach. Qual. Reliab. Eng. Int. n/a (2022). https://doi.org/10.1002/qre.3247
22. Li, H., Yazdi, M.: What are the critical well-drilling blowouts barriers? A progressive DEMATEL-game theory BT. In: Li, H., Yazdi, M. (eds.) Advanced Decision-Making Methods and Applications in System Safety and Reliability Problems: Approaches, Case Studies, Multi-criteria Decision-Making, Multi-objective Decision Making, Fuzzy Risk-Based Models, pp. 29–46. Springer International Publishing, Cham (2022). https://doi.org/10.1007/978-3-031-07430-1_3
23. Kabir, S., Yazdi, M., Aizpurua, J.I., Papadopoulos, Y.: Uncertainty-aware dynamic reliability analysis framework for complex systems. IEEE Access. **6**, 29499–29515 (2018). https://doi.org/10.1109/ACCESS.2018.2843166
24. Cheraghi, M., Eslami Baladeh, A., Khakzad, N.: A fuzzy multi-attribute HAZOP technique (FMA-HAZOP): application to gas wellhead facilities. Saf. Sci. **114**, 12–22 (2019). https://doi.org/10.1016/j.ssci.2018.12.024
25. Duan, Y., Zhao, J., Chen, J., Bai, G.: A risk matrix analysis method based on potential risk influence: a case study on cryogenic liquid hydrogen filling system. Process. Saf. Environ. Prot. **102**, 277–287 (2016). https://doi.org/10.1016/j.psep.2016.03.022
26. Rausand, M.: Risk Assessment: Theory, Methods, and Applications. Wiley (2011)
27. Adesina, K.A., Nedjati, A., Yazdi, M.: A short communication improving marine safety management system by addressing common safety program. Res. Mar. Sci. **5**, 671–680 (2020)
28. Zarreh, A., Wan, H., Lee, Y., Saygin, C., Al Janahi, R.: Risk assessment for cyber security of manufacturing systems: a game theory approach. Procedia Manuf. **38**, 605–612 (2019). https://doi.org/10.1016/j.promfg.2020.01.077
29. Liu, H., Liu, L., Li, P.: Failure mode and effects analysis using intuitionistic fuzzy hybrid weighted Euclidean distance operator. Int. J. Syst. Sci. **45**, 2012–2030 (2014). https://doi.org/10.1080/00207721.2012.760669
30. Liu, H.C., You, J.X., You, X.Y.: Evaluating the risk of healthcare failure modes using interval 2-tuple hybrid weighted distance measure. Comput. Ind. Eng. **78**, 249–258 (2014). https://doi.org/10.1016/j.cie.2014.07.018
31. Lin, Q.L., Wang, D.J., Lin, W.G., Liu, H.C.: Human reliability assessment for medical devices based on failure mode and effects analysis and fuzzy linguistic theory. Saf. Sci. **62**, 248–256 (2014). https://doi.org/10.1016/j.ssci.2013.08.022

32. Liu, Z., Mou, X., Liu, H.C., Zhang, L.: Failure mode and effect analysis based on probabilistic linguistic preference relations and gained and lost dominance score method. IEEE Trans. Cybern. 1–12 (2021). https://doi.org/10.1109/TCYB.2021.3105742
33. Liu, H.C., Chen, X.Q., Duan, C.Y., Wang, Y.M.: Failure mode and effect analysis using multicriteria decision making methods: a systematic literature review. Comput. Ind. Eng. **135**, 881–897 (2019). https://doi.org/10.1016/j.cie.2019.06.055
34. Liu, H.C., Fan, X.J., Li, P., Chen, Y.Z.: Evaluating the risk of failure modes with extended MULTIMOORA method under fuzzy environment. Eng. Appl. Artif. Intell. **34**, 168–177 (2014). https://doi.org/10.1016/j.engappai.2014.04.011
35. Liu, H.-C.: Improved FMEA Methods for Proactive Healthcare Risk Analysis, 1st edn. Springer (2019). https://doi.org/10.1007/978-981-13-6366-5
36. Li, H., Yazdi, M.: Advanced Decision-Making Methods and Applications in System Safety and Reliability Problems. Springer Cham (2022). https://link.springer.com/book/9783031074295
37. Zhou, Q., Thai, V.V.: Fuzzy and grey theories in failure mode and effect analysis for tanker equipment failure prediction. Saf. Sci. **83**, 74–79 (2016). https://doi.org/10.1016/j.ssci.2015.11.013
38. Yazdi, M., Daneshvar, S., Setareh, H.: An extension to fuzzy developed failure mode and effects analysis (FDFMEA) application for aircraft landing system. Saf. Sci. **98**, 113–123 (2017). https://doi.org/10.1016/j.ssci.2017.06.009
39. Fattahi, R., Khalilzadeh, M.: Risk evaluation using a novel hybrid method based on FMEA, extended MULTIMOORA, and AHP methods under fuzzy environment. Saf. Sci. **102**, 290–300 (2018). https://doi.org/10.1016/j.ssci.2017.10.018
40. Li, H., Díaz, H., Guedes Soares, C.: A failure analysis of floating offshore wind turbines using AHP-FMEA methodology. Ocean Eng. **234**, 109261 (2021). https://doi.org/10.1016/j.oceaneng.2021.109261
41. Li, H., Soares, C.G.: Reliability Analysis of Floating Offshore Wind Turbines Support Structure Using Hierarchical Bayesian Network (2019). https://doi.org/10.3850/978-981-11-2724-3
42. Johansen, I.L., Rausand, M.: Ambiguity in risk assessment. Saf. Sci. **80**, 243–251 (2015). https://doi.org/10.1016/j.ssci.2015.07.028
43. Yazdi, M.: Ignorance-aware safety and reliability analysis: a heuristic approach. Qual. Reliab. Eng. Int. **36** (2020). https://doi.org/10.1002/qre.2597
44. Helvacioglu, S., Ozen, E.: Fuzzy based failure modes and effect analysis for yacht system design. Ocean Eng. **79**, 131–141 (2014). https://doi.org/10.1016/j.oceaneng.2013.12.015
45. Liu, H.: FMEA Using Uncertainty Theories and MCDM Methods (2016). https://doi.org/10.1007/978-981-10-1466-6
46. Daneshvar, S., Yazdi, M., Adesina, K.A.: Fuzzy smart failure modes and effects analysis to improve safety performance of system: case study of an aircraft landing system. Qual. Reliab. Eng. Int. 1–20 (2020). https://doi.org/10.1002/qre.2607
47. Yazdi, M., Golilarz, N.A., Nedjati, A., Adesina, K.A.: An improved lasso regression model for evaluating the efficiency of intervention actions in a system reliability analysis. Neural Comput. Appl. (2021). https://doi.org/10.1007/s00521-020-05537-8
48. Momen, S., Tavakkoli-Moghaddam, R., Ghasemkhani, A., Shahnejat-Bushehri, S., Tavakkoli-Moghaddam, H.: Prioritizing surgical cancellation factors based on a fuzzy best-worst method: a case study. IFAC-PapersOnLine **52**, 112–117 (2019). https://doi.org/10.1016/j.ifacol.2019.11.161
49. Mi, C., Chen, Y., Zhou, Z., Lin, C.-T.: Product redesign evaluation: an improved quality function deployment model based on failure modes and effects analysis and 2-tuple linguistic. Adv. Mech. Eng. **10**, 1687814018811227 (2018). https://doi.org/10.1177/1687814018811227
50. Liu, H.C., Lin, Q.L., Mao, L.X., Zhang, Z.Y.: Dynamic adaptive fuzzy petri nets for knowledge representation and reasoning. IEEE Trans. Syst. Man Cybern. Syst. **43**, 1399–1410 (2013). https://doi.org/10.1109/TSMC.2013.2256125
51. Chai, K.C., Jong, C.H., Tay, K.M., Lim, C.P.: A perceptual computing-based method to prioritize failure modes in failure mode and effect analysis and its application to edible bird nest farming. Appl. Soft Comput. J. **49**, 734–747 (2016). https://doi.org/10.1016/j.asoc.2016.08.043

52. Yazdi, M., Kabir, S., Walker, M.: Uncertainty handling in fault tree based risk assessment: state of the art and future perspectives. Process. Saf. Environ. Prot. **131**, 89–104 (2019). https://doi.org/10.1016/j.psep.2019.09.003
53. Nedjati, A., Yazdi, M., Abbassi, R.: A Sustainable Perspective of Optimal Site Selection of Giant Air-Purifiers in Large Metropolitan Areas. Springer Netherlands (2021). https://doi.org/10.1007/s10668-021-01807-0
54. Yazdi, M., Khan, F., Abbassi, R.: Microbiologically influenced corrosion (MIC) management using Bayesian inference. Ocean Eng (2021). https://doi.org/10.1016/j.oceaneng.2021.108852
55. Adumene, S., Okwu, M., Yazdi, M., Afenyo, M., Islam, R., Orji, C.U., Obeng, F., Goerlandt, F.: Dynamic logistics disruption risk model for offshore supply vessel operations in Arctic waters. Marit. Transp. Res. **2**, 100039 (2021). https://doi.org/10.1016/j.martra.2021.100039
56. Kutlu, A.C., Ekmekçioğlu, M.: Fuzzy failure modes and effects analysis by using fuzzy TOPSIS-based fuzzy AHP. Expert Syst. Appl. **39**, 61–67 (2012). https://doi.org/10.1016/j.eswa.2011.06.044
57. Xin Nie, R., Peng Tian, Z., Kang Wang, X., Qiang Wang, J., Li Wang, T.: Risk evaluation by FMEA of supercritical water gasification system using multi-granular linguistic distribution assessment. Knowl.-Based Syst. **162**, 185–201 (2018). https://doi.org/10.1016/j.knosys.2018.05.030
58. Wu, J., Liang, L., Chen, Y.: DEA game cross-efficiency approach to Olympic rankings. Omega **37**, 909–918 (2009). https://doi.org/10.1016/j.omega.2008.07.001
59. Madani, K., Hipel, K.W.: Non-cooperative stability definitions for strategic analysis of generic water resources conflicts. Water Resour. Manag. **25**, 1949–1977 (2011). https://doi.org/10.1007/s11269-011-9783-4
60. Ding, X.F., Liu, H.C.: A new approach for emergency decision-making based on zero-sum game with Pythagorean fuzzy uncertain linguistic variables. Int. J. Intell. Syst. **34**, 1667–1684 (2019). https://doi.org/10.1002/int.22113
61. Dikeos, J., Haas, J., Maddaloni, J., Owen, T., Smolak, T., Songprakorp, R., Sodouri, P., St.Germain, L., Tura, A.: Hydrogen Refueling Station, Design Proposal. British Columbia. http://www.hydrogencontest.org/pdf/2004/uvictoria_2004.pdf (2004)
62. Yoo, B.H., Wilailak, S., Bae, S.H., Gye, H.R., Lee, C.J.: Comparative risk assessment of liquefied and gaseous hydrogen refueling stations. Int. J. Hydrog. Energy **46**, 35511–35524 (2021). https://doi.org/10.1016/j.ijhydene.2021.08.073
63. Casamirra, M., Castiglia, F., Giardina, M., Lombardo, C.: Safety studies of a hydrogen refuelling station: determination of the occurrence frequency of the accidental scenarios. Int. J. Hydrog. Energy **34**, 5846–5854 (2009). https://doi.org/10.1016/j.ijhydene.2009.01.096
64. Yazdi, M.: The application of Bow-Tie method in hydrogen sulfide risk management using layer of protection analysis (LOPA). J. Fail. Anal. Prev. **17**, 291–303 (2017). https://doi.org/10.1007/s11668-017-0247-x
65. Li, H., Yazdi, M.: Integration of the Bayesian network approach and interval Type-2 fuzzy sets for developing sustainable hydrogen storage technology in large metropolitan areas BT. In: Li, H., Yazdi, M. (eds.) Advanced Decision-Making Methods and Applications in System Safety and Reliability Problem, pp. 69–85. Springer International Publishing, Cham (2022). https://doi.org/10.1007/978-3-031-07430-1_5
66. Bahrebar, S., Blaabjerg, F., Wang, H., Vafamand, N., Khooban, M.-H., Rastayesh, S., Zhou, D.: A novel Type-2 fuzzy logic for improved risk analysis of proton exchange membrane fuel cells in marine power systems application. Energies **11** (2018). https://doi.org/10.3390/en11040721
67. Hirayama, M., Shinozaki, H., Kasai, N., Otaki, T.: Comparative risk study of hydrogen and gasoline dispensers for vehicles. Int. J. Hydrog. Energy. **43**, 12584–12594 (2018). https://doi.org/10.1016/j.ijhydene.2018.05.003
68. Park, S., Ahn, J., Choi, K., Kim, S., Yoo, Y., Kang, S., Chang, D.: Fuzzy-inference-based failure mode and effects analysis of the hydrogen production process using Thermococcus onnurineus NA1. Int. J. Hydrog. Energy. **44**, 13135–13146 (2019). https://doi.org/10.1016/j.ijhydene.2019.03.227

Practice and Enlightenment of General Quality Characteristic Management of Us Military Equipment

Dewei Li, Qingjun Meng, Shuaichao Guo, and Jian Zhou

Abstract The impeccable Equipment Quality Management System instituted by the United States Army stands as a pivotal pillar in bolstering the enhancement of combat capabilities. Its exemplary approach to overseeing the general quality of equipment serves as a valuable source of insights and knowledge. This paper delves into the methodologies and accomplishments of the General Quality Characteristic Management within the US military's equipment domain, analyzing it through the lenses of four crucial facets: demand demonstration, incentive mechanisms, test evaluation, and information infrastructure. Moreover, the paper presents a series of strategic recommendations aimed at augmenting the comprehensive management of general quality characteristics in our own military equipment sector. These recommendations are structured around four key areas: foregrounding the importance of early-stage considerations, reshaping the parameters of operational paradigms, fortifying testing proficiency, and enhancing the foundations of information architecture. By embracing these insights and taking proactive measures, our military equipment management can be propelled to new heights of efficiency and effectiveness.

Keywords US military equipment · General quality characteristics · Operation management

1 Introduction

The overall quality characteristics of equipment are intrinsically linked to the specific quality attributes that define the equipment itself or encompass its tactical and technical performance [1–3]. Examples of these particular attributes include a gun's firing

D. Li
MilitaryTraining Center of Naval Staff, Beijing, China

Q. Meng (✉) · S. Guo · J. Zhou
Beijing Headquarters of Jiangsu Automation Research Institute, Beijing, China
e-mail: mengqingjun95@163.com

M. Yazdi (ed.), *Progressive Decision-Making Tools and Applications in Project and Operation Management*, Studies in Systems, Decision and Control 518,
https://doi.org/10.1007/978-3-031-51719-8_4

range, the vehicle's velocity, an aircraft's altitude capabilities, and similar parameters. In China, these broad quality characteristics are denoted by terms like reliability, maintainability, supportability, testability, safety, and environmental adaptability, collectively called the "Six Characteristics."

It is worth noting that the United States military diverges in its approach, as it lacks general quality characteristics akin to those in China. Instead, the U.S. military's framework aligns with the concept of operational applicability [4]. Operational suitability in the U.S. context encompasses a range of factors such as availability, reliability, maintainability, supportability, testability, safety, electromagnetic compatibility, transportability, environmental adaptability, and human–machine adaptability. It is notable that, unlike China's focus on the "Six Characteristics," the U.S. military emphasizes a more expansive set of attributes that contribute to operational effectiveness.

Furthermore, within the U.S. military framework, the term "equipment RMS" (Reliability, Maintainability, and Supportability) holds prominence as a shared quality characteristic [5, 6]. RMS encapsulates the critical aspects of reliability, maintainability, and supportability that are central to the performance and longevity of equipment. This underscores the importance of these attributes in the U.S. military's evaluation and approach to equipment quality.

Recently, our military's dedicated efforts towards equipment development have reached new heights, marked by a noticeable intensification in construction initiatives, a substantial uptick in research-oriented equipment, and a commendable acceleration in the modernization of our armed forces [7]. These collective strides have yielded significant advancements in tactical and technical performance, effectively reducing the disparity that once existed with more formidable adversaries. In several domains, we have not only managed to close the gap but also surpassed certain technologically advanced nations on the global stage.

However, a comprehensive assessment of the situation requires an unwavering recognition of the challenges that arise post-equipment deployment. Despite the strides made in its development, the equipment has encountered formidable hurdles upon integration into the troops. These challenges manifest in various forms, such as an alarming frequency of breakdowns, intricacies in maintenance processes, and formidable obstacles in terms of support infrastructure. These issues serve as a poignant reminder of the substantial discrepancy in our overall quality characteristic management when juxtaposed with international standards.

This discrepancy, in turn, casts a significant shadow over the formation of the equipment's natural combat prowess and support capabilities [8–10]. It becomes abundantly clear that, despite the advancements achieved in specialized attributes and performance benchmarks, the broader quality characteristics that determine reliability, maintainability, supportability, and safety still need to catch up to the standards set by foreign counterparts.

While we have achieved commendable strides in enhancing tactical and technical performance, our focus must now shift towards a comprehensive approach to quality characteristic management. By addressing the issues of breakdowns, maintenance challenges, and support difficulties, we can bridge the existing gap and ensure that our equipment's combat effectiveness and support capabilities align with the global standards we aspire to reach [8–10]. This concerted effort will be instrumental in propelling our armed forces toward an elevated operational readiness and effectiveness, thus solidifying our position on the international stage.

The comprehensive quality management approach adopted by the U.S. military for its equipment reverberates across the entire life cycle of these assets, permeating every facet of equipment development, deployment, and maintenance. This holistic strategy emerges as the culmination of collaborative endeavours involving various departments within the military. In essence, the U.S. military's approach to equipment quality management is underpinned by a well-structured framework encompassing competition, supervision, evaluation, and incentive mechanisms [11]. This multifaceted approach has been pivotal in cultivating the world's most formidable arsenal of weapon systems, accentuating its critical role in global defence capabilities.

The foundation of the U.S. military's quality management strategy is establishing a robust competition mechanism. This mechanism fosters a dynamic environment where manufacturers, researchers, and developers vie to produce cutting-edge equipment. The resultant competition-driven innovation fuels advancements and ensures that only superior technologies are integrated into military assets.

Furthermore, the military's stringent supervision mechanism ensures that every aspect of equipment development adheres to the highest quality and safety standards. This oversight spans design and production to testing and deployment, thus guaranteeing that the equipment's performance aligns with rigorous benchmarks.

The U.S. military's emphasis on evaluation is equally noteworthy. Rigorous assessments are conducted to ascertain the effectiveness and reliability of the equipment. These evaluations serve as a vital feedback loop, facilitating the refinement and enhancement of equipment attributes over time.

Incentive mechanisms within the U.S. military's quality management framework catalyse continuous improvement. Recognizing and rewarding exceptional contributions stimulate innovation, fostering an environment where individuals and teams are motivated to excel in their pursuit of developing superior equipment.

The overarching impact of the U.S. military's comprehensive quality management strategy is undeniable. By cultivating a dynamic competition ecosystem, meticulous supervision, rigorous evaluation, and performance-based incentives, the U.S. military has harnessed the ability to craft a formidable array of weapon systems, solidifying its position as a global defence leader.

Significantly, the applicability of this approach extends beyond national boundaries. As our military endeavours to elevate the overall quality characteristics of our equipment [12], drawing inspiration from the U.S. military's approach could yield practical benefits. Emulating the comprehensive and dynamic strategies underlying the U.S. model can significantly enhance our military equipment's general quality characteristic management, enabling us to attain new heights of operational effectiveness internationally.

2 Make Clear Requirements for General Quality Characteristics

Quality characteristics such as reliability, maintainability and supportability is a basic prerequisite for the quality management of equipment system [13]. The Air Force regulation "Establishes logistical support and integrity requirements," which states that "The best time to influence the design of a equipment system is in the early stages of system acquisition [14]; the most effective way for the user to influence the design is to specify basic qualitative and quantitative integrity and logistical support requirements in the Use Plan and Requirements Document."

The US Army began its research on reliability, maintainability and supportability in the 1950s. After 70 years of development, it has gradually developed and perfected the concept of life-cycle quality management. They pay attention to quality management "Pass forward". The research results of concurrent engineering and equipment integrated support engineering have been widely used in equipment development. The equipment is reliable and durable, easy to maintain, easy to support and low cost. The Department of Defense clearly requires that the design and analysis of reliability, maintainability and supportability must be carried out in the process of equipment development. The military also has clear requirements on reliability, maintainability and supportability, including military requirements, design and analysis methods, as well as assessment and acceptance conditions. In determining the reliability, maintainability and supportability requirements for weapons and equipment, the military has developed a set of requirements determination procedures and methods: from the top to the bottom, from the overall, macro to thc specific, detailed, and with the development of equipment, requirements and refinement requirements are constantly clear. For example, the F-22, the 4th generation fighter of the US Army, has gone through several stages, such as pre-project, scheme exploration, verification and verification, et al., the general quality characteristic index is defined and refined step by step.

3 Using PBL to Stimulate the Growth of General Quality Characteristics

Performance Based Logistics (PBL) is also often translated as "Performance Based Support" in China [15]. In the PBL Assurance: A Project Manager's Guide to Product Assurance issued by the Defense Acquisition University, PBL is defined as a long-term assurance agreement with clear authority and responsibility, safeguards are purchased as a comprehensive, affordable "Performance package" to optimize the performance and operational readiness of equipment systems. Simply put, PBL buys the ultimate guaranteed performance result, not the product or service. PBL is the preferred product support strategy of the US Department of Defense, which requires the lowest cost and the smallest logistics scale, while achieving the best availability of the whole system [16].

In short, the goal of PBL is to buy and secure "Performance outcomes" rather than the traditional simple purchase of products, parts, or repair activities. So it's important to know what "Performance" consists of. The project manager representing the military works with the user/combat unit to determine the performance requirements of the system and then with the product assurance provider, these requirements are met by documenting their requirements, including the corresponding performance parameters, in a performance-based protocol [17]. The effective implementation of PBL depends on top-level performance parameters that accurately reflect user needs and effectively measure the performance of the provider [18].

The top-level performance parameters of PBL generally refer to: availability of use, reliability of use, cost of use per unit, logistics scale, logistics response time, et al. Among them: usability is the percentage of time the system is able to perform tasks or the ability to maintain the operational tempo; usability is a measure of the system's success in achieving mission objectives. Depending on the system, the mission objective may be launch, cruise, launch, destination, or other specific parameters; the cost per unit of use is the total cost of a given system divided by the corresponding unit of measure. Depending on the system, the units of measurement may be flight hours, work hours, launch times, mileage, or other specific parameters of services and systems; Logistics scale refers to the size or "Strength" of the logistics support required by the government/contractor to deploy, maintain, and move systems. Measurable elements include inventory/equipment, personnel, facilities, transportation assets, and real estate; logistics response time is the time it takes to signal logistics requirements and meet those requirements, "Logistics requirements" refer to the systems, components, or resources necessary to support the logistics of a system, including personnel and support services. Of course, in the process of PBL implementation, these top-level performance parameters should be tailored according to the combat mission of equipment system, and be sure that the top-level performance parameters are in accordance with the scope of responsibility of the assurance provider.

In essence, PBL is an innovative user-driven institutional arrangement in which the user (the user of the equipment system) does not have to tell the safeguard provider what to do, but only the required safeguards performance results, the support provider will, according to the requirements of the performance-based agreement (PBA), go all out to improve the general quality characteristics of the equipment, complete the work related to the equipment support, and achieve satisfactory support performance results as far as possible, in order to benefit from it.

4 Emphasizing the Role of Test and Evaluation in the Evolution of General Quality Characteristics

Test and Evaluation is an important content of equipment system life-cycle management, and also an important means to ensure the quality and reliability. The equipment system Test and Evaluation of the U. S. Army includes Development Test and Evaluation (DTE) and Operation Test and Evaluation (OTE), the latter is divided into the initial and subsequent two stages. The testing of general quality characteristics is an important part of DTE and OTE, and more and more attention is paid to it [19].

General quality characteristic test is also called "Logistics test" in the development test stage. It mainly tests the weapon system around the multiple elements of integrated logistics support. These factors include: availability, compatibility, transportability, interoperability, reliability, wartime utilization, maintainability, safety, human factors, human support, logistics support, natural environmental effects and impacts, documentation, and training requirements. Through the test, the design problems affecting the quality can be found early, and the design can be improved to optimize the performance and support characteristics of the weapon system. The 450 logistics test squadron was built at the Edwards Air Force Base Air Force flight test center. B-2, C-17 aircraft in the DTE process have been carried out logistics tests, and there are many successful examples. Air Force maintenance support personnel have been involved from the start of the test flight, along with the contractor's personnel, in the aircraft's test support. Through the actual maintenance operation, they take part in the logistics test and evaluation, especially to verify the maintainability and reliability of each system of the aircraft, and to verify whether the contents and steps of the technical regulations are correct. In addition to professional knowledge and work experience, these personnel should also dare to ask questions and suggestions [20].

OTE is carried out independently by the Center for Operational Test and Evaluation, with the aim of evaluating the operational effectiveness and applicability of equipment systems under as realistic operational conditions as possible [21]. For example, the parameters of the C-17 transport aircraft test include: complete mission rate, complete mission rate, complete mission probability, average inherent repair interval flight time, average repair interval flight time, average disassembly interval flight time, maintenance man time per flight hour and average repair time, etc. In addition, under the terms of the development contract, a special 30-day reliability, maintainability and availability test was conducted on 12 C-17 aircraft. In addition to the above-mentioned parameters, the indexes of fault detection rate, fault isolation rate and fault false alarm are also evaluated. This was followed by a three-year follow-up operational test and evaluation of the C-17 aircraft, with a focus on generic quality characteristics.

5 Exerting the Supporting Role of Information System in the Improvement of Equipment General Quality Characteristics

Continuous Acquisition and Life-cycle Support (CALS) was introduced by the US Army in the 1980s. The purpose of implementing CALS project is to speed up the acquisition process of weapons and equipment, reduce the life-cycle cost, and improve the equipment quality and maintenance support level. The core content of CALS project is the Digitization, standardization and network integration of technical data of weapon system, product or engineering system [22].

The implementation of CALS provides information support for the whole system and whole life management of weapon equipment, that is, in the whole life course of the design, development, production, training, use, support and decommissioning of weapon equipment, using network, database, multimedia and other information technology to build integrated and shared data environment. The information and data needed can be stored, managed and used conveniently by the manufacturers and users concerned.

The US Department of Defense, as well as all branches of the military and related companies, have piloted the CALS program in more than 100 programs, it is proved that CALS project can bring benefits of shortening working cycle, reducing life-cycle cost and improving product quality. In the CALS environment, in recent years, the U. S. military increased investment efforts: update and improve the use of weapons and equipment, maintenance and support of information systems; We will strengthen information transmission between the military and the R & D and production departments, between military departments at all levels, between the service departments and the maintenance departments, and strengthen information management of equipment usage, reliability, and maintenance support. To provide information support for equipment to form combat effectiveness [23, 24].

Under the concept of CALS, the typical information systems that the US military has developed and put into use include logistics support command system, transportation and supply automation information system and maintenance automation information system. These management systems play an active role in ensuring the delivery of forces and their equipment worldwide. Efficient information collection and transmission capability is the basis and means to improve equipment support effectiveness. Making full use of various database resources, the security information obtained through advanced sensors and other means of reconnaissance and surveillance will be compiled and distributed in a timely manner, ensure that decision-makers of equipment support are provided with accurate, near-real-time information.

The US military equipment maintenance information system is divided into three levels: the lowest level is the army's maintenance information collection system, which is responsible for collecting basic maintenance data at the grass-roots level; and the middle level is the military level maintenance management information system, it is managed by the service department in charge of maintenance, and the top layer is the maintenance management information system of the Ministry of Defense, which mainly receives the maintenance information and data reported by various military services. The Ministry of Defence compiles these data into 14 reports (such as total costs, cost statistics for major military repair plants, statistics on repair costs by weapon and system, etc.) for use by the leading authorities. In recent years, with the rapid development of automation and information technology, the US military has developed a new generation of maintenance and support management information systems. The common features of these systems are: comprehensive content, perfect function, real-time data processing, automatic data exchange between systems and intelligent applications.

6 The Enlightenment to Our Military General Quality Characteristics Management

In 2014, a pivotal step was taken within our army as it introduced the "General Quality Characteristics of Equipment Management Provisions." These regulations underscore the imperative of managing general quality characteristics across the entire equipment life cycle—encompassing stages such as equipment demonstration, development, shaping, and deployment.

However, in a comparative evaluation of the practices of the US Army, our military's general quality management endeavours appear to be somewhat lagging. Several discernible issues include outdated perspectives, regulatory gaps, operational mechanism inefficiencies, and inadequate support capabilities.

Nonetheless, there has been notable progress in recent years. Various segments within our military, including equipment demonstration units, research and development entities, production facilities, and user units, are increasingly focusing on general quality characteristics. This surge in attention has yielded promising results, elevating the influence of general quality characteristics within our military operations. Nevertheless, when juxtaposed against the proficiency of formidable foreign counterparts, a substantial gap still warrants attention.

To bridge this divide and fortify our military's capabilities, it's imperative to draw from the experiences of foreign nations. These lessons can be harnessed to bolster our approach in the following key domains:

Enhanced Regulatory Framework: Strengthening laws and regulations that govern general quality characteristics can serve as a cornerstone. Updating and aligning our regulatory infrastructure with contemporary best practices and international standards will provide a solid foundation for effective management.

Operational Mechanism Refinement: Augmenting the operational mechanisms involved in managing general quality characteristics is essential. Streamlining processes, enhancing communication and collaboration among various units, and fostering a culture of accountability can lead to more effective outcomes.

Investment in Training and Education: Promoting a culture of continuous learning within our military can significantly impact general quality management. Comprehensive training and education programs that imbue personnel with the latest methodologies, techniques, and insights are crucial.

Integration of Technological Advancements: Embracing cutting-edge technologies can significantly improve our military's general quality management practices. Incorporating tools like advanced data analytics, predictive modelling, and automation can enhance the accuracy and efficiency of our operations.

International Collaboration: Establishing collaborations and partnerships with foreign military entities renowned for their advanced general quality management practices can facilitate knowledge exchange and accelerate our progress.

Elevating Support Capabilities: Bolstering the support infrastructure—ranging from maintenance and logistics to supply chain management—is essential. A robust support system ensures the longevity and optimal performance of military equipment.

Drawing on the experience of foreign countries, our army can also improve in the following aspects.

6.1 Pay Attention to the Early Stage

The strategic evolution of equipment general quality characteristics necessitates a proactive shift—reorienting from a reactive "After the event" approach to a forward-thinking "Pre-prevention" mode. While it might be tempting to address issues like equipment difficulties, usability, repairability, and support challenges solely during usage, the root of these concerns typically lies in the early stages, particularly during the demonstration phase.

It is imperative to recognize that challenges faced by the military, such as equipment's challenging usability, repair complexities, and support intricacies, trace their origins back to the early stages, particularly the demonstration phase. Therefore, the concept of lifecycle quality management for military equipment should take centre stage to reshape the fulcrum of equipment's general quality characteristics. This approach underscores the importance of thorough Six Characteristics and verification methods demonstrations during the early equipment stages.

The Six Characteristics indicators have been incorporated into general requirements development and the overall testing program. However, the emphasis on performance indicators beyond reliability within the demonstration framework must be increased. Furthermore, the verification methodologies for these general quality characteristics must be delineated within the Test Master Plan to ensure a systematic and comprehensive evaluation.

Concurrently, a crucial strategy involves implementing a continuous assessment paradigm across three pivotal stages: Performance Test, Operation Test, and In-Service Assessment. This strategic triad provides a comprehensive assessment approach that spans the entire equipment lifecycle. It offers invaluable guidance for embedding and realizing the full spectrum of general quality characteristic indicators.

In essence, the transformation from a reactionary standpoint to a proactive "Pre-prevention" mode is paramount. By doing so, the underlying complexities of equipment usage, maintenance, and support can be effectively addressed. Shifting the focus to the early stages, notably during equipment demonstration, offers the potential to identify and rectify issues preemptively. The lifecycle quality management concept is anchored within this approach, demonstrating Six Characteristics indicators gains prominence. While progress has been made in embedding reliability metrics, a more comprehensive demonstration framework is warranted. Concurrently, the strategic implementation of continuous assessment across the three pivotal stages furnishes the means to comprehensively assess and bolster general quality characteristics, aligning with enhancing military equipment efficacy, reliability, and supportability.

6.2 Change the Rules of the Game

At the institutional level of design and arrangement, a paramount goal emerges—to harness the full potential of an entity's general quality characteristics. This imperative perspective is rooted in the premise that disregarding these characteristics can pave the way for the entity's struggle for survival. Borrowing insights from the US Army's experience with Performance-Based Logistics (PBL), this system is a foundation for refinement and advancement.

Within this strategic framework, the initial step entails the definition of top-level performance parameters and indices. This is achieved by placing the user at the nucleus and the equipment system as the focal point. Parameters like availability, reliability, unit cost, logistics scale, and response time form the cornerstone of this endeavour. These parameters and indicators act as the linchpin for PBL execution, encompassing critical roles such as the project manager, product support manager, product support integrator, and product support provider (support source). Among these, the project and product support managers operate within the military management domain. In contrast, product support integrators and providers are responsible for the army or civilian industrial sector.

The project manager prescribes the requisites, the product support integrator executes them, and the subsequent step entails establishing performance-based agreements or contracts founded upon objective metrics. This shift from decentralized transactions—traditionally centred around product and service procurement—towards procuring assurance performance outcomes represents a pivotal transformation. Notably, performance-based agreements typically encompass long-term durations, fostering an environment conducive to bolstering security capabilities, maximizing long-term benefits, and evading short-sighted behaviours. The incentive lies in rendering efficient assurance services, culminating in a synergistic win–win scenario between military users and enterprises.

This symbiotic approach engenders a manifold of advantages. It effectively amplifies the readiness of weapons and equipment, concurrently diminishing the life cycle costs and enriching the user experience. This dynamic also nurtures a collaborative and enduring partnership between the assurance provider and the user. This relationship thrives on a foundation of perpetual enhancement in general quality characteristics, where the assurance provider is incentivized to secure a stable and consistent income stream.

In summation, designing and orchestrating an entity's system that capitalizes on general quality characteristics requires meticulous attention. Drawing inspiration from the US Army's experience with Performance-Based Logistics, the strategic transformation pivots around top-level performance parameters and performance-based agreements. This shift towards an outcomes-oriented approach bolsters the overall security capability and forges lasting partnerships and continuous improvements. Ultimately, this synchronized synergy cultivates a climate of mutual benefit for users and providers, fostering a holistic enhancement of general quality characteristics and sustainable growth.

6.3 Enhance the Test Capability

To comprehensively address the challenges associated with equipment performance, it has become imperative to conduct combat-oriented experiments before and after equipment deployment. This approach facilitates the meticulous scrutiny and resolution of general quality characteristic issues inherent to the equipment, ultimately enhancing army personnel's operational experiences.

In the contemporary landscape, our Army has initiated a progressive three-stage framework encompassing Performance Tests, Operation Test, and In-Service Assessments. The evolution of the Test and Evaluation organization further underlines the commitment to refining equipment assessment methodologies. Nevertheless, certain constraints persist, stemming from test costs, time limitations, and resource availability during testing phases. Notably, a significant discrepancy exists in evaluating our military equipment's maintainability, testability, and supportability prior to its service. Moreover, the current extent of In-Service Assessment and the channels for feedback and associated limitations warrant fortification.

Herein, an opportunity arises to draw inspiration from the management model of test identification practised by the US Army. By adopting certain principles, we can elevate the prominence of Operation Test departments and infuse a more robust investment cycle into the testing process. Accelerating the formulation of comprehensive test standards and norms is crucial. This concerted effort aligns with the overarching objective of nurturing our Army's prowess in equipment test identification.

To elaborate, embedding combat-oriented experiments as a prelude and postlude to equipment service augment our capacity to detect and rectify general quality characteristic concerns preemptively. By subjecting equipment to a comprehensive evaluation, we empower our forces to surmount operational challenges and glean invaluable insights from experiential learning. The formulation of the three-stage framework underscores a strategic approach, yet challenges remain, especially in resource allocation and In-Service Assessment's reach.

Learning from the US Army's methodology, we can strategically bolster our Operation Test units and allocate resources more effectively. Moreover, investing in a robust cycle of test funds expedites the refinement of test standards and norms. These efforts galvanize our military's prowess in identifying equipment-related issues, ensuring a resilient arsenal of functional, reliable, and effective equipment.

In conclusion, the amalgamation of combat-oriented experiments before and after equipment deployment is vital in addressing general quality concerns. The tri-stage assessment framework establishes a framework for progress yet must be underpinned by fortified resources and assessment reach. By emulating the US Army's strategic management, we advance our operational capacities, amplify the influence of Operation Test, and expedite our journey towards establishing a cutting-edge, adaptable, and responsive equipment assessment framework.

6.4 Strengthening Information Construction

The trajectory of our military's equipment development has evolved significantly, transitioning from the initial phases of merely addressing existence and performance gaps to now embracing the paramount stage of "High-quality development." This progression signifies a shift in focus from mere problem-solving and performance catch-up to a concentrated emphasis on attaining and maintaining high standards of excellence. The primary challenge impeding our army's equipment combat capability has transformed. It is no longer rooted in the deficiency of combat technique performance; instead, the pressing issue lies in the imperative enhancement of general quality characteristics. The pursuit of equipment that is not only proficient but also pragmatic, resilient, and productive has become the aspiration of our military. Our armed forces fervently desire such attributes and are instrumental in significantly augmenting our equipment's combat effectiveness and indispensable operational support.

Currently, analyzing and applying information about weapon equipment quality presents notable areas for improvement, particularly in the utilization of general quality information analysis and the subsequent interpretation of its findings. Regrettably, the requisites for available quality information regarding weapon equipment remain insufficiently defined. A conspicuous lack of clarity persists regarding selecting pertinent information to collect and what information should be excluded, contributing to a discord between the information demand and the aspiration for "Systematization and standardization."

Within the existing landscape, the availability of general quality information is beset by "Quality problems." These issues manifest during the compilation of information by troops throughout the equipment's operational lifecycle. The data captured in this process is frequently characterized by incompleteness, untimeliness, lack of rationale, and arbitrary inclusion. Moreover, the landscape of general quality information analysis must be more uniform and systematic. Each branch of the military and armament wields its distinct quality management framework, often incorporating elements of quality information analysis. However, this mosaic of investigations operates as separate "Information Islands," leading to incongruent findings and challenges in their practical application.

Addressing these challenges mandates a two-pronged approach. First, there is a pressing need to reinforce the "Depth" and "Breadth" of general quality information analysis. This entails a proactive engagement with cutting-edge information analysis techniques and methodologies, fostering a comprehensive exploration of the latent value encapsulated within the information corpus. Concurrently, the subsequent stage involves a robust research endeavor focused on enhancing the practical applications of quality information analysis outcomes. The lack of thorough investigation into the avenues and methodologies for applying these outcomes necessitates a concentrated effort to amass knowledge and discern practical implementation strategies.

In essence, rectifying these challenges mandates a concerted endeavor. As we navigate this terrain, it becomes evident that refining the collection and application of weapon equipment quality information is paramount. By meticulously honing the information analysis process, we can surmount the existing barriers, harness the potential of advanced analysis techniques, and unlock the latent value within our data. Integrating consistent analytical approaches and the thoughtful exploration of application pathways will undoubtedly contribute to a more efficient, effective, and informed quality management system for our weapon equipment.

7 Conclusion

The trajectory of our military's equipment development has evolved significantly, transitioning from the initial phases of merely addressing existence and performance gaps to now embracing the paramount stage of "High-quality development." This progression signifies a shift in focus from mere problem-solving and performance catch-up to a concentrated emphasis on attaining and maintaining high standards of excellence. The primary challenge impeding our army's equipment combat capability has transformed. It is no longer rooted in the deficiency of combat technique performance; instead, the pressing issue lies in the imperative enhancement of general quality characteristics. The pursuit of equipment that is not only proficient but also pragmatic, resilient, and productive has become the aspiration of our military. Our armed forces fervently desire such attributes and are instrumental in significantly augmenting our equipment's combat effectiveness and indispensable operational support.

This paper meticulously distils the commendable practices employed by the US Army in general quality management, delineating insights from four pivotal angles. Drawing inspiration from these practices, it crafts four illuminating concepts that hold the potential to reshape the landscape of our army's equipment quality construction. By selectively incorporating the valuable successes of the US Army, we can glean pragmatic guidance for the reform and evolution of our military equipment's general quality characteristics management. This strategic referencing facilitates the circumvention of potential pitfalls and expedites the trajectory towards improved equipment quality. Through this mindful approach, our armed forces can navigate their journey with foresight, harnessing the swiftness of progress while steering clear of unnecessary detours.

References

1. Zhu, H.Y., Li, J.L., Sun, L.Q.: Overview of weapon equipment general quality characteristics management. Eng. Autom. **40**(2), 13–18 (2021)
2. Yazdi, M., Khan, F., Abbassi, R., Rusli, R.: Improved DEMATEL methodology for effective safety management decision- making. Saf. Sci. **127**, 104705 (2020)
3. Li, J.H.: Analysis on coping strategies of military equipment quality management in the new era. Mod. Manuf. Technol. Equip. **5**, 214–217 (2020)
4. Meng, Q.J., Guo, Q.S., Cao, Y.K.: Evaluation index system of equipment in service. Armored Force Eng. Inst. **32**(1), 24–28 (2018)
5. Liu, J., Shao, J.H., Yang, C.M.: Study on quality supervision of equipment with "Six characteristics." Aviat. Stand. Qual. **2**, 33–36 (2016)
6. Huang, C., Gu, C.C., Zhang, L.G.: Engineering capability evaluation method for general quality characteristics of weapon equipment development. Qual. Reliab. **4**, 26–30 (2018)
7. Xu, W., Zhai, F.Y., Niu, Y.: Study on quality control mechanism of American military weapons and equipment. Stand. Sci. **6**, 42–46 (2009)
8. Zhang, Z.J., Gu, H.Q., Yang, L.L., et al.: The practice and enlightenment of battlefield equipment management in American Army. J. Mil. Transp. Univ. **22**(12), 42–44 (2020)
9. Li, Y.Z., Yang, X.D., Guo, R.P.: Equipment quality management single risk assessment model. Ordnance Ind. Autom. **32**(11), 34–36 (2013)
10. Guo, Q.S., Yao, Z.J., Yan, Y.D., et al.: Research on test engineering of weapon systems. Armored Force Eng. Inst. **27**(2), 1–5 (2013)
11. Li, J., Zhang, L.Z., Shen, L.S.: Research on the connotation and mode of weapon system of systems testing. Coll. Armament **26**(5), 110–113 (2015)
12. Liu, C.B., Tang, X.J., Liao, J.P.: Connotation and framework analysis of equipment requirements assessment in joint operations. Mil. Oper. Syst. Eng. **36**(1), 51–54 (2022)
13. Yang, D.C., Ma, Y.Z., Song, K.: Research on demand demonstration of weapon and equipment system. Equip. Eng. China **6**, 250–251 (2021)
14. Wang, J.X., Liu, J., Shi, J., et al.: A summary of the research on equipment demand demonstration. Fire Control. Command. Control. **47**(5), 9–13 (2022)
15. Zuo, L.: Research and discussion of PBL strategy in our Army's equipment support. Inf. Secur. Commun. Secrecy **8**, 109–114 (2020)
16. Yang, J.Z.: Present situation and development trend of electronic information equipment support. Aerosp. Electron. Warf. **35**(3), 53–55 (2019)
17. Zhang, B.Z.: The past, present and future of product support for us military materials. Aviat. Maint. Eng. **8**, 16–19 (2015)
18. Gong, L.D., Zhang, B.Z.: The US military implements performance-based support (PBL)—problems in the process and improvement counter measures. Aviat. Maint. Eng. **9**, 36–38 (2015)
19. Li, Y.Z., Li, D.W.: Analysis of the development process of US military equipment test and evaluation and its enlightenment. Natl. Def. Sci. Technol. **42**(2), 47–49 (2021)

20. Shi, S., Cao, Y.H.: Development status and enlightenment of weapon system of systems test and appraisal in US Army. Mil. Oper. Syst. Eng. **29**(3), 46–49 (2015)
21. Ling, C., Wang, K.: Study on operation test scheme design of weapon system of equipment. Coll. Armament **27**(6), 112–116 (2016)
22. Zhang, Z.G., Shang, Q., Tian, X.Y., et al.: Development trend and enlightenment of integrated joint test quality management of US military space equipment. Qual. Reliab. **4**, 47–50 (2021)
23. Deng, C.J., Liu, H.J., Lv, Z.W.: Application of internet of things technology in equipment quality management information collection. Technol. Innov. Appl. **2**, 141–143 (2018)
24. Li, H., Yazdi, M. (eds.): Stochastic Game Theory Approach to Solve System Safety and Reliability Decision-Making Problem Under Uncertainty BT—Advanced Decision-Making Methods and Applications in System Safety and Reliability Problems: Approaches, Case Studies, Multi-criteria De, pp. 127–151. Springer International Publishing, Cham (2022). https://doi.org/10.1007/978-3-031-07430-1_8
25. Li, H., Yazdi, M. (eds.): Integration of the Bayesian Network Approach and Interval Type-2 Fuzzy Sets for Developing Sustainable Hydrogen Storage Technology in Large Metropolitan Areas BT—Advanced Decision-Making Methods and Applications in System Safety and Reliability Problem, pp. 69–85. Springer International Publishing, Cham (2022). https://doi.org/10.1007/978-3-031-07430-1_5
26. Li, H., Yazdi, M.: Advanced Decision-Making Methods and Applications in System Safety and Reliability Problems. Springer, Cham (2022)

Bi-Objective Portfolio Optimization with Mean-CVaR Model: An Ideal and Anti-Ideal Compromise Programming Approach

AmirMohammad Larni-Fooeik, Hossein Ghanbari, Mostafa Shabani, and Emran Mohammadi

Abstract Portfolio optimization is an important topic in the financial management literature. This problem consists of managing wealth across available assets to gain maximum profit with the least amount of risk. In this paper, a novel approach for mean conditional value at risk (Mean-CVaR) as a bi-objective portfolio optimization problem is proposed. In order to address mean-CVaR, we utilized an ideal and anti-ideal compromise programming approach. Compared to compromise programming and goal programming, this approach is more efficient for solving the portfolio optimization problem, as it aims to increase the distance between solutions and anti-ideal criteria. This research utilizes a real-life case study from the Tehran Stock Exchange (TSE) to demonstrate the effectiveness of the proposed approach. The computational results determined the effectiveness of the solution methodology and it could be applied quite reliably in other engineering contexts without a significant degradation in performance.

Keywords Portfolio optimization · Mean-CVaR · Bi-objective · Ideal and anti-ideal compromise programming approach

1 Introduction

Portfolio optimization is the selection of an optimal combination of assets to minimize risk for a given level of expected return or maximize return for a given level of expected risk which was first introduced by Markowitz's in 1952 [1]. Portfolio optimization aims to develop a diversified portfolio that offers the best risk-return tradeoff. Despite its great significance, the Markowitz mean–variance model is not suitable for addressing data in a real-life situation because it requires quadratic programming

A. Larni-Fooeik · H. Ghanbari · M. Shabani · E. Mohammadi (✉)
Faculty of Industrial Engineering, Iran University of Science and Technology, Tehran, Iran
e-mail: e_mohammadi@iust.ac.ir

M. Yazdi (ed.), *Progressive Decision-Making Tools and Applications in Project and Operation Management*, Studies in Systems, Decision and Control 518,
https://doi.org/10.1007/978-3-031-51719-8_5

which is a complex process to calculate [2]. As a result of research carried out in the 1990s, downside risk measures have been developed to calculate and evaluate risk measures more accurately. A measure based on percentiles that is generally considered to be an appropriate measure for assessing risk is Value at Risk (VaR) [3]. Despite its positive mathematical features, VaR has some bad mathematical characteristics, which attracted Rockafellar and Uryasev [4] to introduce a new risk measure which is called Conditional Value at Risk (CVaR). The risk can be captured more effectively with CVaR than with other traditional risk measures like standard deviation or Value at Risk (VaR). By estimating the expected loss when the portfolio's returns fall below a threshold, CVaR offers a more comprehensive understanding of the tail risk of a portfolio [4]. For these reasons, CVaR is used as a measure of risk in this paper. For enhancing the outcomes of optimization, A wide range of papers about portfolio optimization utilize the impact of decision-making techniques as well as considering several criteria to improve the efficiency of the problem [5, 6]. A variety of decision-making techniques have been employed to formulate portfolio optimization problems [7–10]. An approach to portfolio optimization is proposed that uses bi-objective programming, this type of optimization is based on multi-objective optimization, a field of mathematics that addresses problems involving multiple conflicting objectives [11, 12]. Ideally, the goal of bi-objective optimization would be to find a portfolio that offers an optimal balance between risk and return that is efficient or Pareto-optimal. By applying Ideal and Anti-Ideal Compromise Programming, we can establish a logical mathematical relationship between the two contradictions [13, 14]. Anti-ideal solutions represent the worst possible values for each objective individually, and the objective is to maximize the distance between all objectives and the anti-ideal values [15, 16]. As a first contribution in this paper, we develop a two-criteria portfolio optimization model. These two criteria conflict with each other. The second contribution to this paper relates to a problem-solving approach; As compared to other methods of solving portfolio optimization problems, this solution method works better. By applying this approach, the solution and anti-ideal criteria are distanced from one another while the solution and ideal criteria are separated from one another. In portfolio optimization problems, risk is the anti-ideal factor that decision-makers try to distance themselves from, and return is the ideal factor that decision-makers try to distance themselves from. The remainder of this paper is structured as follows. Section 2 presents the mean-CVaR model as a bi-objective portfolio optimization problem. In the third section, a compromise model is proposed using ideal and anti-ideal compromise programming, and in the fourth section, results from historical data are examined. Finally, the conclusions and suggestions for possible future research are presented in the final section of the paper.

2 Bi-Objective Portfolio Optimization Model

In this section, mean-CVaR is defined as a bi-objective model, and the theoretical constraints of our portfolio optimization model are presented. Next, the final model is described in detail. The mean-CVaR evaluates the entire distribution of returns below the specified threshold, rather than just focusing on a particular percentile. Furthermore, extreme or catastrophic events that may have a low probability of occurrence but a high impact can be assessed and managed more effectively in the mean-CVaR model [4]. The mean-CVaR model based on Rockafellar and Uryasev [4] is as follow:

$$Min\,CVaR \tag{1}$$

Subjected to

$$\sum_{i=1}^{n} \mu_i x_i = \mu_p \tag{2}$$

$$\sum_{i=1}^{n} x_i = 1 \tag{3}$$

$$x_i \geq 0, i = 1, 2, \ldots, n \tag{4}$$

The bi-objective mean-CVaR portfolio optimization problem can be expressed as below.

$$Max \sum_{i=1}^{n} \mu_i x_i \tag{5}$$

$$Min\,CVaR \tag{6}$$

Subjected to

$$\sum_{i=1}^{n} x_i = 1 \tag{7}$$

$$x_i \geq 0, i = 1, 2, \ldots, n \tag{8}$$

The above formulation can also be represented as:

$$Max \sum_{i=1}^{n} \mu_i x_i \tag{9}$$

$$MinCVaR = \lambda + \frac{1}{(1-a)T}\sum_{t=1}^{T}(y_t) \tag{10}$$

Subjected to

$$y_t \geq \mu_p - \sum_{i=1}^{n} r_{it}x_i - \lambda \tag{11}$$

$$y_t \geq 0 \tag{12}$$

$$\sum_{i=1}^{n} x_i = 1 \tag{13}$$

$$x_i \geq 0, i = 1, 2, \ldots, n \tag{14}$$

2.1 Constraints

In real-life portfolio optimization, additional constraints are usually incorporated into the model to improve its realistic aspects [17]. In order to evaluate the feasibility of the proposed portfolio optimization model, this study examines the following additional practical constraints.

2.1.1 Cardinality Constraints

The cardinality constraint restricts the number of assets held in a portfolio. In this constraint, the binary variable Z_i indicates whether an asset has been selected or not; therefore, the cardinality constraint can be determined as follows:

$$\sum_{i=1}^{N} Z_i = K \tag{15}$$

$$Z_i \in \{0, 1\}, i = 1, 2, \ldots, n \tag{16}$$

2.1.2 Quantity Constraints

A quantity constraint specifies the minimum and maximum amount that can be invested in each asset within a portfolio. Quantity constraints can be determined as below:

$$l_i Z_i \leq x_i \leq u_i Z_i, i = 1, 2, ..., n \tag{17}$$

$$0 \leq l_i \leq u_i \leq 1 \tag{18}$$

2.2 *The Proposed Bi-Objective Portfolio Optimization Model*

The proposed bi-objective portfolio optimization model with cardinality and quantity constraints is formulated as follows:

Parameters:

n : The number of available assets.
μ_i : Expected return of asset i
K : Desired number of assets to be hold in the portfolio.
l_i : Minimum fraction of the budget invested in the asset i
u_i : Maximum fraction of the budget invested in the asset i

Variables:

x_i : the fraction of the budget invested in the asset i $(i = 1, 2, \ldots, n)$

$$Z_i = \begin{cases} 1\, If\, asset\, i\, is\, chosen\, to\, be\, held \\ 0\, If\, asset\, i\, is\, not\, chosen\, to\, be\, held \end{cases}$$

$$Max \sum_{i=1}^{n} \mu_i x_i \tag{19}$$

$$Min\, CVaR = \lambda + \frac{1}{(1-a)T} \sum_{t=1}^{T} (y_t) \tag{20}$$

Subjected to

$$y_t \geq \mu_p - \sum_{i=1}^{n} r_{it} x_i - \lambda \tag{21}$$

$$y_t \geq 0 \tag{22}$$

$$\sum_{i=1}^{n} x_i = 1 \tag{23}$$

$$\sum_{i=1}^{N} Z_i = K \tag{24}$$

$$l_i Z_i \le x_i \le u_i Z_i \tag{25}$$

$$0 \le l_i \le u_i \le 1 \tag{26}$$

$$Z_i \in \{0, 1\} \tag{27}$$

$$x_i \ge 0, i = 1, 2, \ldots, n \tag{28}$$

3 Programming an Ideal and an Anti-Ideal Compromise Solution for Selecting Portfolios Based on Mean CVaR

This section presents a method for optimizing the mean-CVaR model using L-P metrics. Our approach to solving the model involves a combination of ideal and anti-ideal compromise programming techniques [13, 18]. There are two competing objective functions in the model: Maximizing return and minimizing risk. This multi-objective problem-solving approach called utopia was introduced by Triantaphyllou [19]. According to this approach, the model's solution should be as close as possible to the utopia point based on L-P metrics. For each objective function, $f_{i(x)}$ represents the solution to the problem [20–23].

$$min\, or\, max f_{i(x)} \tag{29}$$

Subjected to

$$g_{i(x)} \le 0 \tag{30}$$

$$x \in E^n \tag{31}$$

The optimal solution for a bi-objective problem is $\{f(x^{*1}), f(x^{*2})\}$, but this situation does not usually occur, so the decision maker tries to minimize the distance between the ideal solution and the solution of the model as follow:

$$L - P = \left\{ \sum_{j=1}^{k=2} \varphi_j \left| f_j\left(x^{*j}\right) - f_j(x) \right|^p \right\}^{\frac{1}{p}} \tag{32}$$

In the mentioned equation the φ_j is the weight of each objective function. In order to find ideal and anti-ideal we should solve two single objective problems at first. In our problem these equations can be determined as follow:

$$Max \sum_{i=1}^{n} \mu_i x_i = f_1(x) \tag{33}$$

Subjected to

$$\sum_{i=1}^{n} x_i = 1 \tag{34}$$

$$\sum_{i=1}^{N} Z_i = K \tag{35}$$

$$l_i Z_i \leq x_i \leq u_i Z_i \tag{36}$$

$$0 \leq l_i \leq u_i \leq 1 \tag{37}$$

$$Z_i \in \{0, 1\} \tag{38}$$

$$x_i \geq 0, i = 1, 2, \ldots, n \tag{39}$$

and

$$Min\, \lambda + \frac{1}{(1-a)T} \sum_{t=1}^{T} (y_t) = f_2(x) \tag{40}$$

Subjected to

$$y_t \geq \mu_p - \sum_{i=1}^{n} r_{it} x_i - \lambda \tag{41}$$

$$y_t \geq 0 \tag{42}$$

$$\sum_{i=1}^{n} x_i = 1 \tag{43}$$

$$\sum_{i=1}^{N} Z_i = K \tag{44}$$

$$l_i Z_i \le x_i \le u_i Z_i \tag{45}$$

$$0 \le l_i \le u_i \le 1 \tag{46}$$

$$Z_i \in \{0, 1\} \tag{47}$$

$$x_i \ge 0, i = 1, 2, \ldots, n \tag{48}$$

Then we should solve the following model:

$$Min\varphi_1\left[f_1\left(x^*\right) - f_1(x)\right]^p + \varphi_2\left[f_2(x) - f_2\left(x^*\right)\right]^p \tag{49}$$

Subjected to

$$y_t \ge \mu_p - \sum_{i=1}^{n} r_{it} x_i - \lambda \tag{50}$$

$$y_t \ge 0 \tag{51}$$

$$\sum_{i=1}^{n} x_i = 1 \tag{52}$$

$$\sum_{i=1}^{N} Z_i = K \tag{53}$$

$$l_i Z_i \le x_i \le u_i Z_i \tag{54}$$

$$0 \le l_i \le u_i \le 1 \tag{55}$$

$$Z_i \in \{0, 1\} \tag{56}$$

$$x_i \ge 0, i = 1, 2, \ldots, n \tag{57}$$

Above formulation is based on the ideal and anti-ideal compromise programming that consider different objective function.

4 Computational Results

In this section, the ideal compromise result is illustrated along with the anti-ideal compromise result. To develop the technique, we use data from the Tehran Stock Exchange (TSE). A total of 50 periods and 14 stocks are covered by the data monthly. The descriptive statistics of the selected assets are presented in Table 1.

Two models are derived from this paper's methods, and the following results are presented:

Based on these numerical results, a 95% confidence level ($\alpha = 0.95$) was applied to the CVaR model. In this case study, we also set the lower bound $l_i = 0.1$ and the upper bound $u_i = 0.6$ for each asset, and the cardinality of the portfolio is set to $k = 3$. The list of the results that are obtained after solving the CVaR model with $p = 1$ is shown in Table 2.

Table 1 Descriptive statistics of the selected assets

	Mean	SD	Variance	Min	Max
PSHZ	0.2337	−0.2266	0.0132	0.0126	0.1140
PARS	0.2406	−0.2577	0.0202	0.0179	0.1421
PARS	0.2308	−0.4199	0.0216	0.0201	0.1471
TPEZ	0.2243	−0.4411	0.0316	−0.0203	0.1777
PFSP	−0.0006	−0.2222	0.0039	−0.0861	0.0622
PFAN	0.2272	−0.2634	0.0176	0.0156	0.1327
NORI	0.2975	−0.2636	0.0218	0.0385	0.1477
GARN	0.2863	−0.3045	0.016	−0.0147	0.1266
SSIN	0.2325	−0.3593	0.0201	−0.0302	0.1410
PTAP	0.1951	−0.2431	0.0156	−0.0328	0.1251
PETR	0.3731	−0.262	0.0236	−0.0361	0.1537
PKLJ	0.2194	−0.2854	0.0205	−0.0084	0.1431
SHFS	0.6212	−0.4169	0.0649	−0.0208	0.2549
GTSH	0.1665	−0.2851	0.0175	−0.0646	0.1322

Table 2 Summary of the optimal solution

Utility function	x_2	x_4	x_{14}
0.00188935	0.6	0.3	0.1

5 Conclusions

The portfolio optimization purpose is to ensure investors manage risk and maximize returns, thereby aligning their portfolios with their financial goals. This paper proposed a novel optimization method to solve the mean-CVaR portfolio optimization model. This new approach uses ideal and anti-ideal compromising programming to solve the model, which decreases the distance between ideal and anti-ideal criteria and increases the distance between anti-ideal criteria and the solution. In this paper, a real-life case study from the Tehran Stock Exchange (TSE) is utilized to demonstrate the effectiveness of the proposed approach. The computational results determined the effectiveness of the solution methodology and it could be applied quite reliably in other engineering contexts without a significant degradation in performance.

In future research, the approach described above can be used for other risk measures such as semi-variance (SV) and conditional drawdown at risk (CDaR). Furthermore, it might be interesting to consider the uncertainty of parameters in modeling.

6 Declaration of Conflicting Interests

The authors declared no potential conflicts of interest concerning the research, authorship, and/or publication of this article.

References

1. Markowitz, H.: Portfolio selection. J. Finance **7**(1), 77–91 (1952). https://doi.org/10.1111/j.1540-6261.1952.tb01525.x
2. Markowitz, H.M., Perold, A.F.: Portfolio analysis with factors and scenarios. J. Finance **36**(4), 871–877 (1981). https://doi.org/10.1111/j.1540-6261.1981.tb04889.x
3. Jorion, P.: Value at risk: a new benchmark for measuring derivatives risk. Irvin Professional Publisher (1996)
4. Tyrrell Rockafellar, R., Uryasev, S.: Optimization of conditional value-at-risk (2001). http://www.gloriamundi.org/
5. García, M.J.R.: Financial education and behavioral finance: new insights into the role of information in financial decisions. J. Econ. Surv. **27**(2), 297–315 (2013). https://doi.org/10.1111/j.1467-6419.2011.00705.x
6. Kumar, P.C.: Applications and implementation conflict resolution in investment decisions: implementation of goal programming methodology for dual-purpose funds* (1979)
7. Wang, Z., Werner, J.: Portfolio characterization of risk aversion (1994)
8. Lwin, K.T., Qu, R., MacCarthy, B.L.: Mean-VaR portfolio optimization: a nonparametric approach. Eur. J. Oper. Res. **260**(2), 751–766 (2017). https://doi.org/10.1016/j.ejor.2017.01.005
9. Ferreira, F.G.D.C., Cardoso, R.T.N.: Mean-CVaR portfolio optimization approaches with variable cardinality constraint and rebalancing process. Arch. Comput. Methods Eng. **28**(5), 3703–3720 (2021). https://doi.org/10.1007/s11831-020-09522-1

10. Papahristodoulou, C., Dotzauer, E.: Optimal portfolios using linear programming models. J. Oper. Res. Soc. **55**(11), 1169–1177 (2004). https://doi.org/10.1057/palgrave.jors.2601765
11. Alexander, G.J., Resnick, B.G.: Using linear and goal programming to immunize bond portfolios* (1985)
12. Ghahtarani, A., Najafi, A.A.: Robust goal programming for multi-objective portfolio selection problem. Econ. Model. **33**, 588–592 (2013). https://doi.org/10.1016/j.econmod.2013.05.006
13. Ballestero, E., Romero, C.: Portfolio selection: a compromise programming solution (1996). www.jstor.org
14. Amiri, M., Ekhtiari, M., Yazdani, M.: Nadir compromise programming: a model for optimization of multi-objective portfolio problem. Expert Syst. Appl. **38**(6), 7222–7226 (2011). https://doi.org/10.1016/j.eswa.2010.12.061
15. Bilbao-Terol, A., Pérez-Gladish, B., Arenas-Parra, M., Rodríguez-Uría, M.V.: Fuzzy compromise programming for portfolio selection. Appl. Math. Comput. **173**(1), 251–264 (2006). https://doi.org/10.1016/j.amc.2005.04.003
16. Soyster, A.L.: Technical note—convex programming with set-inclusive constraints and applications to inexact linear programming. Oper. Res. **21**(5), 1154–1157 (1973). https://doi.org/10.1287/opre.21.5.1154
17. Jin, Y., Qu, R., Atkin, J.: Constrained portfolio optimisation: the state-of-the-art Markowitz models. In: ICORES 2016—Proceedings of the 5th International Conference on Operations Research and Enterprise Systems, pp. 388–395 (2016). https://doi.org/10.5220/0005758303880395
18. Mansini, R., Speranza, M.G.: LP solvable models for portfolio optimization: a classification and computational comparison (2003). http://imaman.oxfordjournals.org/
19. Triantaphyllou E.: 6 3) MCDM methods: a comparative study, by (2000)
20. Jiménez, M., Arenas, M., Bilbao, A., Rodríguez, M.V.: Linear programming with fuzzy parameters: an interactive method resolution. Eur. J. Oper. Res. **177**(3), 1599–1609 (2007). https://doi.org/10.1016/j.ejor.2005.10.002
21. Zhang, Z.: Intelligent optimization of the financial sharing path based on accounting big data. Math. Probl. Eng. (2022). https://doi.org/10.1155/2022/1310994
22. Golilarz, N.A., Gao, H., Pirasteh, S., Yazdi, M., Zhou, J., Fu, Y.: Satellite multispectral and hyperspectral image de-noising with enhanced adaptive generalized gaussian distribution threshold in the wavelet domain. Remote Sens. **13** (2021). https://doi.org/10.3390/rs13010101
23. Golilarz, N.A., Mirmozaffari, M., Gashteroodkhani, T.A., Ali, L., Dolatsara, H.A., Boskabadi, A., et al.: Optimized wavelet-based satellite image de-noising with multi-population differential evolution-assisted harris hawks optimization algorithm. IEEE Access **8**, 133076–133085 (2020). https://doi.org/10.1109/ACCESS.2020.3010127

Developing Alternative Multilinear Regression-Based Intelligence Hybrid Model

Kehinde Adewale Adesina and Mohammad Yazdi

Abstract The study focuses on improving quantitative structure–property relationships (QSPR) simulation for mycotoxins by employing multi-oriented data-intelligent approaches. While previous methods only considered the molecular structure, this study aims to overcome this limitation using four data-driven model approaches based on mycotoxins' retention time (tR). The goal is to leverage the correlation between various molecular properties and develop a hybrid intelligence-based model. The three hybrid intelligence models examined in this study are MLR-SVM, MLR-ANN, and MLR-ANFIS. Four molecular properties, namely retention index, peak symmetry, mono isotopic mass, and relative sensitivity factor, are selected as input variables for modelling. The relationships between the input and output parameters are measured using Spearman Pearson correlation. Several metrics are employed to evaluate the fitness, performance, and adequacy of the models. These include the coefficient of determination, root mean square error, mean square error and correlation coefficient. The results indicate that the ANFIS model outperformed the other three single models (MLR, SVM, and ANN). Furthermore, the hybrid models outperformed the single models, with the hybrid model MLR-ANFIS being the adequate hybrid model. Based on these findings, the hybrid MLR-ANFIS intelligence model could be an alternative high-performance model for simulating QSPR in mycotoxins. This highlights the potential of using hybrid intelligence approaches that combine multiple techniques to enhance the accuracy and effectiveness of QSPR simulations.

K. A. Adesina
Independent Researcher, 32 Radburn Close, Harlow, Essex, UK

M. Yazdi (✉)
School of Computing, Engineering & Physical Sciences, University of the West of Scotland (UWS), London, UK
e-mail: mohammad.yazdi@mq.edu.au

Faculty of Science and Engineering, Macquarie University, Sydney, Australia

M. Yazdi (ed.), *Progressive Decision-Making Tools and Applications in Project and Operation Management*, Studies in Systems, Decision and Control 518,
https://doi.org/10.1007/978-3-031-51719-8_6

Keywords Quantitative structure–property relationship (QSPR) · Operations management · Data-driven single based models · Hybrid MLR-based intelligence models

1 Introduction

Generally, fungi are considered as one of the major disease causative agents in plants and animals [1]. However, they are not a vital agent that causes diseases in vertebrates due to the fact that the number of fungi that are of medical significance is relatively small compared to other organisms such as virus and bacteria. The growth of fungi on animal hosts lead to the production of diseases and disorders. These are collectively called mycoses; while dermal, respiratory, and dietary and others exposure to harmful fungal metabolites and precursors lead to the production of diseases collectively known as mycotoxicoses [2, 3]. These mycotoxicoses are generally analogous to pathologies that are caused through natural poisoning produced through exposure to heavy metals scums and pesticides [4]. The intensity of mycotoxin poisoning may be attributed to factors such as deficiency of a certain vitamin, health status, abuse of chemical such as alcohol, drugs as well as caloric deprivation. Consecutively, mycotoxicoses may lead to microbial diseases, increase the synergistic effects of other harmful and toxic substances and increase the negative impact of malnutrition [5].

There are many other microbes that do result in the deterioration of foods and feeds. Studies such as that of Adesina [6] had revealed various microbes in honey produced and store at different locations. However, mycotoxins are known with the tendency to form secondary metabolites of various fungi that can easily colonize food produce in the farm or in stores after harvesting [7]. Therefore, it results in potential threat to animals and human health via ingestion of the food products prepared from such commodities. Some of these contaminations occur at both pre- and post-harvest [6]. Mycotoxins have the ability of thriving in both tropical and temperate regions due to the variants of the fungi species involved [8]. The main food crops affected are beans, dried peas, oil seeds, spices, cocoa, coffee, dried fruit, nuts and cereals; some have been reported in wine and beer [9]. They have the ability of entering human food chain through meat as well as other animal products like cheese, milk and eggs due to the fact that the livestock have ingested a food infested by the mycotoxins [10].

Aflatoxins are considered as widely related and closed group of mycotoxins, which are produced from fungi *A. parasiticus* and *A. flavus*. This group of mycotoxins were first understood following the death of almost 100,000 Turks in England on a poultry farm, which are related to the ingestion of various Brazilian peanut meals. Aflatoxins are said to be regarded as difuranocoumarin derivatives, which are synthesized via polyketide mechanism through various strains of *A. parasiticus* and *A. flavus* [11]. Based on mycological points of view, there are quantitative and qualitative differences in the toxigenic capabilities shown by various strains found in each aflatoxin specie

[12]. For instance, about 50% of *A. flavus* are involved in the production of aflatoxins. The four major classes of aflatoxins are G1, G2, B1, and B2, which are generally classified based on their UV-properties as well as chromatographic behaviour [13].

Ochratoxins was discovered in the year 1965, which is considered as a fungal metabolite, which showed harmful behaviour to animals [12], and containing of species that thrives widely in different environmental and biological conditions [14]. Ochratoxins is generally found in cereals, other starch sources, dried fruits, spices and coffee. Mostly the concentration of Ochratoxins is about some part per billion (ppb), the biological properties of this mycotoxin are well reported [15]. Some of its carcinogenic, mutagenic, fertility inhibition, teratogenic and immune-suppressive effects have widely reported [16]. Recent studies have described that various mycotoxin composed of some carcinogenic characteristics. Most of them are DNA-reactive while some are not; however, when the endpoint is carcinogenic, in-vivo or in-vitro studies might be a constructive way of determining possible gene expression, molecular modified tumor suppressor and/or proto-oncogenes genes and genomic-based instability. These estimators would provide indicators for cancer process and for portraying various hazards [17].

The Mycotoxins that exist in food materials are usually secondary metabolites of various filamentous fungi that have the ability of contaminating the food via food chain. Even though, different classes of fungal toxins were known limited numbers of them are considered vital in the area of food safety (see [6] for brief descriptions of some of these mycotoxins). Therefore, based on these, a wide range of analytical tools have been used to designed methods for their elucidation and determination [6]. The diversity of the chemical nature of some micro-fungi can be found in the sample of extract either in the natural or derivative form [18]. Most of the methods adopted by these previous techniques require the use of water-acetonitrile gradient elution system using reversed-phase carbon-18 and carbon-8 as stationary contactors in chromatographic experiments [19, 20]. Accordingly, due to the high chemical diversity nature of these micro-fungi, the native and/or derivatives extract would comprise of kinds of chemical structure ranging from acids, alcohols, ketones, alkaloids, antraquinones and cylic-peptides [1].

Generally, different methods have been used to determine, evaluate and quantify mycotoxins. Analytical methods such as spectroscopy (nuclear magnetic resonance, spectroscopy and UV-spectroscopy), spectrometry method (mass spectrometry) and chromatographic method (HPLC, GC–MS and GC-FID) are currently being used in the elucidation of mycotoxins [21]. The high performance liquid chromatography (HPLC) or sometimes-called high-pressure liquid chromatography (HPLC) which is one the classes of liquid chromatographic methods involve the use of a liquid mobile phase [22–26]. Mass spectrometry is considered as one of the most informative analytical instruments, owing to its wide range of application. Its application composed of determination of isotopic elements, elucidation of quantitative and qualitative composition of complex mixtures, various elemental determination of samples and structural determinations of biological, organic and inorganic samples [27–29].

Many studies have been carried out to elucidate these structures—chemical diversities in mycotoxins. Khosronkhavar et al. [1] reported these techniques were based on reversed-phase liquid chromatography employing electrospray ionization (ESI), diode array detection (DAD), atmospheric pressure chemical ionization (APCI) mass spectrometry (MS) amongst others. Nielsen and Smedsgaard, [30] adopted the aforementioned techniques for fungal metabolite screening of 474 mycotoxins and fungal metabolites using standardized liquid chromatography-mass spectrometry-ultra violet (UV) spectroscopy combined with the high resolution orthogonal time-of-flight accurate mass spectrometry (oaTOF-MS) basically for evaluating the retention time. The conclusion put the methods as being fast and powerful for the dereplicating of the secondary metabolites and for highlighting metabolites groups that can be appraised by LC-DAD and/or ESI–MS techniques. Earlier, Eckers et al. [31] and Hopfgartner and Vilbois [32] had reported that combining these techniques enhances the mass precision and permits regular formula evaluation.

However, the major drawbacks of these approaches are that only the molecular formula is either utilized or provided and little efforts have been made in this direction to assess the relationship between the molecular structure and its parameters with the dynamics of the retention time. Furthermore, shortages of the approach also include the need for gaseous sample for the MS and the suitability of the choice of chromatography techniques adopted. The challenge is in the selection of either LC or GC that can adequately vapourize the excess solvent [33, 34]. The major application of LC–MS method is that it gives the molecular fingerprint of a certain elucidate without much relying on the retention time, which is contrary in the case of the classical HPLC instrument. This combination equally gives information on the molecular mass, exert quantitative result as well as structural information [34].

To properly study the properties of various mycotoxins, their chemical behaviour and structure needs to be understood. The high costs associated with instrumentation, reagents, conducting laboratory experiment amongst others has led to limited studies on the determination of mycotoxins using these chromatographic methods. More often, computational approaches involved in the prediction of the retention behaviour can be categorized into two classes [35, 36]. There is a mathematical equation approach mainly for correlating retention time with the molecular parameters. The second is the use of an experientially based quantitative structure–property relationship (QSPR) approach where multiple linear regression (MLR) and support vector machine (SVM) models are employed. According to the study of Adesina, 2021, although various mathematical models could be employed for simulating mycotoxins of which the computational method of this study offers some advantages over some trivial mathematical equations.

Interestingly, most reported chromatographic techniques in the field have primarily focused on data-driven machine learning techniques such as neural networks and other forms of artificial intelligence (AI). However, the utilization of hybrid intelligence-based approaches still needs to be improved. Therefore, this study aims to employ AI-based models to investigate mycotoxins' quantitative structure–property relationship (QSPR) based on various molecular properties elucidated through chromatography experiments, explicitly focusing on retention behaviour. Despite the growing adoption of AI-based models, inherent drawbacks and weaknesses still need to be addressed. For instance, artificial neural networks (ANN) often suffer from lower accuracy due to overfitting. Moreover, the non-linearity in the input data of molecular descriptors poses a challenge for single-model approaches. To overcome these limitations, a hybrid approach based on multiple linear regression (MLR) and intelligence models is proposed and examined in this study, which adds further credibility to the research. The objective of this work is to introduce a high-performance hybrid MLR-intelligence approach. By integrating the strengths of the classical linear MLR and AI-based models, this approach is expected to enhance the model's overall performance and effectively address the inadequacies present in each direction.

For this study, the hybrid techniques are of two steps; the first stage involves the training of the MLR model in order to have the best linear fitted values. Subsequently, since MLR is a single model that can capture only the linear behaviour of the dataset, therefore the residuals of MLR result will be taken, which contains the information regarding the non-linearity properties of the datasets [37]. In the second step, according to Nourani et al. [68], is merging of the MLR with the fitted values of the non-linear models. This study, we aim at harnessing the precision prowess of the intelligent hybrid model from these data driven models. Moreover, studies on this dimension employing hybrid intelligence-based linear and non-linear for modelling the qualitative properties of mycotoxins are scarce and have not been widely reported. We considered multiple linear regression (MLR), support vector machine (SVM), artificial neural network (ANN) and adaptive neuro-fuzzy inference systems (ANFIS). This is with view of determining an adequate new hybrid MLR-based intelligent model for the quantitative structure–property relationship (QSPR). The retention factor (time) as the dependent variable is predicted using retention index, mono isotopic mass, relative sensitivity factor and peak symmetry.

2 Basic Model Considerations

Artificial neural networks (ANNs) non-linear model consists of information processing unit known as neurons incorporated with advance tools which mimicked the human biological neurons [38]. The structure of ANN generally composed of three layers, whereby each layer perform different and special tasks; The layer is the one responsible for processing the information, the input layer distributes the data set into the network, while the output layer processes each of the input vector [39]. The neurons are considered as the smallest units which aid in processing the networks [15]. ANN have different models such as Feed forward neural network (FFNN), Back propagation neural network (BPNN) and multi-layer perceptron. In this study the multilayer perceptron (MLP) is adopted which is one of the commonest ANNs applied nowadays [40]. MLP is a vital technique that has the capacity of simulating the non-linear relationships between dependent and independent variables [40]. Just like the normal ANN, MLP also composed of input layer, a minimum of one hidden layer as well as an output layer.

3 Adaptive-Neuro Fuzzy Inference System

ANFIS is one of the artificial intelligences (AI)-based machine learning tools commonly regarded as the ANN-based Takagi–Sugeno Fuzzy Inference System (FIS). ANFIS integrates neural networks and Fuzzy Logic principles and can take advantage of both within a single framework. This is a non-linear data-driven models leverage the advantage of both fuzzy logic and neural networks [41]. Basically, there are three models of ANFIS namely Mamdani, Sugeno and Tsumoto; however, Sugeno has been found to be the most efficient. Defuzzifier and Fuzzier are main components of the fuzzy logic systems [42]. The fuzzy logic works through converting the input data into fuzzy (crisp) values by using the membership functions (MFs). The nodes here act as the membership function which converts the data into values from 0 to 1. In the model, the reasoning engine comprises of two states viz fuzzy logic and neural network. Generally we have four classes of MFs, namely; trapezoidal, Gaussian, Sigmoid and triangular [43]. If we assume that the FIS is composed of 2 inputs namely 'x' and 'y' having a single output "f". The Sugeno fuzzy first order will then have these rules.

$$\text{Rule 1: if } \mu(x)\, is\, A_1 \text{ and } \mu(y)\, B_1 \text{ then } \mathrm{f}_1 = p_1 x + q_1 y + r_1 \tag{1}$$

$$\text{Rule 2: if } \mu(x)\, is\, A_2 \text{ and } \mu(y)\, B_2 \text{ then } \mathrm{f}_2 = p_2 x + q_2 y + r_2 \tag{2}$$

A_1, B_1, A_2, B_2 parameters are membership functions for x and y inputs.

$p_1, q_1, r_{1,} p_2, q_2, r_2$ are outlet function parameters.

The structure and formulation of ANFIS follows a five-layer neural network arrangement. Refer to [44] for more information about ANFIS.

4 Support Vector Machine (SVM)

The first idea of SVM was in 1995 and it is built on the theory of statistical learning that has been widely employed for pattern recognition, classification and regression problems [45]. Basically, SVM has also been applied for minimization of structural risk [46]. SVM is generally classified into two classes viz-a-viz the linear and the non-linear SVM. Moreover, for the regression problems, SVM is also referred as SVR [47] Moreover, the learning machine determines the regression by employing a hyper-plane in a higher-dimensional space as the linear function. To address the challenges of expected error and over-fitting, the risk function is utilized to minimize these issues conscientiously. In Support Vector Regression (SVR), the weights are fitted into the data, and subsequently, the output transition to a non-linear kernel. This allows for the examination of non-linear behaviour within the dataset. SVR is a supervised learning algorithm primarily used similarly to Support Vector Machines

(SVM) for predicting discrete values. Equation 8 below demonstrates the general equation of SVM [48].

$$y = f(x) = w\phi(x_i) + b \tag{3}$$

where $\varphi(x_i)$ indicates feature spaces, nonlinearly mapped from input vector x. Here, SVR ensures the best fit represents the hyperplanes having the highest number of points. Sharghi et al. [49] had attributed the efficacy of SVM as its capability to reduce errors, complexity and to enhance performance generalization ability of the network. These abilities distinguish SVM from ANFIS and ANN.

5 Multi-Linear Regression (MLR)

In general, the regression models are used in predicting the correlation that exists between the dependent and the independent parameters. The mechanism of linear regression is based on the least square method [50]. The linear regression is classified into 2 main groups: the simple and multiple linear regressions. Simple linear regression (SLR) involved the modeling of a single dependent variable using a single independent variable [51]. Conversely, a multiple linear regression (MLR) model involves the use of two or more independent variables toward the prediction and modeling of a single dependent variable [52].

MLR is the commonest applied linear model used nowadays in various fields of study such as engineering, science and social science [53]. It is very important to note that MLR displays its correlation based on a straight line (linear) pattern that can be used in estimating the entire data points of both the dependent and independent variables [39]. MLR can be best described using Eqs. 2–4 below

$$y = b_0 + b_1x_1 + b_2x_2 + \ldots b_ix_i \tag{4}$$

where by x_1, is the value of the *i*th output, b_0 is the regression constant, and b_i is the coefficient of the *i*th output.

6 Hybrid-Based Intelligent Model Concepts

The performance accuracy of the simple models is mostly related to various conditions such as the time scale, input determination and configuration of the model [54]. These parameters were sum-up and tagged influencers of the chromatographic conditions [55]. Researchers have adopted various data-driven approaches for modelling the properties of mycotoxins using different inputs and outputs. Khosrokhavar et al. 2010 [1] employed MLR and SVM methods for the prediction of retention behaviour of different mycotoxins using data obtained through the liquid chromatography with

UV and MS detector. A strong agreement between the experimental and the simulated values was reported using the R^2 and Q^2. Similarly Stangierski et al. [5] study on the comparative performance of MLR and ANN models for the prediction of the overall quality of cheese reported the robustness of ANN over MLR through the higher R^2 and lower RMSE values. The major drawback of some of these single models is the inability to cope with linear and nonlinear data simultaneously [56]. A novel techniques hybrid model could be poised to capably cope with this shortage by measuring both the non-linear and linear co-relations that exists between the independent and dependent variables. As aforesaid, the non-linear models such as SVM, ANFIS and ANN) have shown propensity for both regression and classification [57].

Some data driven algorithms such as SVM, principal component regression, partial least squares (PLS) and PCA-SVM technique have been employed. According to Marrero-Ponce et al. 2018 [55], both linear and non-linear models have demonstrated success in classification and regression. Furthermore, the "no free lunch" theorem" has opined that there is no single existing model can be used decipher issues in all dataset [58]. As such the properties of data such as size, normality and linearity are militating against the performance of various learning algorithms [59]. Moreover, different works have illustrated that even when the same data set is used in modelling the performance of various properties, the performance of various models differs [60]. An investigation was launched by Ge et al. 2016 to detect the presence of aflatoxin B1 in 160 samples using terahertz (THZ) spectral method with different range of concentration. The results revealed the reliability of the non-linear SVM and PCA-SVM models and higher performance accuracy were observed. Another study to determine the mechanism of ocean water system, [61] applied hybrid models to boost the performance of the classical MLR model for the simulation of Chlorophyll using various physical and biological properties; the hybrid MLR-ANN outperformed the MLR. The results indicated that the proposed techniques have the ability of improving the performance by reducing the errors generated and by increasing the correlation amongst variables.

$$f(y_t) = q_t + r_t \tag{5}$$

where q_t represents the linear phase and r_t represents the non-linear phase. These two phases have to be estimated from the data. Let ϵ denote the residual at time t from the linear model. Then

$$\epsilon_t = y_t - \hat{q}_t \tag{6}$$

where $\hat{q}_t$ is the forecast value for time t.

From the estimated correlation and by modelling residuals using AI-based models, nonlinear relationships can be discovered; the overall AI model of the residuals for n input nodes will be:

$$\epsilon_t = f(\epsilon_{t-1}, \epsilon_{t-2}, \ldots, \epsilon_{t-n}) + \varepsilon_t \tag{7}$$

where f is a nonlinear function determined by the AI models (ANN, ANFIS and SVM) and ε_t is the random error.

7 Model Conceptualization

7.1 Single Model

First, the prediction of the mycotoxins is done using single models as SVM, ANFIS, ANN and MLR. This method is conducted using the MS Excel 2016 using the normalized data. The regression is done using single output variable retention time and input variables. The MLR model follows the least square method culminates in Eq. 8.

$$y = b_0 + b_1 x_1 + b_2 x_2 + \cdots b_i x_i \tag{8}$$

The ANN topology adopted consists of three layers viz; input layer, out layer and hidden layer. The numbers of neurons at the hidden layer are selected according to the complexity of the parameter settings as detailed in the study of Daneshvar and Adesina [62].

You are describing a study that used various artificial neural network (ANN) models, as well as other techniques like the Adaptive Neuro-fuzzy Inference System (ANFIS) and Support Vector Machine (SVM) for analysis. Here is a summary of the information you provided: Artificial Neural Networks (ANN): The study employed several ANN models named M1, M2, M3, M4, and M5. These models had different numbers of neurons, specifically 3, 4, 5, and 4 neurons, respectively. The Multilayer Perceptron (MLP) topology was used for simulation, and the Tan-sigmoid transfer function was applied. Adaptive Neuro-fuzzy Inference System (ANFIS): The ANFIS technique was incorporated into the study and involved two stages. In the first stage, the normalised data were imported into MATLAB and divided into training and testing sets. The sub-clustering Membership function (MP) was used throughout the analysis, with the number of input variables adjusted as needed. The grid-partitioning method optimised the training Fuzzy Inference System (FIS), with an error tolerance of 0.0005 and 60 epochs set for the analysis. Support Vector Machine (SVM): The study also utilised the SVM technique using a cubic model. The normalised data was imported into the working space and transferred to the regression learner. The data was trained and tested with various parameter settings to obtain the analysis results.

It's important to note that with specific details or context about the study or its objectives, it's easier to analyse the results or their implications comprehensively.

7.2 Hybrid MLR-Based Intelligence Model

The hybrid data-driven approach combines the strengths of both linear and non-linear models to achieve more accurate predictions. In this approach, the residuals obtained from multiple linear regression (MLR) models, representing the difference between the MLR model's predicted and actual experimental values, are incorporated into the best-performing non-linear single model. Integrating these residual values enhances the predictive power of the non-linear model. To carry out the studies, three MLR-based intelligence models are utilized as follows.

(a) Hybrid I: MLR-ANN-the MLR residuals are the added to the best performing ANN single model.
(b) Hybrid II: MLR-ANFIS-the residuals are further added to the best performing result of the ANFIS single model.
(c) Hybrid III: MLR-SVM-the residuals are further added to the best performing result of the SVM model.

These simulation techniques are conducted on MATLAB 9.3 (R2019a).

7.3 Performance Evaluation

The performances of the models are evaluated by comparing the experimental results with the predicted. Four performance indicators [63–65] as mean square error (MSE), root mean square error (RMSE), correlation co-efficient (R) and determination co-efficient (R^2) expressed respectively in Eq. 9–12 as

$$MSE = \frac{1}{N}\sum_{i=1}^{N}(Y_{obsi} - Y_{comi})^2 \tag{9}$$

$$RMSE = \sqrt{\frac{\sum_{i=1}^{N}(Y_{obsi} - Y_{comi})^2}{N}} \tag{10}$$

$$R = \frac{\sum_{i=1}^{N}(Y_{obs} - \overline{Y}_{obs})(Y_{com} - \overline{Y}_{com})}{\sqrt{\sum_{i=1}^{N}(Y_{obs} - \overline{Y}_{obs})^2 \sum_{i=1}^{N}(Y_{com} - \overline{Y}_{com})^2}} \tag{11}$$

$$R^2 = 1 - \frac{\sum_{j=1}^{N}\left[(Y)_{obs,j} - (Y)_{com,j}\right]^2}{\sum_{j=1}^{N}\left[(Y)_{obs,j} - (\bar{Y})_{obs,j}\right]^2} \tag{12}$$

8 Model Demonstration—Case Study

Phase 1: Data Acquisition and Correlation Analysis

To illustrate the propensity of the study, a set of 150 dataset of seven different groups of mycotoxins are adopted from the reported by Nielsen and Smedsgaard [30] (see Table 1). The mycotoxin QSPRs considered is retention index, peak symmetry, relative sensitivity factor, mono isotopic mass and retention time.

Phase 2: Correlation and Statistical Analysis

The Spearman correlation and descriptive statistical analysis are conducted using the Microsoft Excel 2016. The correlation analysis was done with 95% confidence limit. These checking are necessary to enhance the fitness of the data, reduce over-fitting and under-fitting according to [66].

Phase 3: Model Development Using Correlation Results

Based on the output of the correlation analysis, model QSRPs settings are developed and are set as different alternatives for the simulation. The data is divided into training and test data; with 100 (67%) as training data set data and 50 (33%) as test data set.

Phase 4: Data Normalization

The data set employed are normalized to minimize the redundancy of the data and to reduce the larger numerical errors using the method of Pfeifer et al. [67] expressed as:

$$y = \left(\left(\frac{x - x_{min}}{x_{max} - x_{min}}\right)\right) \tag{13}$$

Phase 5: Model Implementation

See Table 1.

9 Results and Discussion

9.1 *Statistical and Correlation Analysis*

The results of the pre-analysis of the dataset describes an adequate QSRP input variables for the modelling the chemical and physical behaviour of various mycotoxins (Table 2). As opined by Elkiran et al. [47], higher skewness has the ability of impacting extensively on the ANN model whereas lower skewness reduces the predictability of the model. It can be observed that the retention index possesses the highest skewness. Complementarily, the output of the Spearman Correlation

Table 1 The historical data used for simulation of mycotoxins in the study

Class	Metabolites	Retention index	Peak symmetry	Mono isotopic mass	Relative sensitivity factor	Retention time (tR) (Min)
	Aflatoxicol I	880	1.1	314.08	0.1	12.45
	Aflatoxin B1	859	1.1	312.06	0.2	11.50
	Aflatoxin B2	834	1.1	314.08	0.2	10.33
	Aflatoxin B2α	753	1.4	330.074	0	6.6
	Aflatoxin G1	830	1.1	328.06	0.2	10.16
	Aflatoxin G2	804	1.1	330.07	0.2	8.97
Aflatoxins and their Precursors	Aflatoxin G2α	718	1.4	346.07	0	5
	Aflatoxin M1	781	1.3	328.06	0.3	7.21
	Austocystin A	1140	1.3	372.04	0.2	21.57
	Averufin	1289	1.5	368.09	0.07	25.65
	5-Methoxysterigmatocystin	1028	0.9	354.07	0.2	18.02
	Dihydroxysterigmatocystin	1018	0.9	326.08	0.1	17.7
	Methoxysterigmatocystin	944	1.1	338.08	0.05	15.03
	Sterigmatocystin	1055	0.9	324.06	0.7	18.91
	Norsolorinic acid	1514	2.3	370.105	0.0002	31.08
	Parasiticol	842	1.3	302.08	0.1	10.73
	Nivalenol	638	0.9	312.12	0.002	1.27
	Fusarenone X	661	1.5	354.13	0.005	2.35
	Deoxynivalenol	644	1.2	296.13	0.003	1.54
	3-Acetyldeoxynivalenol	723	1.3	338.14	0.011	5.21

(continued)

Table 1 (continued)

Class	Metabolites	Retention index	Peak symmetry	Mono isotopic mass	Relative sensitivity factor	Retention time (tR) (Min)
	15-O-Acetyl-4-deoxynivalenol	721	1.4	338.14	0.02	5.1
	Scirpentriol	650	1.2	282.1467	0.03	1.82
	15-Acetoxyscirpenol	770	1.4	324.16	0.01	7.4
	Diacetoxyscirpenol	854	1.6	366.17	0.015	11.28
	3α-Acetyldiacetoxyscirpenol	958	1.5	408.18	0.025	15.56
	Neosolaniol	679	1.4	382.2	0.02	3.19
Trichothecenes	T-2 Triol	841	1.4	382.199	0.005	10.66
	HT-2 Toxin	908	1.1	424.2097	0.01	13.69
	T-2 Toxin	999	1.4	466.2203	0.001	17.06
	Iso-T-2 toxin	1015	1.8	466.2203	0.02	17.61
	Acetyl-T- 2toxin	1125	1.6	508.23	0.04	21.12
	Trichodermin	974	1.3	292.17	0.04	16.13
	Trichodermol	820	1	250.16	0.005	9.69
	7-α-Hydroxytrichodermol	666	1.4	266.16	0.03	2.59
	Verrucarol	673	1.5	266.16	0.003	2.89
	4,15 Diacetylverrucarol	920	1.2	350.17	0.015	14.15
	Trichothecin	978	1.5	332.1624	0.001	16.29
	Trichothecolone	689	1.6	264.1362	0.005	3.63
	Isosatratoxin F	960	1.4	542.23	0.2	15.63
	Roridin A	978	1.4	532.27	0.003	16.29
	RoridinE	1107	1.7	514.27	0.3	20.6

(continued)

Table 1 (continued)

Class	Metabolites	Retention index	Peak symmetry	Mono isotopic mass	Relative sensitivity factor	Retention time (tR) (Min)
	Roridin H	1238	1.8	512.24	0.07	24.33
	Roridin L-2	900	1.4	530.25	0.05	13.37
Macrocyclic Trichothecens and Precursors	Satratoxin G	901	1.5	544.2	0.08	13.43
	Satratoxin H	920	1.4	528.24	0.02	14.14
	Trichoverrin A	922	1.3	532.27	0.2	14.2
	Trichoverin B	919	1.3	532.26	0.1	14.1
	Trichoverrol A	830	1.4	420.22	0.1	10.16
	Trichoverrol B	823	1.4	420.22	0.2	9.85
	Verrucarin A	981	1.4	502.22	0.005	16.4

Table 2 The statistical analysis of the experimental variables

Variables	Retention index	Peak symmetry[a]	Mono isotopic Mass	Relative sensitivity factor [b]	Retention time (tR) (min)
Mean	1066.8	2.7	432.2	0.3	16.3
Median	981.0	1.4	416.7	0.1	16.5
Mode	1055.0	1.4	481.3	0.2	6.6
Standard Deviation	769.7	3.1	189.8	0.6	7.6
Kurtosis	124.3	6.3	29.2	21.9	0.1
Skewness	10.7	2.6	4.1	4.4	0.2
Range	9371.0	17.0	1963.1	4.0	37.4
Minimum	638.0	0.0	0.0	0.0	1.3
Maximum	10,009.0	17.0	1963.1	4.0	38.6

conducted to ascertain the extent of the quantity of the relationships and the direction reveals that retention time correlated positively retention index, peak symmetry, relative sensitivity factor and negatively with mono isotopic mass. The values of the correlation are +0.7810, +0.6105. +0.1336 and −0.1454 respectively (see Table 3).

Therefore, based on the correlation analysis, five different simulation alternatives are arranged as M1–M5 for the studies (see Table 4).

Table 3 The correlation analysis matrix

Variables	Retention index	Peak symmetry	Mono isotopic Mass	Relative sensitivity factor	Retention time (tR) (min.)
Retention index	1.0000				
Peak symmetry	−0.0947	1.0000			
Mono isotopic Mass	0.4669	−0.1329	1.0000		
Relative sensitivity factor	0.1029	0.3690	0.0715	1.0000	
Retention time (tR)	0.7810	0.6105	−0.1454	0.1336	1.0000

Table 4 Input variable parameter settings

	Model input combination
M1	Mono isotopic mass + relative sensitivity factor
M2	Mono isotopic mass + peak symmetry + relative sensitivity factor
M3	Retention index + peak symmetry + relative sensitivity factor + Mono isotopic mass
M4	Retention index + peak symmetry + relative sensitivity factor
M5	Retention index + peak symmetry + Mono isotopic mass

9.2 Single Models—SVM, ANN, ANFIS and MLR

The performances the single models are presented in Table 5. Nourani et al. [68] had reported determination co-efficient (R^2) of 80% as the acceptable minimum performance required for an adequate data driven model. In line with the aforesaid, the following alternatives: MLR-M3 to MLR-M5 (R^2 range 81–87%), ANN-M2 to ANN-M5 (R^2 range 85–98%), SVM-M3 to SVM-M5 (R^2 range 92–95%), and ANFIS-M2 to ANFIS-M5 (R^2 range 82–99%) are considered acceptable and adequate single models.

Furthermore, the results show that M1 of all the single models could not attain the minimum acceptable determination co-efficient (R^2). This observation is not unconnected with the quantity and direction of the correlations that exist between the retention time and Mono isotopic mass and relative sensitivity factor. Similarly, observations across all the models reveal that M4 of all the models performed better than M, M2, M3 and M5 of all the models. The reason for these trends could be alluded to the correlation as discussed. Intuitively, the settings of the simulation alternatives could also be attributed to the trends witnessed. At this time, further analyses focused on the M4's shows the superiority of ANFIS-M4 over other models—SVM-M4, ANN-M4 and MLR-M4 in all the performance measurement employed.

The performance of the M4 single models is illustrated with the scatter plots to ascertain the adequacy of the models. Based on the plots (Figs. 1, 2, 3, 4), a good agreement exists between the predicted and the experimental results of all the M4 models. Interestingly, the scatter plot of ANFIS-M4 shows that it is most adequate with the highest predictability (see Fig. 3).

The predictive performance tendencies of the single models are demonstrated using the performance error of the models: the higher the error, the lower the performance of the model and vice-versa. In Fig. 5, the performance error using RMSE-values as presented. The essence of depicting this graph is to clearly demonstrate the performance of each model using a bar chart for better understanding. It is interesting to note that ANFIS-M4 have the highest performance tendency. This further gives credence to the adequacy of the ANFIS-M4 model for predicting the retention time of various classes of mycotoxin through the inputs considered in the study.

Moreover, the capability of this model in capturing the chaotic and complex properties of the data is also revealed. These findings are in agreement with [69–71].

Table 5 Performance indicator evaluation for the single models

	Training				Testing			
Models	R^2	RMSE	MSE	CC	R^2	RMSE	MSE	CC
MLR-M1	0.4126183	0.126091	0.015899	0.642354	0.459973	0.11278	0.012719	0.678213
MLR-M2	0.5245308	0.113445	0.01287	0.724245	0.607718	0.096122	0.00924	0.779563
MLR-M3	0.8137542	0.100138	0.010028	0.902083	0.815155	0.065983	0.004354	0.902859
MLR-M4	0.8727467	0.058689	0.003444	0.934209	0.876753	0.053878	0.002903	0.936351
MLR-M5	0.8460088	0.064561	0.004168	0.919787	0.812934	0.066378	0.004406	0.901628
ANN-M1	0.7854856	0.076199	0.005806	0.886276	0.559203	0.101893	0.010382	0.747799
ANN-M2	0.8462503	0.064511	0.004162	0.919919	0.725962	0.08034	0.006454	0.852034
ANN-M3	0.9342788	0.042177	0.001779	0.966581	0.909113	0.046268	0.002141	0.953474
ANN-M4	0.9750963	0.025963	0.000674	0.98747	0.978962	0.02226	0.000496	0.989425
ANN-M5	0.9648481	0.030846	0.000951	0.982267	0.926096	0.041722	0.001741	0.962339
SVM-M1	0.5722622	0.1076	0.011578	0.75648	0.562944	0.10146	0.010294	0.750296
SVM-M2	0.6378167	0.099012	0.009803	0.798634	0.6792	0.12293	0.015112	0.824136
SVM-M3	0.9332619	0.042502	0.001806	0.966055	0.910133	0.046007	0.002117	0.954009
SVM-M4	0.9463757	0.038098	0.001451	0.972818	0.928889	0.040926	0.001675	0.963789
SVM-M5	0.9267449	0.044529	0.001983	0.962676	0.929217	0.040831	0.001667	0.963959
ANFIS-M1	0.667417	0.09488	0.009002	0.816956	0.67565	0.087404	0.00764	0.821979
ANFIS-M2	0.8230232	0.069212	0.00479	0.907206	0.822984	0.06457	0.004169	0.907185
ANFIS-M3	0.9878777	0.018114	0.000328	0.99392	0.958831	0.03114	0.00097	0.979199
ANFIS-M4	0.9893603	0.01697	0.000288	0.994666	0.995069	0.010777	0.000116	0.997531
ANFIS-M5	0.9605229	0.032689	0.001069	0.980063	0.930701	0.040401	0.001632	0.964728

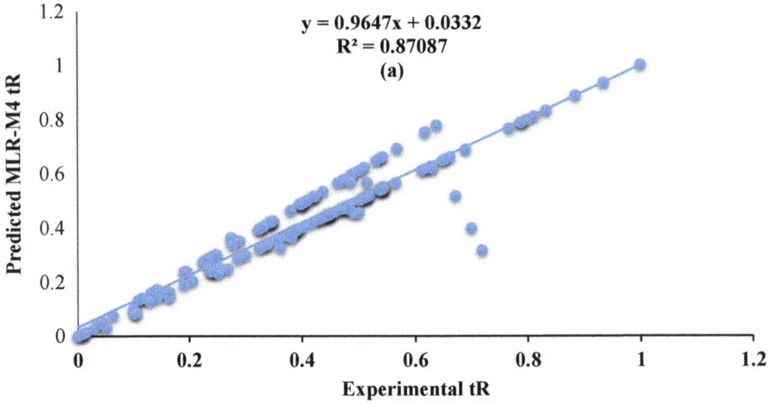

Fig. 1 Experimental and predicted values fitted scatter plot for MLR-M4 single model

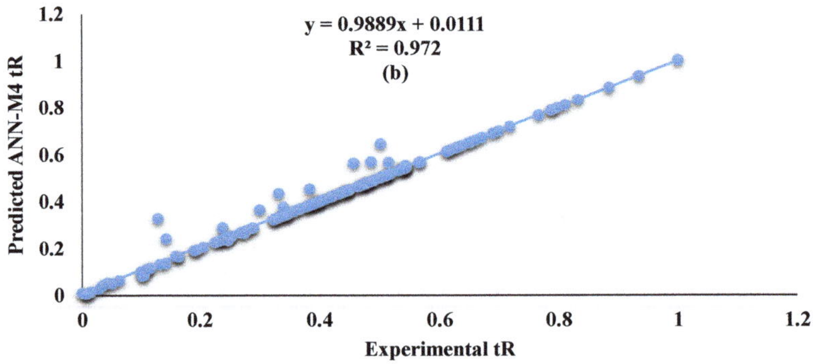

Fig. 2 Experimental and predicted values fitted scatter plot for ANN-M4 single model, First one

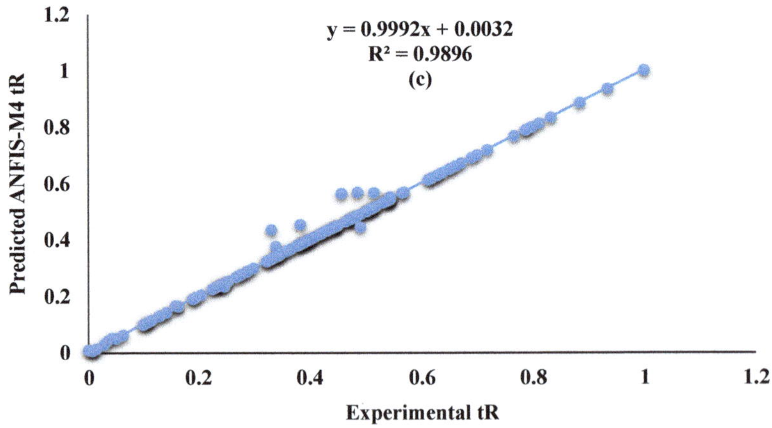

Fig. 3 Experimental and predicted values fitted scatter plot for ANFIS-M4 single model, Second one

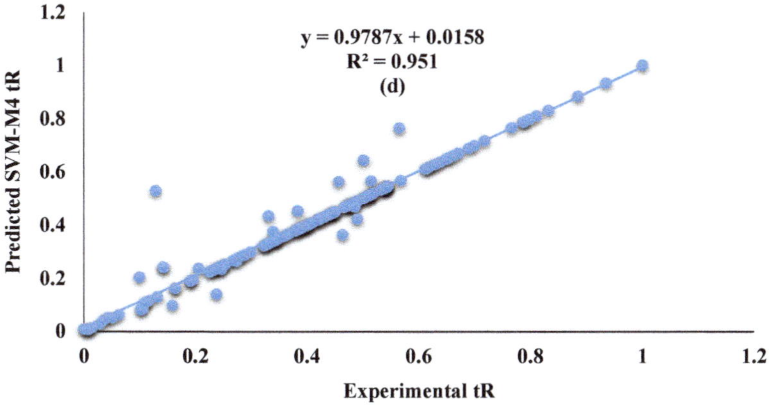

Fig. 4 Experimental and predicted values fitted scatter plot for MLR-M4 single model

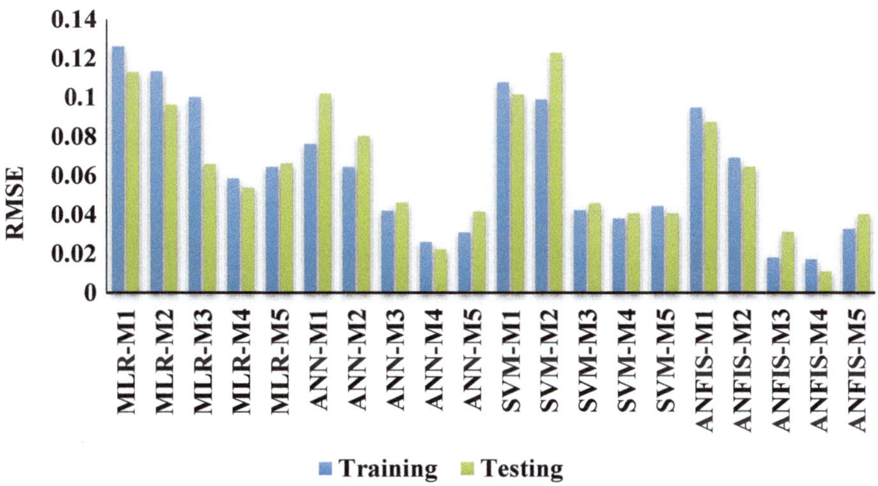

Fig. 5 RMSE performance comparison of the single models

10 Hybrid Intelligence Model Approach

The hybrid intelligence model approach is developed to enhance the adequacy and predictability of the model for studying the QSPR properties of mycotoxins aimed at elucidating and predicting for the purpose of developing methods of prevention and control. Table 6 presents the performance efficiency of the hybrid models MLR-ANN, MLR-ANFIS and MLR-SVM. The hybrid involved the integration of the adequate single models SVM-M4, ANFIS-M4 and ANN-M4 with the linear model MLR. This integration is also aimed at dealing with the variability that may be influencing the accuracy of the estimated performance of the model.

Table 6 The hybrid intelligence models

	Training				Testing			
Models	R2	RMSE	MSE	CC	R2	RMSE	MSE	CC
MLR-ANN	0.9823	0.0322	0.0010	0.9911	0.9828	0.0201	0.0004	0.9914
MLR-SVM	0.9736	0.0394	0.0016	0.9867	0.9741	0.0247	0.0006	0.9870
MLR-ANFIS	0.99996	0.0015	2E-06	0.9999	0.9899	0.0154	0.0002	0.995

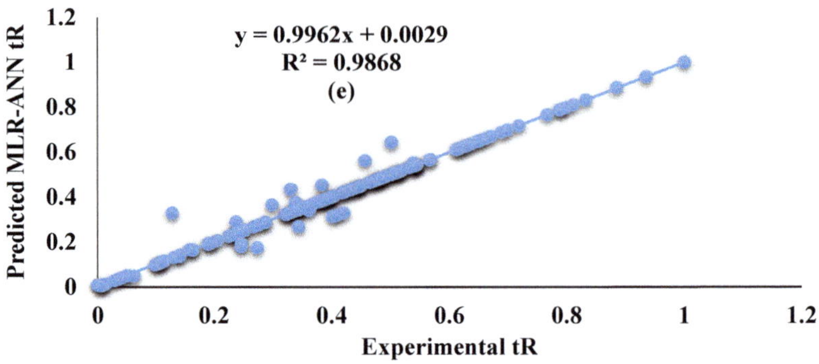

Fig. 6 Experimental and predicted values fitted scatter plot of the MLR-ANN hybrid intelligence model

Table 6 presents the adequacy of the hybrid-intelligence models over the classical single model approaches currently being used for the prediction. The hybrid-intelligence MLR-ANFIS model with highest values of CC and R^2 and lowest MSE and RMSE shows the highest performance among other hybrid models. These measures obtained are in agreement with the conclusion of Abdullahi et al. [72]. The performances of the single models are enhanced by the hybridization of ANFIS single model. This is vividly demonstrated in the R^2-value; the highest of the single models is of 0.9893603 in the training phases while MLR-ANFIS hybrid model shows the highest performance with R^2-value of 0.99996 also in the same phase.

The correlation co-efficient (CC) demonstrated in Table 6 indicates the adequacy of the hybrid-intelligence models in modelling the performance efficiency of the various mycotoxins classes using HPLC-QSPR data. The adequacy and predictability of the hybrid-intelligence models MLR-ANN, MLR-ANFIS and MLR-SVM are further demonstrated with the best fitted scatter plots of experimental and predicted values (Figs. 6, 7, 8).

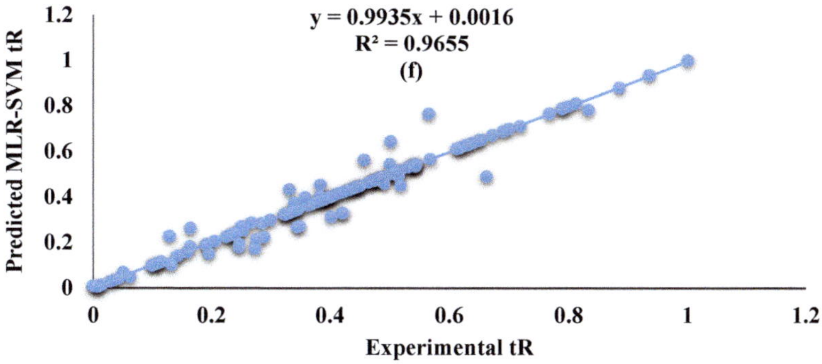

Fig. 7 Experimental and predicted values fitted scatter plot of the MLR-SVM hybrid intelligence model

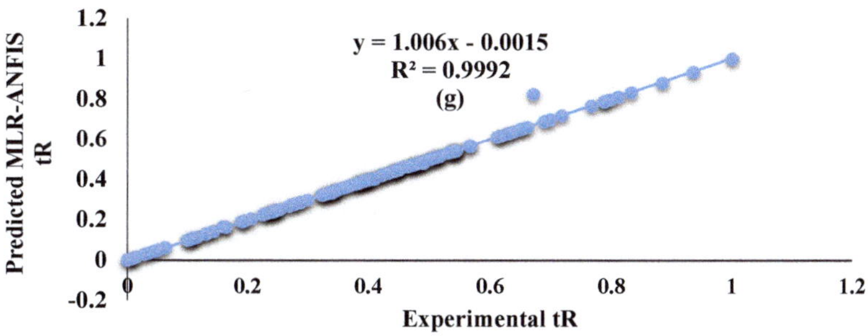

Fig. 8 Experimental and predicted values fitted scatter plot of the MLR-ANFIS hybrid model

The MSE of the hybrid models at calibration and verification phases are found to be within the acceptable range (see Fig. 9) showing conformity with the correlations of the different alternatives presented and have observed in similar studies. The results are equally in concordance with the assertions made [73], which reported the application of hybrid ANFIS model in the simulation of clostridium from various patients. The application of the potential of hybrid models in modelling rainfall reported Pham et al. [60] is also corroborated by this present study.

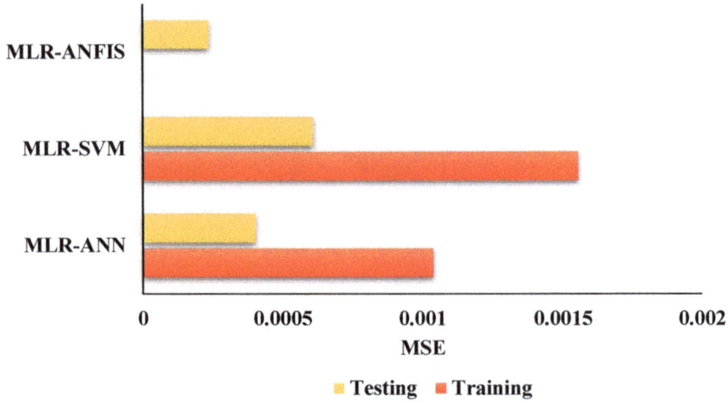

Fig. 9 MSE performance measures of Training and testing phases for the hybrid models

11 Conclusion

The major objective of this research is to assess the potential applications of data driven approaches consisting of both single and hybrid modelling methods. These techniques are employed in order to determine the adequacy and performance of hybrid-intelligence models for modeling and predicting the QSPR of various classes of mycotoxins using data obtained by HPLC technique. The mycotoxin properties considered in this are study retention index, peak symmetry, and mono isotopic mass and relative sensitivity as inputs while the modeling is hinged on the retention behaviour as the output. Four data driven models are employed as SVM, ANFIS, ANN and MLR with the single model ANFIS-M4 having the best performance in all the performance metrics of RMSE, CC, MSE and R^2 employed. The hybrid models developed to improve the performance of the single models are MLR-ANN, MLR-ANFIS and MLR-SVM. It is found that ANFIS based the single and hybrid techniques performed better that ANN, SVM and MLR based models. Conclusively, hybrid-intelligence MLR-ANFIS show superiority and outperformed other single and hybrid models employed in this current study.

Finally, there is the need to extend the study on the proposed hybrid MLR-based intelligence model with other machine learning methods. The integration of single model pproaches such as Hammerstein Wiener (HW), step-wise linear regression (SWLR), principal component analysis (PCA), and extreme learning machine (ELM) needs to be investigated. Similarly, other hybrid techniques such as SWLR-HW, SWLR-ELM and SWLR-PCA could also provide more insight into the best fitted hybrid-intelligence model most adequate for the prediction of the QSPR of various mycotoxins using date from HPLC experimentation. The consideration of uncertainty due to the relationships between various input–output variable of the studies needs to the smoothened for a more reliable results. The application of robust models such as data envelopment analysis Taguchi robust parameter techniques (see Adesina and Daneshvar [62]) could be considered, besides of other advance optimization models [74, 75].

References

1. Khosrokhavar, R., Ghasemi, J.B., Shiri, F.: 2D Quantitative Structure-property relationship study of mycotoxins by multiple linear regression and support vector machine. Int. J. Mol. Sci. **11**, 3052–3068 (2010). https://doi.org/10.3390/ijms11093052
2. Magan, N., Aldred, D.: Post-harvest control strategies: minimizing mycotoxins in the food chain. Int. J. Food Microbiol. **119**, 131–139 (2007). https://doi.org/10.1016/j.ijfoodmicro.2007.07.034
3. Magan, M., Olsen, N.: Mycotoxins in food: detection and control woodhead publishing series in food science (1994)
4. El-Nezami, H., Kankaanpaa, P., Salminen, S., Ahokas, J.: Ability of dairy strains of lactic acid bacteria to bind a common food carcinogen, aflatoxin B1. Food Chem. Toxicol. **36**, 321–326 (1998). https://doi.org/10.1016/S0278-6915(97)00160-9
5. Stangierski, J., Weiss, D., Kaczmarek, A.: Multiple regression models and Artificial Neural Network (ANN) as prediction tools of changes in overall quality during the storage of spreadable processed Gouda cheese. Eur. Food Res. Technol. **245**, 2539–2547 (2019). https://doi.org/10.1007/s00217-019-03369-y
6. Adesina, K.: Mycotoxins-causes, prevention and control: mathematical modeling strategies. Fungal Territ. **4** (2021). https://doi.org/10.36547/ft.337
7. Afsah-Hejri, L., Hajeb, P., Ehsani, R.J.: Application of ozone for degradation of mycotoxins in food: a review. Compr. Rev. Food Sci. Food Saf. **19**, 1777–1808 (2020). https://doi.org/10.1111/1541-4337.12594
8. Ayofemi Olalekan Adeyeye, S.: Aflatoxigenic fungi and mycotoxins in food: a review. Crit. Rev. Food Sci. Nutr. **60**, 709–721 (2020). https://doi.org/10.1080/10408398.2018.1548429
9. Foerster, C., Muñoz, K., Delgado-Rivera, L., Rivera, A., Cortés, S., Müller, A., Arriagada, G., Ferreccio, C., Rios, G.: Occurrence of relevant mycotoxins in food commodities consumed in Chile. Mycotoxin Res. **36**, 63–72 (2020). https://doi.org/10.1007/s12550-019-00369-5
10. Colombo, R., Papetti, A.: Pre-concentration and analysis of mycotoxins in food samples by capillary electrophoresis. Molecules **25**, 1–19 (2020). https://doi.org/10.3390/molecules25153441
11. Bauchet, J., Prieto, S., Ricker-Gilbert, J.: Improved drying and storage practices that reduce aflatoxins in stored maize: experimental evidence from smallholders in Senegal. Am. J. Agric. Econ. **00**, 1–21 (2020). https://doi.org/10.1111/ajae.12106
12. Abdelhaliem, E., Al-Otaibi, H.: Modulation of ochratoxins a-induced genotoxicity in phaseolus vulgaris by multi-walled carbon nanotubes evaluated by phenotypic, cellular ultrastructure, and nuclear dna bioassays. Genet. Mol. Res. **19**, 1–33 (2020). https://doi.org/10.4238/gmr18509
13. Roohi, R., Hashemi, S.M.B., Mousavi Khaneghah, A.: Kinetics and thermodynamic modelling of the aflatoxins decontamination: a review. Int. J. Food Sci. Technol. 1–8 (2020). https://doi.org/10.1111/ijfs.14689
14. El-Shahir, A.A., Abdel-Sater, M.A., Yassein, A.S.: Lipolysis activity of aflatoxins and ochratoxins producing fungal strains contaminating fresh cow meat. Biosci. Biotechnol. Res. Asia **17**, 163–172 (2020). https://doi.org/10.13005/bbra/2821
15. Veenaas, C., Linusson, A., Haglund, P.: Retention-time prediction in comprehensive two-dimensional gas chromatography to aid identification of unknown contaminants. Anal. Bioanal. Chem. **410**, 7931–7941 (2018). https://doi.org/10.1007/s00216-018-1415-x

16. Dankovich, R.S.: Clinical sings and pathomorphologic changes of pigs by the spontaneous ochratoxicosis. Sci. Messenger LNU Vet. Med. Biotechnol. **21**, 75–80 (2019). https://doi.org/10.32718/nvlvet9613
17. Bull, A.T., Ward, A.C., Goodfellow, M.: Search and discovery strategies for biotechnology: the paradigm shift. Microbiol. Mol. Biol. Rev. **64**, 573–606 (2000). https://doi.org/10.1128/mmbr.64.3.573-606.2000
18. Martin, L., White, M.P., Hunt, A., Richardson, M., Pahl, S., Burt, J.: Nature contact, nature connectedness and associations with health, wellbeing and pro-environmental behaviours. J. Environ. Psychol. **68**, 101389 (2020). https://doi.org/10.1016/j.jenvp.2020.101389
19. Corley, D.G., Miller-Wideman, M., Durley, R.C.: Isolation and structure of Harzianum A: a new trichothecene from *Trichoderma harzianum*. J. Nat. Prod. **57**, 422–425 (1994). https://doi.org/10.1021/np50105a019
20. Dorner, J.W., Cole, R.J.: Effect of application of nontoxigenic strains of *Aspergillus flavus* and *A. parasiticus* on subsequent aflatoxin contamination of peanuts in storage. J. Stored Prod. Res. **38**, 329–339 (2002). https://doi.org/10.1016/S0022-474X(01)00035-2
21. Xiong, Y., Zhang, P., Warner, R.D., Shen, S., Johnson, S., Fang, Z.: Comprehensive profiling of phenolic compounds by HPLC-DAD-ESI-QTOF-MS/MS to reveal their location and form of presence in different sorghum grain genotypes. Food Res. Int. **137**, 109671 (2020). https://doi.org/10.1016/j.foodres.2020.109671
22. Ostertag, F., Schmidt, C.M., Berensmeier, S., Hinrichs, J.: Development and validation of an RP-HPLC DAD method for the simultaneous quantification of minor and major whey proteins. Food Chem. 128176 (2020). https://doi.org/10.1016/j.foodchem.2020.128176
23. Sparkman, O.D.: GC/MS: a practical user's guide, 2nd edn Marvin McMaster. J. Am. Soc. Mass Spectrom. **19**, R1–R5 (2008). https://doi.org/10.1016/j.jasms.2008.05.001
24. Adamson, G.E., Lazarus, S.A., Mitchell, A.E., Prior, R.L., Cao, G., Jacobs, P.H., Kremers, B.G., Hammerstone, J.F., Rucker, R.B., Ritter, K.A., Schmitz, H.H.: HPLC method for the quantification of procyanidins in cocoa and chocolate samples and correlation to total antioxidant capacity. J. Agric. Food Chem. **47**, 4184–4188 (1999). https://doi.org/10.1021/jf990317m
25. Biswas, A.K., Rao, G.S., Kondaiah, N., Anjaneyulu, A.S.R., Mendiratta, S.K., Prasad, R., Malik, J.K.: A simple multi-residue method for determination of oxytetracycline, tetracycline and chlortetracycline in export buffalo meat by HPLC-photodiode array detector. J. Food Drug Anal. **15**, 278–284 (2007). https://doi.org/10.38212/2224-6614.2419
26. Zapata, M., Rodríguez, F., Garrido, J.L.: Separation of chlorophylls and carotenoids from marine phytoplankton: a new HPLC method using a reversed phase C8 column and pyridine-containing mobile phases. Mar. Ecol. Prog. Ser. **195**, 29–45 (2000). https://doi.org/10.3354/meps195029
27. Wang, M., Jarmusch, A.K., Vargas, F., Aksenov, A.A., Gauglitz, M., Weldon, K., Petras, D., Silva, R., Quinn, R., Alexey, V., Van Der Hooft, J.J.J., Mauricio, A., Rodríguez, C., Felix, L., Aceves, C.M., Panitchpakdi, M., Brown, E., Di, F., Sikora, N., Elijah, E.O., Labarta-bajo, L., Gentry, E.C.: HHS Public Access **38**, 23–26 (2020). https://doi.org/10.1038/s41587-019-0375-9.Mass
28. Cui, J.J., Wang, L.Y., Tan, Z.R., Zhou, H.H., Zhan, X., Yin, J.Y.: Mass spectrometry-based personalized drug therapy. Mass Spectrom. Rev. **39**, 523–552 (2020). https://doi.org/10.1002/mas.21620
29. Liu, X.R., Zhang, M.M., Gross, M.L.: Mass spectrometry-based protein footprinting for higher-order structure analysis: fundamentals and applications. Chem. Rev. (2020). https://doi.org/10.1021/acs.chemrev.9b00815
30. Nielsen, K.F., Smedsgaard, J.: Fungal metabolite screening: database of 474 mycotoxins and fungal metabolites for dereplication by standardised liquid chromatography–UV–mass spectrometry methodology. J. Chromatogr. A. **1002**, 111–136 (2003). https://doi.org/10.1016/S0021-9673(03)00490-4
31. Eckers, C., Wolff, J.-C., Haskins, N.J., Sage, A.B., Giles, K., Bateman, R.: Accurate mass liquid chromatography/mass spectrometry on orthogonal acceleration time-of-flight mass analyzers using switching between separate sample and reference sprays. 1. Proof of Concept, Anal. Chem. **72**, 3683–3688 (2000). https://doi.org/10.1021/ac000448i

32. Hopfgartner, G., Vilbois, F.: The impact of accurate mass measurements using quadrupole/time-of-flight mass spectrometry on the characterisation and screening of drug metabolites. Analusis **28**, 906–914 (2000). https://doi.org/10.1051/analusis:2000280906
33. Xuan, Q., Zheng, F., Yu, D., Ouyang, Y., Zhao, X., Hu, C., Xu, G.: Rapid lipidomic profiling based on ultra-high performance liquid chromatography–mass spectrometry and its application in diabetic retinopathy. Anal. Bioanal. Chem. **412**, 3585–3594 (2020). https://doi.org/10.1007/s00216-020-02632-6
34. Yoshikawa, K., Furuno, M., Tanaka, N., Fukusaki, E.: Fast enantiomeric separation of amino acids using liquid chromatography/mass spectrometry on a chiral crown ether stationary phase. J. Biosci. Bioeng. **130**, 437–442 (2020). https://doi.org/10.1016/j.jbiosc.2020.05.007
35. Wang, X.: Magnetic bead-based immunoassays for Aflatoxin B1 using biofunctionalized gold nanoparticles (2015)
36. Wang, Y., Zhang, C., Wang, J., Knopp, D.: Recent progress in rapid determination of mycotoxins based on emerging biorecognition molecules: a review. Toxins (Basel) **14** (2022). https://doi.org/10.3390/toxins14020073
37. Solgi, A., Pourhaghi, A., Bahmani, R., Zarei, H.: Improving SVR and ANFIS performance using wavelet transform and PCA algorithm for modeling and predicting biochemical oxygen demand (BOD). Ecohydrol. Hydrobiol. **17**, 164–175 (2017). https://doi.org/10.1016/j.ecohyd.2017.02.002
38. Ma, J., Cai, J., Lin, G., Chen, H., Wang, X., Wang, X., Hu, L.: Development of LC-MS determination method and back-propagation ANN pharmacokinetic model of corynoxeine in rat. J. Chromatogr. B Anal. Technol. Biomed. Life Sci. **959**, 10–15 (2014). https://doi.org/10.1016/j.jchromb.2014.03.024
39. D'Archivio, A.A., Giannitto, A., Maggi, M.A.: Cross-column prediction of gas-chromatographic retention of polybrominated diphenyl ethers. J. Chromatogr. A **1298**, 118–131 (2013). https://doi.org/10.1016/j.chroma.2013.05.018
40. Lotfi, E., Akbarzadeh-T., M.R.: Adaptive brain emotional decayed learning for online prediction of geomagnetic activity indices. Neurocomputing **126**, 188–196 (2014). https://doi.org/10.1016/j.neucom.2013.02.040
41. Ahmed, A.A.M., Mustakim, S., Shah, A.: Application of adaptive neuro-fuzzy inference system (ANFIS) to estimate the biochemical oxygen demand (BOD) of Surma River. J. King Saud Univ. Eng. Sci. **29**, 237–243 (2017). https://doi.org/10.1016/j.jksues.2015.02.001
42. Tao, Y., Wang, Y., Pan, M., Zhong, S., Wu, Y., Yang, R., Han, Y., Zhou, J.: Combined ANFIS and numerical methods to simulate ultrasound-assisted extraction of phenolics from chokeberry cultivated in China and analysis of phenolic composition. Sep. Purif. Technol. **178**, 178–188 (2017). https://doi.org/10.1016/j.seppur.2017.01.012
43. Saini, R., Kumar, P.: Optimization of chlorpyrifos degradation by Fenton oxidation using CCD and ANFIS computing technique. J. Environ. Chem. Eng. **4**, 2952–2963 (2016). https://doi.org/10.1016/j.jece.2016.06.003
44. Gaya, M.S., Abdul Wahab, N., Sam, Y.M., Samsudin, S.I.: ANFIS modelling of carbon and nitrogen removal in domestic wastewater treatment plant. J. Teknol. **67** (2014). https://doi.org/10.11113/jt.v67.2839
45. Vapnik, V.: The nature of statistical learning theory. **188** (1995). https://doi.org/10.1007/978-1-4757-2440-0
46. Haghiabi, A.H., Azamathulla, H.M., Parsaie, A.: Prediction of head loss on cascade weir using ANN and SVM. ISH J. Hydraul. Eng. **23**, 102–110 (2017). https://doi.org/10.1080/09715010.2016.1241724
47. Elkiran, G., Nourani, V., Abba, S.I.: Multi-step ahead modelling of river water quality parameters using ensemble artificial intelligence-based approach. J. Hydrol. **577**, 123962 (2019). https://doi.org/10.1016/j.jhydrol.2019.123962
48. Su, M., Zhang, Z., Zhu, Y., Zha, D.: Data-driven natural gas spot price forecasting with least squares regression boosting algorithm. Energies **12** (2019). https://doi.org/10.3390/en12061094

49. Sharghi, E., Nourani, V., Behfar, N.: Earthfill dam seepage analysis using ensemble artificial intelligence based modeling. J. Hydroinformatics. **20**, 1071–1084 (2018). https://doi.org/10.2166/hydro.2018.151
50. Lee, J.K., Han, W.S., Lee, J.S., Yoon, C.N.: A novel computational method for biomedical binary data analysis: development of a thyroid disease index using a brute-force search with MLR analysis. Bull. Korean Chem. Soc. **38**, 1392–1397 (2017). https://doi.org/10.1002/bkcs.11308
51. Li, H., Yazdi, M. (eds.): Advanced Decision-Making Neutrosophic Fuzzy Evidence-Based Best–Worst Method BT—Advanced Decision-Making Methods and Applications in System Safety and Reliability Problems: Approaches, Case Studies, Multi-criteria Decision-Making, Multi-objective Decision, pp. 153–184. Springer International Publishing, Cham (2022). https://doi.org/10.1007/978-3-031-07430-1_9
52. Khademi, F., Behfarnia, K.: Evaluation of concrete compressive strength using artificial neural network and multiple linear regression models. Iust **6**, 423–432 (2016)
53. Kazemi, P., Khalid, M.H., Szlek, J., Mirtič, A., Reynolds, G.K., Jachowicz, R., Mendyk, A.: Computational intelligence modeling of granule size distribution for oscillating milling. Powder Technol. **301**, 1252–1258 (2016). https://doi.org/10.1016/j.powtec.2016.07.046
54. Ghaedi, M., Hosaininia, R., Ghaedi, A.M., Vafaei, A., Taghizadeh, F.: Adaptive neuro-fuzzy inference system model for adsorption of 1,3,4-thiadiazole-2,5-dithiol onto gold nanoparticales-activated carbon, Spectrochim. Acta—Part A Mol. Biomol. Spectrosc. **131**, 606–614 (2014). https://doi.org/10.1016/j.saa.2014.03.055
55. Marrero-Ponce, Y., Barigye, S.J., Jorge-Rodríguez, M.E., Tran-Thi-Thu, T.: QSRR prediction of gas chromatography retention indices of essential oil components. Chem. Pap. **72**, 57–69 (2018). https://doi.org/10.1007/s11696-017-0257-x
56. Park, S.H., Haddad, P.R., Talebi, M., Tyteca, E., Amos, R.I.J., Szucs, R., Dolan, J.W., Pohl, C.A.: Retention prediction of low molecular weight anions in ion chromatography based on quantitative structure-retention relationships applied to the linear solvent strength model. J. Chromatogr. A **1486**, 68–75 (2017). https://doi.org/10.1016/j.chroma.2016.12.048
57. Chandwani, V., Vyas, S.K., Agrawal, V., Sharma, G.: Soft computing approach for rainfall-runoff modelling: a review. Aquat. Procedia. **4**, 1054–1061 (2015). https://doi.org/10.1016/j.aqpro.2015.02.133
58. Yaseen, Z.M., Ehteram, M., Hossain, M.S., Fai, C.M., Koting, S.B., Mohd, N.S., Jaafar, W.Z.B., Afan, H.A., Hin, L.S., Zaini, N., Ahmed, A.N., El-Shafie, A.: A novel hybrid evolutionary data-intelligence algorithm for irrigation and power production management: application to multi-purpose reservoir systems. Sustainability **11** (2019). https://doi.org/10.3390/su11071953
59. Yaseen, Z.M., Ghareb, M.I., Ebtehaj, I., Bonakdari, H., Siddique, R., Heddam, S., Yusif, A.A., Deo, R.: Rainfall pattern forecasting using novel hybrid intelligent model based ANFIS-FFA. Water Resour. Manag. **32**, 105–122 (2018). https://doi.org/10.1007/s11269-017-1797-0
60. Pham, Q.B., Abba, S.I., Usman, A.G., Thi, N., Linh, T.: Potential of hybrid data-intelligence algorithms for multi-station modelling of rainfall (2019)
61. Lola, M.S., Noor, M., Ramlee, A., Gunalan, G.S., Zainuddin, N.H., Zakariya, R., Idris, M., Khalil, I.: Improved the prediction of multiple linear regression model performance using the hybrid approach: a case study of chlorophyll-a at the offshore Kuala Terengganu. Terengganu (2016). https://doi.org/10.4236/ojs.2016.65065
62. Daneshvar, S., Adesina, K.A.: Modified variable return to scale back-propagation neural network robust parameter optimization procedure for multi-quality processes. Eng. Optim. **51**, 1352–1369 (2019). https://doi.org/10.1080/0305215X.2018.1524463
63. Yazdi, M., Golilarz, N.A., Nedjati, A., Adesina, K.A.: An improved lasso regression model for evaluating the efficiency of intervention actions in a system reliability analysis. Neural Comput. Appl. (2021). https://doi.org/10.1007/s00521-020-05537-8
64. Huang, C.-G., Men, C., Yazdi, M., Han, Y., Peng, W.: Transfer fault prognostic for rolling bearings across different working conditions: a domain adversarial perspective. Int. J. Adv. Manuf. Technol. (2022). https://doi.org/10.1007/s00170-022-09452-1

65. Li, H., Peng, W., Adumene, S., Yazdi, M.: Intelligent Reliability and Maintainability of Energy Infrastructure Assets. Springer Nature Switzerland (2023)
66. Gözen, D., Çaka, S.Y., Beşirik, S.A., Perk, Y.: First bathing time of newborn infants after birth: a comparative analysis. J. Spec. Pediatr. Nurs. **24**, e12239 (2019). https://doi.org/10.1111/jspn.12239
67. Pfeifer, J.H., Lieberman, M.D., Dapretto, M.: "I know you are but what am i?!": neural bases of self- and social knowledge retrieval in children and adults. J. Cogn. Neurosci. **19**, 1323–1337 (2007). https://doi.org/10.1162/jocn.2007.19.8.1323
68. Nourani, V., Molajou, A., Uzelaltinbulat, S., Sadikoglu, F.: Emotional artificial neural networks (EANNs) for multi-step ahead prediction of monthly precipitation; case study: northern Cyprus. Theor. Appl. Climatol. **138**, 1419–1434 (2019). https://doi.org/10.1007/s00704-019-02904-x
69. Ghali Muhammad, U., Alhosen, M., Alsharksi Çankırı Karatekin Üniversitesi, A., Hoti, Q., Muhammad Ghali, U., Alhosen Ali Degm, M., Nouri Alsharksi, A., Garba Usman, A.: Development of computational intelligence algorithms for modelling the performance of humanin and its derivatives in HPLC optimization method development simulation of various active ingredients in HPLC method development using artificial intelligence models view project development of computational intelligence algorithms for modelling the performance of humanin and its derivatives in HPLC optimization method development. Artic. Int. J. Sci. Technol. Res. 9:110–117 (2020). www.ijstr.org
70. Abba, S.I., Usman, A.G., IŞIK, S.: Simulation for response surface in the HPLC optimization method development using artificial intelligence models: a data-driven approach. Chemom. Intell. Lab. Syst. **201**, 104007 (2020). https://doi.org/10.1016/j.chemolab.2020.104007
71. Kobayashi, Y., Yoshida, K.: Automated retention time prediction of new psychoactive substances in gas chromatography. Procedia Comput. Sci. **207**, 654–663 (2022). https://doi.org/10.1016/j.procs.2022.09.120
72. Abdullahi, H.U., Usman, A.G., Abba, S.I.: Modelling the absorbance of a bioactive compound in HPLC method using artificial neural network and multilinear regression. Methods **6**, 362–371 (2020)
73. Alsharksi, A.N., Danmaraya, Y.A., Abdullahi, H.U., Ghali, U.M., Usman, A.G.: Potential of hybrid adaptive neuro fuzzy model in simulating clostridium difficile infection status. 1–6 (2020). https://doi.org/10.35940/ijbsac.A0191.073120
74. Yazdi, M., Nedjati, A., Zarei, E., Abbassi, R.: Chapter 6—application of multi-criteria decision-making tools for a site analysis of offshore wind turbines. In: Asadnia, M., Razmjou, A., Beheshti, ABT-AI and DS in ES (eds.) Cognitive Data Science in Sustainable Computing, pp. 109–127. Academic Press (2022). https://doi.org/10.1016/B978-0-323-90508-4.00008-3
75. Nedjati, A., Yazdi, M., Abbassi, R.: A Sustainable Perspective of Optimal Site Selection of Giant Air—Purifiers in Large Metropolitan Areas. Springer Netherlands (2021). https://doi.org/10.1007/s10668-021-01807-0
76. Golilarz, N.A., Gao, H., Pirasteh, S., Yazdi, M., Zhou, J., Fu, Y.: Satellite multispectral and hyperspectral image de-noising with enhanced adaptive generalized gaussian distribution threshold in the wavelet domain. Remote Sens. **13** (2021). https://doi.org/10.3390/rs13010101

Empowering Emergency Operations Management: A Pride Day

Nankai Cheng, Li Yang, Rosita Moradi, and Mohammad Yazdi

Abstract In recent years, the importance of inclusivity and diversity in emergency operations management has gained significant recognition. This chapter introduces "Pride Day," a novel initiative aimed at empowering emergency operations management by celebrating and including LGBTQ + communities. Pride Day serves as a platform to raise awareness, promote understanding, and foster a culture of inclusivity within emergency response organisations. It underlines the need for diverse perspectives and experiences in emergency operations management, highlighting the benefits of embracing LGBTQ + individuals and their unique contributions. It explores the challenges faced by LGBTQ + communities during emergencies, such as discrimination, lack of tailored support, and barriers to accessing essential services. Pride Day is proposed as an annual event that encourages emergency response agencies to engage in various activities centred around LGBTQ + awareness and inclusion. These activities may include training sessions on LGBTQ + cultural competence, policy reviews to ensure non-discrimination and equality, and fostering partnerships with local LGBTQ + organisations. Emergency response organisations can enhance their effectiveness and responsiveness to diverse communities by actively promoting inclusivity. Pride Day also creates a safe and supportive environment for LGBTQ + employees, promoting employee morale, well-being, and retention within emergency response agencies. It should be noted that Pride Day is not limited to a single day but

N. Cheng
School of Media Design, College of Technology and Design, University of Economics Ho Chi Minh City, Ho Chi Minh City, Vietnam

L. Yang
Department of Design, Faculty of Architecture, University of Lisbon, Lisbon, Portugal

R. Moradi
University of Wollongong, Northfields Avenue, Wollongong, NSW 2522, Australia

M. Yazdi (✉)
School of Computing, Engineering & Physical Sciences, University of the West of Scotland (UWS), London, UK
e-mail: mohammad.yazdi@mq.edu.au

Faculty of Science and Engineering, Macquarie University, Sydney, Australia

M. Yazdi (ed.), *Progressive Decision-Making Tools and Applications in Project and Operation Management*, Studies in Systems, Decision and Control 518,
https://doi.org/10.1007/978-3-031-51719-8_7

serves as a catalyst for long-term changes within emergency operations management. It fosters ongoing efforts to create an inclusive and resilient emergency management system that effectively addresses the needs of all individuals, regardless of their sexual orientation or gender identity.

Keywords Emergency operations management · Inclusivity · Diversity · LGBTQ + · Pride Day · Awareness · Cultural competence · Non-discrimination · Equality · Resilience

1 Introduction

In recent years, Pride events have gained significant momentum worldwide as celebrations of diversity, inclusion, and equality for the LGBTQ + community. These vibrant and empowering gatherings bring together individuals from all walks of life to commemorate the progress made and advocate for the rights of sexual and gender minorities. However, with the scale and complicatedness of such events, it is crucial to prioritize the safety and well-being of all participants, requiring meticulous emergency operations management [1–4].

Emergency operations management is vital in ensuring the smooth execution of large-scale events, including Pride celebrations [5–9]. While Pride events are primarily focused on fostering a sense of community and celebration, it is essential to address potential risks and emergencies during the festivities proactively. This chapter aims to present a comprehensive approach to empowering emergency operations management in the context of Pride Day, with a specific focus on enhancing safety and inclusivity.

The primary objective of this chapter is to provide event organizers, emergency responders, and relevant stakeholders with a framework encompassing pre-event planning, risk assessment, emergency response protocols, and post-event evaluations. By adopting this holistic approach, organizers can effectively mitigate potential hazards, strengthen emergency response capabilities, and ensure that the Pride Day experience remains secure, accessible, and inclusive for all attendees.

Throughout this chapter, we will explore critical components of empowering emergency operations management, including:

- Pre-event planning: This stage involves identifying potential risks, establishing communication channels with relevant authorities, designing a comprehensive emergency response plan, and establishing robust coordination mechanisms between event organizers, local emergency services, and community stakeholders.
- Risk assessment and mitigation: Through a thorough evaluation of potential risks, including crowd management, medical emergencies, and public safety concerns, event organizers can implement proactive measures such as traffic control, crowd flow optimization, and contingency plans, ensuring the safety of participants and reducing the likelihood of emergencies.

- Communication and information dissemination: Effective communication strategies are crucial in emergencies. This chapter will address the importance of clear and accessible information channels during the event and in the lead-up to keep attendees informed about safety protocols, emergency exits, medical services, and relevant contact points.
- Training and capacity building: The capacity of event staff, security personnel, and volunteers is pivotal to effective emergency response. This chapter will explore the significance of specialized training programs, inclusive approaches to emergency response, and collaboration with local emergency services to enhance the overall preparedness for the event.
- Post-event evaluation and lessons learned: Conducting comprehensive evaluations following Pride Day events allows for identifying strengths, weaknesses, and areas of improvement. This chapter will emphasize the importance of post-event assessments, feedback collection, and knowledge sharing to inform future emergency operations management strategies and ensure continuous enhancement of safety measures.

By empowering emergency operations management within the context of Pride Day events, this chapter aims to contribute to developing inclusive and secure celebrations for the LGBTQ + community and their allies. Recognizing the unique challenges and opportunities associated with these gatherings, this comprehensive approach will enable event organizers to proactively address potential risks, ensuring that Pride celebrations remain safe, welcoming, and empowering spaces for all participants.

In this chapter, Sydney Mardi Gras as an example. A brief introduction to this event is provided as the following. Sydney Mardi Gras is an annual festival that celebrates diversity, inclusion, and the LGBTQ + community in the vibrant city of Sydney, AU. It is one of the most significant and iconic LGBTQ + events in the world, drawing thousands of participants and spectators from across the globe.

The origins of Sydney Mardi Gras can be traced back to June 24, 1978, when the first protest march called the "Sydney Gay Solidarity Celebrations" took place. What began as a demonstration of gay rights quickly became a celebration of love, identity, and equality? Over the years, the event grew in scale and significance, becoming a symbol of strength, resilience, and the ongoing fight for LGBTQ + rights. Today, Sydney Mardi Gras is a multifaceted festival that spans several weeks, culminating in a grand parade and a series of exciting events. The festival's highlight is the dazzling Mardi Gras Parade, where flamboyantly decorated floats, vibrant costumes, and live performances take over the streets of Sydney. The parade showcases diverse participants, including LGBTQ + community members, allies, advocacy groups, corporations, and cultural organizations. Beyond the parade, Sydney Mardi Gras offers an array of cultural, artistic, and entertainment events. These include parties, art exhibitions, theatre performances, film screenings, and thought-provoking discussions on LGBTQ + issues. The festival creates a platform for self-expression, empowerment, and fostering a sense of belonging within the LGBTQ + community and beyond. Sydney Mardi Gras has become an essential part of Sydney's cultural fabric and has played a significant role in advocating for LGBTQ + rights and promoting social progress. It continues to serve as a reminder of the ongoing struggle for equality while celebrating the achievements and resilience of the LGBTQ + community. Whether we all are a member of the LGBTQ + community, an ally, or simply someone seeking a vibrant and inclusive celebration, Sydney Mardi Gras offers a unique and unforgettable experience that embraces diversity, love, and acceptance.

2 Emergency Operations Management

The Public Messaging Framework is a strategic approach to disseminating critical messages to specific audience segments through various communication channels. Its primary purpose is to facilitate the coordinated management of pedestrian and visitor communications and wayfinding strategies. Public messaging within this framework refers to delivering important messages to audience segments throughout the event, triggered by specific events or stages outlined in the Crowd Load Threshold Table. These messages aim to either set customer expectations or provide redirection information to the audience.

While particular messages may be pre-determined, others may need to be created based on the specific stage in the Crowd Load Threshold Table. The overall objective of the Public Messaging Framework is to support the Strategic Communication and Management Plan by preventing overcrowding in apparent, partially obscured, and obstructed areas.

The Public Messaging Framework places particular emphasis on messaging that accomplishes two primary goals:

1. Setting customer expectations involves informing the audience about anticipated wait times or any other relevant information that can help manage their expectations effectively.
2. Providing redirection information: During periods of high-volume attendance in the event footprint, the messaging aims to guide patrons towards alternative routes or areas to alleviate congestion and ensure a smooth flow of foot traffic.

Several organisations should manage localised crowd issues. Decision-making is done in consultation with the subject matter experts (SMEs), and these decisions are made locally. They are then communicated to the Government Coordination Centre for information purposes, to assess the impact on other areas, and to facilitate public communication if necessary.

Crowd Load Threshold, Various stages of the Crowd Load Threshold Test may lead to a specific location or operational zone reaching a Crowd Load Threshold. At this point, actions are taken to manage the situation, which may require messaging to different audiences. The Public Messaging Framework is designed to support tactical crowd management procedures by diverting crowds at decision-making points away from locations/sectors where the Crowd Load Threshold is above Stage 2. Early notification of increasing Crowd Load Thresholds is crucial to enable the broadcasting of diversion and expectation messages.

Public Messaging Crowd Load Threshold Action Plan, Early notification of increasing Crowd Load Thresholds is vital to ensure the timely broadcast of diversion and/or expectation messages.

Controls (I): Monitor access routes and the flow of patrons entering the area, review circulation, and monitor egress levels. Report back to the SMEs. Actions: The Police and security personnel are responsible for taking action. Further consultation will be conducted regarding the required number of security personnel and their locations.

Controls (II): Implement alternative routes (activate flippers in critical locations) once spatial densities have reached 90% capacity. Actions: The Police and security personnel are responsible for taking action. The media channels will be activated. Wayfinding personnel and variable message signs will be utilised to divert crowds/patrons to other areas that have yet to reach capacity.

Controls (III). No further access is allowed (egress only) to enable crowd numbers and flow to correct themselves. The area can reopen when the number of people drops below 80%. Security personnel and Police are responsible for taking action. The media channels will be activated. Crowds/patrons will be diverted at transport hubs and ingress routes to areas with exceeded capacity. Wayfinding personnel will redirect crowds away from the exceeded area.

3 Emergency Risk Management

Emergency Risk Management (ERM) is a systematic approach to identifying, assessing, and mitigating risks associated with emergencies and disasters [10–12]. It involves a comprehensive process of planning, organizing, coordinating, and implementing measures to reduce the potential impact of crises on individuals, communities, and organizations [13–15].

The primary goal of emergency risk management is to enhance preparedness and response capabilities to deal with emergencies and minimize their negative consequences effectively [8, 16]. It involves understanding and evaluating potential hazards and vulnerabilities, developing strategies to prevent or mitigate risks, and establishing effective response and recovery mechanisms.

Critical components of emergency risk management typically include:

- Risk Assessment [17]: Identifying and analysing potential hazards, vulnerabilities, and risks to determine their likelihood and potential impact. This involves studying natural disasters (such as earthquakes, floods, or hurricanes), technological accidents (such as chemical spills or nuclear incidents), or other emergencies specific to the region or context.

- Prevention and Mitigation [18]: Implementing measures to reduce or eliminate the likelihood and severity of emergencies. This may involve land-use planning, infrastructure development, structural improvements, environmental management, public education, and awareness campaigns.
- Preparedness [19]: Developing plans, procedures, and protocols to ensure emergency readiness. This includes creating emergency response plans, training teams, conducting drills and exercises, establishing communication systems, and stockpiling necessary resources and supplies.
- Response [20]: Activating and executing the planned emergency response activities in a coordinated manner. This involves mobilizing emergency personnel, deploying resources, implementing evacuation plans, providing emergency medical care, managing communication and information dissemination, and coordinating with relevant stakeholders.
- Recovery [21, 22]: Initiating post-emergency actions to restore affected areas and communities to normal functioning. This includes assessing damage, providing relief and support to affected individuals, repairing infrastructure, restoring essential services, and facilitating long-term recovery and reconstruction efforts.

ERM is typically implemented by government agencies, emergency management organizations, non-governmental organizations (NGOs), and other public safety and disaster response entities. It emphasizes a proactive approach to risk management, aiming to minimize the impact of emergencies on lives, infrastructure, and the environment through effective planning, preparedness, and response.

A examples of risk assessment is conducted to show that how to prevent, and mitigate the risk of pride day, in the following format: (I) Hazard; (II) Risk; (III) Safety measures:

- (I) Event conducted in busy public areas; (II) Injury of a public member due to absence of event site isolation and access control. Damage of a member of the public due to being hit by a vehicle; (III) To instruct staff that event vehicle drivers must:- Hold a current unrestricted licence,- Adhere to the relevant state Road Rules,- Don't stop through traffic lanes or Stopping areas,- Travel with walking speed (5 km/h max) while in the pedestrian domain,- Avoid reversing or U-turns. The plan for the load-in and load-out activities so that vehicles are driven in a one-way flow, 2. Any cars driving over the pedestrian domain are to be escorted. Escort to:—Wear high visibility work wear—Stay at least 3 m. clear from the vehicle, and never stand between an immovable object and the car (maintain an escape path),—Maintain eye contact with the driver,—Proactively direct the public away from the vehicle's travel path—using hand gestures and verbal commands, 3—first aid capabilities to be established during load-in and load-out.
- (I) Workers handling electrical equipment—leads/cords and plugging/ unplugging to locally available power outlets, (II) Electric shock to a worker, (III) 1. Power mains (house power or generator) to be turned off (de-energised) and physically unplugged while the electrical grid and appliances are being plugged/unplugged, 2. Mains and portable distribution boards to be RCD-protected and, if practicable,

Overload protected, 3. Crew to test RCDs daily and confirm functionality. Reset with buddy-check immediately after the test, 4. No complicated wiring or any electrical alterations are allowed unless done by a licensed electrician, 5. Only 3-prong leads are to be used, and leads/cords are to be uncoiled when energised and electrical equipment isolated from patron reach; 6—leads and appliances to bear current AS 3760 inspection test tags. RCD to handle recent RCD test tags. Overload Protection Breakers to be considered, 7. Access to power mains and distribution boards to be isolated by keeping doors shut and locked as much as practicable, 8. Workers must be shown how to inspect leads for damage, nine visually. Leads and electrical equipment are to be kept away from water. Consider using devices certified by the manufacturer with high IP (Ingress Protection).

- (I) Float Deck—live/unsecured edges; (II) Fall of the worker from float deck and associated injury; (III) 1. Deck to be erected per stage system specifications—connection of boards, bracing. The stage system should be rated to achieve 5kN/ 1 sq.m, 2. Isolate live edges of presentation stage deck using bollards and tape—delineation to be placed at no less than 1 m from the edges, 3—presence of workers on stages without handrails to be limited, 4. Stage contractors are to mark the edges of the stage with 2 cm. Comprehensive tape line of colour contrasting from the stage deck colour (white colour preferred). Such a line will be applied to any steps, ramps and deck platforms to ensure visibility of platform dimensions, 5. Supervisors of contractor workers to closely monitor adherence to these guidelines—workers not allowed to walk near unsecured edges, 6. Artists' positions on stage to be marked, and presenters are given the opportunity to rehearse and become accustomed to the environment, 7. Avoid shining lights in Artists' faces from perpendicular angles (to have better understanding about float deck, see the Fig. 1).
- (I) Outdoor parade; (II) Outdoor Inclement weather such as rain and wind causing patron injury (e.g., debris blown, slip and fall incidents); (III) 1. Consult with

Fig. 1 An example of float deck in a Pride day

stakeholders and determine specific weather conditions under which the parade group would not participate, e.g., severe wind conditions, heavy rain, lightning storm, 2. To document the agreed criteria and ensure that the requirements are communicated to participants) 3. To discuss the conditions with the participants and allow them to exercise their risk appetite if this is more conservative to the established criteria, e.g. participants not to feel pursued to participate in the parade if the prevailing conditions make them feel unsafe, 4. Designate a team member to check the severe weather warnings and weather forecasts—check in the afternoon and evening before the parade and every 2 h on the day of the parade, 5. Designate authorised persons designated to decide whether to proceed subject to weather conditions. Protocol for decision-making to include aim for clarity and concurrence.
- (I) Inadequate safety communication consultation, coordination and cooperation, (II) Workplace incident and injury/illness attributable to inadequate stakeholder communication and consultation, (III) 1. To engage in proactive communication with all event stakeholders, 2. Seek confirmation on critical matters in writing, 3. Include groups of stakeholders as much as practicable (e-mail) so more parties are included on discussions, 4. Stress time frames for crucial decisions to allow planning time for all relevant parties, 5. Document critical choices in writing and double-check/confirm as required, 6. Encourage all stakeholders to communicate, consult and cooperate on WHS and operational matters—a legal requirement (legislation and approved Code of Practice) and industry best practice.

4 How to Empower It?

Emergency Operations Management involves coordinating and responding to various emergencies, such as natural disasters, public health crises, and other critical incidents. Empowering it entails providing the necessary tools, resources, and strategies to enhance the effectiveness and efficiency of emergency response efforts. Here are several ways to empower Emergency Operations Management:

- Technology Integration: Utilize advanced technologies, such as artificial intelligence (AI), data analytics, and the Internet of Things (IoT), to improve situational awareness, decision-making, and resource allocation. These technologies can help collect, analyze, and visualize data in real-time, enabling faster and more informed emergency responses.
- Communication and Information Management: Establish robust communication systems that facilitate seamless information sharing among emergency responders, government agencies, and the public. This can involve utilizing interoperable communication technologies, social media platforms, and mobile applications to disseminate critical updates, warnings, and instructions to affected populations [23].

- Training and Capacity Building: Invest in training programs and exercises to enhance the skills and capabilities of emergency responders. This includes providing comprehensive training on incident management, leadership, and the use of specialized equipment. Continuous professional development ensures that emergency personnel are well-prepared to handle various emergency scenarios effectively, [24].
- Collaborative Partnerships: Foster collaboration and partnerships among stakeholders, including government agencies, non-profit organizations, private sector entities, and community groups. Engaging all relevant parties in emergency planning, response, and recovery efforts enhances coordination, resource sharing, and the overall effectiveness of emergency operations [25].
- Community Engagement and Empowerment: Involve the local community in emergency preparedness initiatives, encouraging individuals to take an active role in their safety. This can include public awareness campaigns, community training programs, and the establishment of community emergency response teams. Empowering individuals and communities to contribute to emergency response efforts improves overall resilience, like study [26].
- Data-driven Decision Making: Utilize data and analytics to inform emergency decision-making processes. Real-time data on incident status, resource availability, and population demographics can help emergency managers make informed decisions on resource allocation, prioritization, and response strategies. Data-driven insights enable more efficient and effective emergency operations, such as in studies [27, 28].
- Continuous Improvement and Evaluation: Implement mechanisms to review and evaluate emergency response efforts regularly. This includes conducting after-action reviews, collecting stakeholder feedback, and identifying improvement areas. Regular evaluations help identify gaps, refine procedures, and enhance future emergency response capabilities, such as in studies [29, 30].

5 Conclusions

In conclusion, introducing Pride Day as an initiative in emergency operations management is a significant step towards promoting inclusivity and diversity within emergency response organizations. By celebrating and including LGBTQ + communities, Pride Day catalyzes positive change and fosters a culture of acceptance, understanding, and support. The chapter emphasizes embracing diverse perspectives and experiences in emergency operations management. By actively engaging with LGBTQ + individuals, emergency response agencies can gain valuable insights and improve their effectiveness and responsiveness to the needs of all individuals, regardless of their sexual orientation or gender identity.

Furthermore, Pride Day highlights the specific challenges faced by LGBTQ + communities during emergencies, such as discrimination, lack of tailored support, and barriers to accessing essential services. Emergency response organizations can create a safer and more inclusive environment for all individuals by addressing these challenges and implementing measures to ensure non-discrimination and equality. Pride Day is not limited to a single day but catalyses long-term changes within emergency operations management. It encourages ongoing efforts to create an inclusive and resilient emergency management system that effectively addresses the needs of diverse communities. Additionally, Pride Day plays a crucial role in fostering a safe and supportive environment for LGBTQ + employees within emergency response agencies. Pride Day promotes employee morale, well-being, and retention and promotes a more positive and inclusive work culture.

In summary, implementing Pride Day as an annual event and the accompanying activities focused on LGBTQ + awareness, and inclusion can significantly enhance emergency response organizations' effectiveness, responsiveness, and overall ability to serve diverse communities. By actively promoting inclusivity, emergency operations management can work towards a more resilient and equitable future, ensuring that all individuals, regardless of their sexual orientation or gender identity, are fully supported and protected during times of crisis.

References

1. Zali, M.N., Ibrahim, A.M., Venkat, I., Omar, S., Belaton, B.: Crowd management: Simulation modelling and visualization via fuzzy agents. AIP Conf. Proc. **2643**, 40024 (2023). https://doi.org/10.1063/5.0110450
2. Cheng, J., Chen, B., Huang, Z.: Collective-based ad transparency in targeted hotel advertising: Consumers' regulatory focus underlying the crowd safety effect. J. Retail. Consum. Serv. **72**, 103257 (2023). https://doi.org/10.1016/j.jretconser.2023.103257
3. Skipworth, S.: Soccer match tragedy offers lessons for improving crowd management, safety. Campus Leg. Advis. **23**, 4–5 (2023). https://doi.org/10.1002/cala.40998
4. Yazdi, M., Ding, Y., Adumene, S., Shafie, P.: Progressive system safety and reliability analysis: A sustainable game theory approach. Qual. Reliab. Eng. Int. n/a (2022). https://doi.org/10.1002/qre.3247
5. Kang, J., Wang, L., Wang, Z., Zhang, J., Dai, H.: Improving the emergency management of energy infrastructure using scenario construction. Int. J. Hydrogen Energy. **48,** 8731–8742 (2023). https://doi.org/10.1016/j.ijhydene.2022.11.208
6. Adesina, K.A., Yazdi, M., Omidvar, M.: Emergency decision making fuzzy-expert aided disaster management system BT—linguistic methods under fuzzy information in system safety and reliability analysis, in: M. Yazdi (Ed.), Springer International Publishing, Cham, pp. 139–150 (2022). https://doi.org/10.1007/978-3-030-93352-4_6
7. Atkinson, C.L.: Local government emergency management. Encyclopedia **3**, 1–14 (2023). https://doi.org/10.3390/encyclopedia3010001
8. Jiang, G.-J., Chen, H.-X., Sun, H.-H., Yazdi, M., Nedjati, A., Adesina, K.A.: An improved multi-criteria emergency decision-making method in environmental disasters. Soft. Comput. (2021). https://doi.org/10.1007/s00500-021-05826-x
9. Li, H., Guo, J.-Y., Yazdi, M., Nedjati, A., Adesina, K.A.: Supportive emergency decision-making model towards sustainable development with fuzzy expert system. Neural Comput. Appl. **33**, 15619–15637 (2021). https://doi.org/10.1007/s00521-021-06183-4
10. Li, H., Peng, W., Adumene, S., Yazdi, M.: Cutting edge research topics on system safety, reliability, maintainability, and resilience of energy-critical infrastructures BT—Intelligent Reliability and Maintainability of Energy Infrastructure Assets, in: H. Li, W. Peng, S. Adumene, M.

Yazdi (Eds.), Springer Nature Switzerland, Cham, pp. 25–38 (2023). https://doi.org/10.1007/978-3-031-29962-9_2
11. Li, H., Yazdi, M.: A holistic question: is it correct that decision-makers neglect the probability in the risk assessment method? BT —Advanced Decision-Making Methods and Applications in System Safety and Reliability Problems: Approaches, Case Studies, Multi-criteria Deci, in: H. Li, M. Yazdi (Eds.), Springer International Publishing, Cham, pp. 185–189 (2022). https://doi.org/10.1007/978-3-031-07430-1_10
12. Yazdi, M., Golilarz, N.A., Adesina, K.A., Nedjati, A.: Probabilistic risk analysis of process systems considering epistemic and aleatory uncertainties: a comparison study. Int. J. Uncertainty, Fuzziness Knowledge-Based Syst. **29**, 181–207 (2021). https://doi.org/10.1142/S0218488521500098
13. Huang, P., Gu, Y., Li, H., Yazdi, M., Qiu, G.: An optimal tolerance design approach of robot manipulators for positioning accuracy reliability. Reliab. Eng. Syst. Saf. **109347** (2023). https://doi.org/10.1016/j.ress.2023.109347
14. Li, H., Yazdi, M.: Reliability analysis of correlated failure modes by transforming fault tree model to Bayesian network: a case study of the MDS of a CNC Machine Tool BT—Advanced Decision-Making Methods and Applications in System Safety and Reliability Problems: Approac, in: H. Li, M. Yazdi (Eds.), Springer International Publishing, Cham, pp. 15–28 (2022). https://doi.org/10.1007/978-3-031-07430-1_2
15. Li, H., Yazdi, M.: Stochastic game theory approach to solve system safety and reliability decision-making problem under uncertainty BT —Advanced Decision-Making Methods and Applications in System Safety and Reliability Problems: Approaches, Case Studies, Multi-criteria De, in: H. Li, M. Yazdi (Eds.), Springer International Publishing, Cham, pp. 127–151 (2022). https://doi.org/10.1007/978-3-031-07430-1_8
16. Yazdi, M.: Improving failure mode and effect analysis (FMEA) with consideration of uncertainty handling as an interactive approach. Int. J. Interact. Des. Manuf. **13**, 441–458 (2019). https://doi.org/10.1007/s12008-018-0496-2
17. Adesina, K.A., Yazdi, M., Zarei, E., Pouyakian, M.: smart decision fuzzy-based data envelopment model for failure modes and effects analysis BT —Linguistic Methods Under Fuzzy Information in System Safety and Reliability Analysis, in: M. Yazdi (Ed.), Springer International Publishing, Cham, pp. 151–170 (2022). https://doi.org/10.1007/978-3-030-93352-4_7
18. Yazdi, M., Khan, F., Abbassi, R.: A dynamic model for microbiologically influenced corrosion (MIC) integrity risk management of subsea pipelines. Ocean Eng. **269** , 113515 (2023). https://doi.org/10.1016/j.oceaneng.2022.113515
19. Azadeh, A., Salehi, V., Arvan, M., Dolatkhah, M.: Assessment of resilience engineering factors in high-risk environments by fuzzy cognitive maps: a petrochemical plant. Saf. Sci. **68,** 99–107 (2014). https://doi.org/10.1016/j.ssci.2014.03.004
20. Yazdi, M., Adesina, K.A., Korhan, O., Nikfar, F.: Learning from fire accident at Bouali Sina petrochemical complex plant. J. Fail. Anal. Prev. (2019). https://doi.org/10.1007/s11668-019-00769-w
21. Adumene, S., Islam, R., Amin, M.T., Nitonye, S., Yazdi, M., Johnson, K.T.: Advances in nuclear power system design and fault-based condition monitoring towards safety of nuclear-powered ships. Ocean Eng. **251** , 111156 (2022). https://doi.org/10.1016/j.oceaneng.2022.111156
22. Yazdi, M., Khan, F., Abbassi, R., Quddus, N.: Resilience assessment of a subsea pipeline using dynamic Bayesian network. J. Pipeline Sci. Eng. **2**, 100053 (2022). https://doi.org/10.1016/j.jpse.2022.100053
23. Qi, N., Zhang, X.: Optimization design and implementation of shared information management system for industrial design network platform. J. Comb. Optim. **45**, 29 (2022). https://doi.org/10.1007/s10878-022-00956-w
24. Huang, C.-G., Men, C., Yazdi, M., Han, Y., Peng, W.: Transfer fault prognostic for rolling bearings across different working conditions: a domain adversarial perspective. Int. J. Adv. Manuf. Technol. (2022). https://doi.org/10.1007/s00170-022-09452-1

25. Lubell, M.: Collaborative partnerships in complex institutional systems. Curr. Opin. Environ. Sustain. **12**, 41–47 (2015). https://doi.org/10.1016/j.cosust.2014.08.011
26. Frandsen, M., Paton, D., Sakariassen, K.: Fostering community bushfire preparedness through engagement and empowerment. Aust. J. Emerg. Manag. **26,** 23–30 (2011). https://search.informit.org/doi/https://doi.org/10.3316/ielapa.140692861394554
27. Jiang, H., Zhang, H., Zhang, K., Cui, X.: Neurocomputing Data-driven adaptive dynamic programming schemes for non-zero-sum games of unknown discrete-time nonlinear systems. Neurocomputing **275**, 649–658 (2018). https://doi.org/10.1016/j.neucom.2017.09.020
28. Heidary, R., Gabriel, S.A., Modarres, M., Groth, K.M., Vahdati, N.: A review of data-driven oil and gas pipeline pitting corrosion growth models applicable for prognostic and health management. Int. J. Progn. Heal. Manag., 1–13 (2018)
29. Li, H., Yazdi, M.: Developing failure modes and effect analysis on offshore wind turbines using two-stage optimization probabilistic linguistic preference relations BT —Advanced Decision-Making Methods and Applications in System Safety and Reliability Problems: Approaches, Case Studies, Multi-criteria Decision-Making, Multi-objective Decision-Making, Fuzzy Risk-Based Models, in: H. Li, M. Yazdi (Eds.), Springer International Publishing, Cham, pp. 47–68 (2022). https://doi.org/10.1007/978-3-031-07430-1_4
30. Li, H., Yazdi, M.: What are the critical well-drilling blowouts barriers? A Progressive DEMATEL-Game Theory BT —Advanced Decision-Making Methods and Applications in System Safety and Reliability Problems: Approaches, Case Studies, Multi-criteria Decision-Making, Multi-obj, in: H. Li, M. Yazdi (Eds.), Springer International Publishing, Cham, pp. 29–46 (2022). https://doi.org/10.1007/978-3-031-07430-1_3

Enhancing Hydrogen Risk Management: A Modified EDAS-Based FMEA Approach

Hamidreza Seiti, Amir Mohammad Larni-Fooeik, and Ali Elkamel

Abstract Hydrogen emerges as a versatile and environmentally friendly energy carrier, garnering attention due to its potential to drive a low-carbon economy. The diverse methods employed for hydrogen production offer adaptability to varying resources and technological advancements. However, the risks associated with hydrogen leakage demand a comprehensive assessment due to their implications for safety, the environment, and economics. This paper introduces an innovative modification to the EDAS method, complemented by the FMEA technique, to analyze hydrogen leakage risks within a steam-reforming hydrogen generation unit. The study underscores the significance of comprehending failure modes and their underlying causes in ensuring system safety and efficiency. Moreover, the paper addresses the limitations of the EDAS method and introduces a novel similarity measure aimed at mitigating challenges posed by outliers. Through a case study and a systematic exploration of failure modes, this research contributes to informed decision-making, with a strong emphasis on enhancing safety and reliability within hydrogen production systems.

Keywords Hyrogen energy · Risk management · EDAS method · FMEA · Leakage analysis

H. Seiti (✉) · A. M. Larni-Fooeik
Department of Industrial Engineering, Iran University of Science and Technology, Tehran, Iran
e-mail: hamidreza.seiti@srbiau.ac.ir

A. Elkamel
Department of Chemical Engineering, Khalifa University, Abu Dhabi, UAE

Department of Chemical Engineering, University of Waterloo, ON, Canada

M. Yazdi (ed.), *Progressive Decision-Making Tools and Applications in Project and Operation Management*, Studies in Systems, Decision and Control 518,
https://doi.org/10.1007/978-3-031-51719-8_8

1 Introduction

The shift towards a sustainable energy future is paramount in our endeavors to combat the effects of climate change and decrease our reliance on fossil fuels. Within this context, hydrogen has emerged as a clean and adaptable energy carrier, drawing significant attention for its potential to play a role in a low-carbon economy [1]. Hydrogen can be generated from a variety of domestic resources, including natural gas, coal, biomass, solar, wind, and nuclear energy. Numerous methods are employed for producing hydrogen, including steam reforming, gasification, electrolysis, and thermochemical cycles [2]. These diverse production methods provide flexibility to adapt to resource availability and technological progress. Moreover, hydrogen serves various purposes across different sectors, rendering it a remarkably versatile energy source. Its applications range from power generation, heating systems, and transportation to a variety of industrial processes [3]. The utilization of hydrogen involves an array of devices and technologies, including fuel cells, turbines, boilers, engines, and emerging innovations.

However, it's essential to recognize that hydrogen also poses potential risks, particularly related to its potential leakage throughout the entire value chain encompassing production, distribution, and end-use [4, 5]. Hydrogen leakage can lead to adverse safety, environmental, and climatic consequences. Due to its high flammability, hydrogen accumulation in confined spaces can result in fires or explosions [6]. Additionally, hydrogen leakage can impact atmospheric chemistry, contributing to the extended lifespan of methane and other greenhouse gases, thereby exacerbating concerns about climate change [7]. Moreover, such leakage can lead to economic losses through wasted energy and reduced system efficiency. Given these potential risks, a thorough investigation of the sources, rates, and impacts of hydrogen leakage is imperative [8]. This entails examining the entire value chain to identify potential vulnerabilities.

Hydrogen leakage risk management within hydrogen production systems can be approached through two distinct perspectives. The first involves a comprehensive exploration of the underlying causes and modes leading to hydrogen leakage occurrences. The second perspective delves into the potential consequences arising from hydrogen leakage incidents. In this paper, the primary objective is to investigate the failure modes and underlying causes of hydrogen leakage in the context of a hydrogen generation unit. This investigation is carried out using the Failure Modes and Effects Analysis (FMEA) technique, a widely recognized and utilized method for such analysis. FMEA offers a structured approach to understanding failure scenarios and their potential impacts [10].

The failure modes and causes of hydrogen leakage are of paramount concern in maintaining the safety and efficiency of hydrogen production systems. By identifying these failure modes and their root causes, operators and engineers can take proactive measures to mitigate risks and enhance system reliability. FMEA, as a systematic methodology, provides a structured framework for evaluating potential failure scenarios. It involves identifying possible failure modes, understanding their causes, assessing their severity, and determining the effectiveness of existing safeguards or preventive measures [10]. This approach aids in prioritizing corrective actions and allocating resources where they are most needed.

The steam-reforming hydrogen generation unit serves as a specific case study in this paper. Steam reforming is a prevalent method for producing hydrogen from hydrocarbon feedstocks. However, this process carries inherent risks, including the potential for hydrogen leakage. Analyzing the failure modes and causes of hydrogen leakage in a steam-reforming unit allows for a detailed examination of critical points in the production process. This understanding, coupled with the FMEA methodology, empowers decision-makers to develop effective risk mitigation strategies.

In addition to FMEA, Multi-Criteria Decision-Making (MCDM) methods play a pivotal role in handling complex decision problems. Among the MCDM techniques, the EDAS (Evaluation based on Distance from Average Solution) [11] method stands out as one of the most applicable. EDAS provides a systematic approach to evaluating multiple criteria simultaneously, helping decision-makers select the optimal solution from a range of alternatives. By incorporating such methods, this paper aims to offer a comprehensive framework that combines failure mode analysis, risk assessment, and decision-making techniques to enhance the safety and reliability of hydrogen production systems.

However, the EDAS method, while providing a structured approach to decision-making, does exhibit certain limitations in its implementation. The EDAS method relies on the arithmetic mean as a central point of reference for evaluating alternatives, which might not capture the nuanced comparative analysis needed in complex decision scenarios. This reliance on the mean could potentially oversimplify the evaluation process, overlooking crucial variations and interactions among criteria. Additionally, the utilization of a linear relative difference in the EDAS method, while aiding in quantifying differences, could render it sensitive to outliers or extreme values within the dataset.

This paper introduces a novel modification to the EDAS method, designed to tackle the challenge of outliers in decision-making. Our proposed modification incorporates a new similarity measure to enhance decision outcomes by addressing the impact of outliers on results.

Section 2 of the paper provides our proposed modification to the EDAS method, emphasizing its unique similarity measure's potential to mitigate the influence of outliers. The study framework is also introduced in this section. Moving to Sect. 3, a case study delves into the application of the modified EDAS method in a risk management analysis for a hydrogen steam generation unit. Finally, Sect. 4 consolidates the findings, discussing the implications of the modification and suggesting future research avenues.

2 Proposed Method

In this section, we introduce our modified EDAS method. For information about the core EDAS method, please refer to the following source [11].

Consider a scenario involving a decision problem characterized by a selection of m choices and n factors for evaluation. The problem is represented through a decision matrix outlined below:

$$D = \begin{array}{c} \\ a_1 \\ a_2 \\ \vdots \\ a_m \end{array} \begin{array}{c} \begin{array}{cccc} c_1 & c_1 & \ldots & c_n \end{array} \\ \begin{pmatrix} x_{11} & x_{12} & \ldots & x_{1m} \\ x_{21} & x_{22} & \ldots & x_{2m} \\ \vdots & \vdots & \vdots & \vdots \\ x_{n1} & x_{n2} & \ldots & x_{nm} \end{pmatrix} \end{array} \tag{1}$$

In this relation, D stands for the decision matrix, where $a_i(i = 1, \ldots, m)$ and $c_j(j = 1, \ldots, n)$ represent the i-th choice and j-th attribute respectively.

Step 1: *Computing Averages and Evaluating Utility and Loss for Each Option.*

In first step, the utility and loss for each option based on average value can be obtained thorugh following options:

$$\begin{cases} u_{ij}{}^{S} = \frac{x_{ij}-x_{sj}}{x_{sj}} if x_{ij} \geq x_{sj}, otherwise\ u_{ij}{}^{S} = 0 \\ l_{ij}{}^{S} = -\frac{x_{ij}-x_{sj}}{x_{sj}} if x_{ij} < x_{sj}, otherwise\ l_{ij}{}^{S} = 0 \end{cases} \text{for beneficial attributes} \tag{2}$$

$$\begin{cases} u_{ij}{}^{S} = -\frac{x_{ij}-x_{sj}}{x_{sj}} if x_{ij} < x_{sj}, otherwise\ u_{ij}{}^{S} = 0 \\ l_{ij}{}^{S} = \frac{x_{ij}-x_{sj}}{x_{sj}} if x_{ij} \geq x_{sj}, otherwise\ l_{ij}{}^{S} = 0 \end{cases} \text{for non-beneficial attributes} \tag{3}$$

where $x_{sj} = \sum_{i=1}^{m} \frac{x_{ij}}{m}$ and $u_{ij}{}^{S}$ and $l_{ij}{}^{S}$ show the utility or loss associated with the i-th alternative concerning the j-th attribute.

Step 2. *Calculating the relative utility and relative loss of each alternative.*

Now the maximum and minimum utility and loss values are defined as follows:

$$\begin{cases} u_j{}^{Smin} = min_i(u_{ij}{}^{S}) \\ u_j{}^{Smax} = max_i(u_{ij}{}^{S}) \\ l_j{}^{Smin} = min_i(l_{ij}{}^{S}) \\ l_j{}^{Smax} = max_i(l_{ij}{}^{S}) \end{cases} \tag{4}$$

where $u_j{}^{Smin}$ and $u_j{}^{Smax}$ represent the minimum and maximum utility values for each criterion, and $l_j{}^{Smin}$ and $l_j{}^{Smax}$ indicate the minimum and maximum loss values for each attribute. Subsequently, the relative utility and loss of the i-th alternative for each attribute are derived by considering the reference points and the extremal values of utility and loss. This is accomplished using the subsequent closeness (similarity) coefficient function:

$$\begin{cases} u_{ij}{}^{r} = \dfrac{\left(u_{ij}{}^{s}+\left(u_{ij}{}^{s}-u_{j}{}^{smin}\right)\right)}{\left(\left(u_{j}{}^{smax}-u_{ij}{}^{s}\right)+u_{ij}{}^{s}+\left(u_{ij}{}^{s}-u_{j}{}^{smin}\right)\right)} 0 \le u_{ij}{}^{r} \le 1 \\ l_{ij}{}^{r} = \dfrac{\left(l_{ij}{}^{s}+\left(l_{ij}{}^{s}-l_{j}{}^{smin}\right)\right)}{\left(l_{ij}{}^{s}+\left(l_{ij}{}^{s}-l_{j}{}^{smin}\right)+\left(l_{j}{}^{smax}-l_{ij}{}^{s}\right)\right)} 0 \le l_{ij}{}^{r} \le 1 \end{cases} \tag{5}$$

Here, $u_{ij}{}^{r}$ and $l_{ij}{}^{r}$ denote the relative utility and loss, respectively, of the i-th alternative in relation to the j-th attribute.

Step 3. *Evaluating the overall weighted relative utility and relative loss for every alternative.*

During this stage, the cumulative weighted relative utility and loss for the i-th alternative concerning all criteria are computed using the subsequent additive functions:

$$\begin{cases} u_i = \sum_{j=1}^{n} w_j u_{ij}{}^{r} \\ l_i = \sum_{j=1}^{n} w_j l_{ij}{}^{r} \end{cases} \tag{6}$$

where $0 \le u_i \le 1$ and $0 \le l_i \le 1$, since $0 < w_j \le 1$, $0 \le u_{ij}{}^{r} \le 1$ and $0 \le l_{ij}{}^{r} \le 1$.

In Eq. (6), u_i and l_i represent the combined weighted utility and loss for the i-th alternative, respectively, while w_j signifies the weight assigned to the j-th attribute.

Step 4: *Computation of Overall Performance for Each Option and Establishing Rankings.*

In the concluding phase, the objective is to determine the overall performance of each alternative, considering their acquired weighted relative values and losses. To achieve this, it's essential to transform the acquired utility and loss values into a uniform advantageous format, thereby enabling normalization. When utilizing the target-oriented linear normalization approach, the following formula applies:

$$\begin{cases} u_i{}^{n} = \frac{u_i}{max_i u_i} \\ l_i{}^{n} = 1 - \frac{l_i}{max_i l_i} \end{cases} \tag{7}$$

In this relation, $u_i{}^{n}$ and $l_i{}^{n}$ symbolize the normalized combined weighted utility and loss for the i-th alternative. Additionally, $max_i u_i$ and $max_i l_i$ represent the maximum total utility and loss values observed among the alternatives. At this point, the comprehensive performance score for each alternative is computed using the normalized weighted total utility and loss values, as outlined by Eq. (8).

$$P_i = \frac{1}{2}\left(u_i{}^{n} + l_i{}^{n}\right) 0 \le P_i \le 1 \tag{8}$$

Here, P_i represents the overall performance of the i-th choice. Ultimately, the options can be arranged in a descending sequence according to their performance.

2.1 Steps of the FMEA Analsysi for Heyrogen Risk Managemneet

Here are the steps for performing hydrogen leakage risk management using the FMEA method and modified EDAS method:

Step 1. Identify Failure Modes

Identify potential failure modes related to hydrogen leakage.

Step 2. Assess Consequences

Evaluate the consequences of each failure mode in terms of safety, environment, and operational impact.

Step 3. Assign Severity Ratings

Rate the severity of consequences associated with each failure mode.

Step 4. Identify Causes and Detection Methods:

Determine underlying causes and methods for detecting or monitoring failure modes.

Step 5. Assign Occurrence and Detection Ratings

Assign likelihood of occurrence and effectiveness of detection ratings to each failure mode.

Step 6. Calculate Risks

Consider appropriate weights for occurrence, severity, and detection factors. Calculate risks using the modified EDAS method, incorporating multiple criteria and their respective weights for comprehensive risk assessment.

3 Case Study

The examined case study centered on a specific hydrogen generation unit utilizing steam reforming, boasting a capacity of 1000 Nm3/h. This process efficiently converts natural gas and steam into hydrogen and carbon dioxide. Steam reforming, involving hydrocarbon-water reactions, is chiefly catalyzed by natural gas, the prevailing hydrocarbon choice for this procedure. The steam methane reformer is structured into two primary segments. In the initial stage, the hydrocarbon reacts with water within a temperature range of 400 to 800 degrees, yielding H2 and CO. This reforming

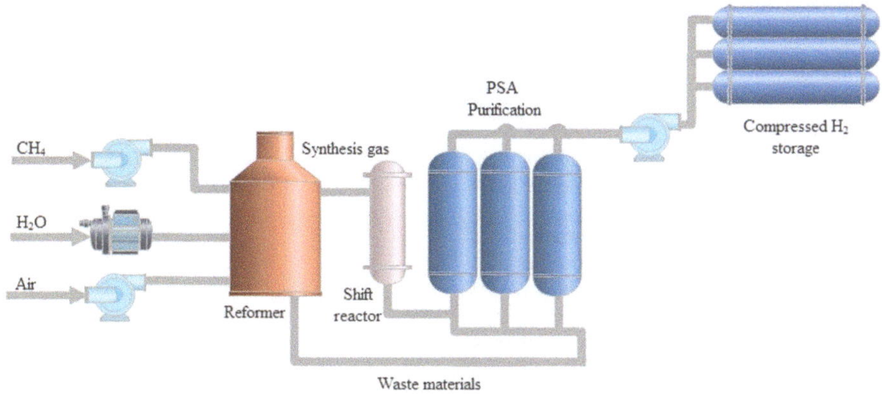

Fig. 1 Hydrogen production process

segment, constituting a crucial component, yields approximately 75% hydrogen and encompasses components such as compressors, desulfurizers, preheat sections, steam reformers, heat recovery systems, and shift reactors. The subsequent phase, pressure swing adsorption (PSA), refines hydrogen to 99% purity and diminishes CO concentration to below five ppm. Illustrated in Fig. 1, the holistic representation of the steam methane reforming process elucidates its overall configuration.

Now our focus lies in identifying and understanding the various failure modes and their underlying causes that may lead to hydrogen leakage within the production system. By scrutinizing these failure modes and causes, we aim to enhance our knowledge of potential vulnerabilities and risks associated with the operation of the hydrogen generation unit. This investigation will enable us to implement effective risk mitigation strategies and bolster system safety, thereby contributing to the overall reliability and sustainability of hydrogen production processes (Table 1).

Using the proposed methodology and following steps 1 to 4, the performance of various failure modes is calculated and subsequently ranked. The ranking outcomes obtained from the modified EDAS approach are presented in Table 2. Additionally, several weight scenarios have been taken into account, and a sensitivity analysis on these weights has been conducted, as illustrated in Fig. 2.

The results indicate that "Corrosion of Piping and Equipment," "Cracking in Piping or Components," and "Inadequate Maintenance" are identified as critical failure modes. Conversely, "Insufficient Heat Recovery" and "Contamination of Feedstock" are considered the least critical failure modes.

To enhance the reliability and consistency of the results, the ranking of failure modes has been evaluated using other established Multiple-Criteria Decision-Making (MCDM) methods, including WASPAS [12], TOPSIS [13], ARAS [14], EDAS [11], and MARCOS [15]. The performance and outcomes for each failure mode are presented in Table 2.

Remarkably, the results derived from our proposed method exhibit strong agreement with the outcomes of the WASPAS method and maintain consistency

Table 1 FMEA evaluations

Index	Failure mode	Causes	Occurrence (O)	Severity (S)	Consequences (C)
			WO = 0.3	WS = 0.4	WD = 0.3
FM1	Corrosion of Piping and Equipment	Exposure to corrosive gases (e.g., sulfur compounds), impurities in feedstock, high-temperature operation, lack of corrosion-resistant materials	3	8	6
FM2	Seal Degradation in Valves and Joints	Wear and tear due to frequent cycling, thermal expansion/contraction, exposure to high-temperature/pressure, inadequate maintenance	4	6	5
FM3	Cracking in Piping or Components	Thermal cycling, mechanical stress from process fluctuations, improper welding practices, material defects	2	9	7
FM4	Improper Reactor Operation	Incorrect temperature control, inadequate pressure regulation, malfunctioning valves, deviations from optimal operational conditions	3	7	4
FM5	Inadequate Feedstock Preparation	Inconsistent feedstock composition, improper mixing, variations in feedstock quality, inadequate quality control	4	5	3

(continued)

Table 1 (continued)

Index	Failure mode	Causes	Occurrence (O)	Severity (S)	Consequences (C)
FM6	Leakage in Seals and Gaskets	Deterioration due to thermal/chemical exposure, mechanical wear from operational stresses, compromised containment	4	6	5
FM7	Insufficient Heat Recovery	Inadequate heat exchanger performance due to fouling or scaling, lack of effective heat recovery mechanisms	3	4	4
FM8	Incorrect Pressure Regulation	Malfunctioning pressure control devices, inconsistencies in pressure measurement, deviations from optimal settings	3	5	4
FM9	Contamination of Feedstock	Introduction of impurities during storage/handling, variations in feedstock sources, inadequate purification steps	2	6	3
FM10	Inadequate Maintenance	Irregular inspection schedules, lack of preventive maintenance planning, deferred repairs due to resource constraints	3	7	6

(continued)

Table 1 (continued)

Index	Failure mode	Causes	Occurrence (O)	Severity (S)	Consequences (C)
FM11	Poor Hydrogen Purification	Inefficient PSA process design, degradation of purification materials over time, deviations from optimal PSA operation	2	8	5
FM12	Control System Malfunction	Sensor inaccuracies, communication failures between control units, software glitches affecting control parameters	2	7	4

Table 2 Ranking results of different approaches

	Proposed method	Rank	TOPSIS	Rank	WASPAS	Rank	Basic EDAS	Rank	Aras	Rank	MARCOS	Rank
FM1	0.91	1	0.708	1	0.836	1	0.307	7	0.98	2	0.732	2
FM2	0.68	4	0.546	5	0.774	4	0.284	9	0.91	5	0.683	4
FM3	0.79	2	0.662	2	0.831	2	0.703	1	1	1	0.743	1
FM4	0.509	7	0.469	7	0.704	7	0.157	12	0.82	7	0.619	7
FM5	0.305	8	0.375	9	0.631	8	0.612	2	0.75	8	0.569	8
FM6	0.68	5	0.546	4	0.774	5	0.284	10	0.91	4	0.683	5
FM7	0.14	11	0.241	12	0.567	11	0.437	4	0.67	11	0.502	11
FM8	0.209	10	0.291	10	0.615	10	0.305	8	0.72	10	0.541	10
FM9	0	12	0.247	11	0.540	12	0.5	3	0.63	12	0.477	12
FM10	0.78	3	0.623	3	0.792	3	0.205	11	0.93	3	0.694	3
FM11	0.56	6	0.527	6	0.710	6	0.390	5	0.84	6	0.629	6
FM12	0.27	9	0.390	8	0.626	9	0.345	6	0.73	9	0.553	9

with TOPSIS, ARAS, and MARCOS methods. In contrast, the results from the conventional EDAS approach show significant inconsistency and divergence when compared to the proposed method and other methodologies. This discrepancy might be attributed to inherent limitations within classic EDAS, as discussed in the Introduction.

Consequently, our proposed methodology holds potential for practical application by decision-makers across diverse sectors. Based on the derived results, organizations can devise plans to address critical failure modes and mitigate various risk factors encompassing occurrence, severity, and detection. For instance, for addressing the risk associated with "Corrosion of Piping and Equipment," mitigation strategies such

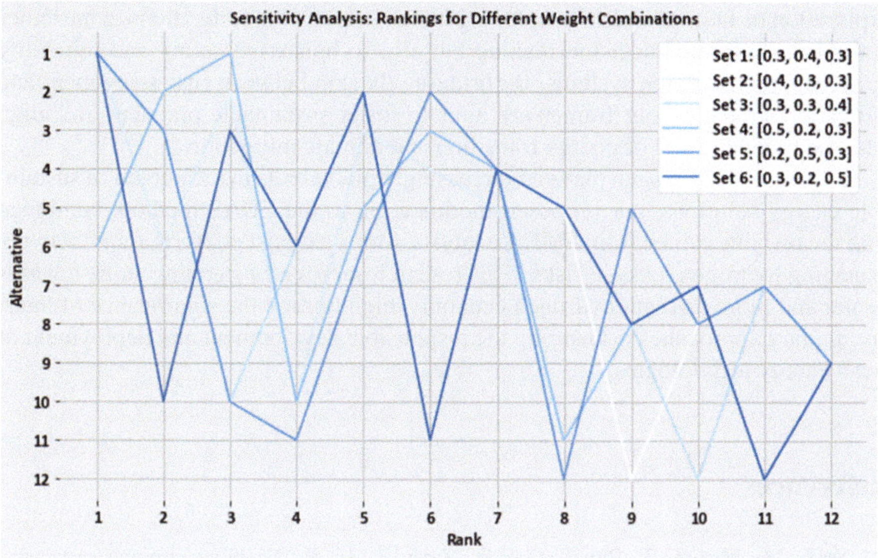

Fig. 2 Sensitivity analysis

as implementing Coatings and Linings, employing Cathodic Protection measures, and more can be carefully considered and tailored accordingly.

4 Conclusion

In conclusion, the journey towards a sustainable energy future hinges on harnessing the potential of hydrogen as a versatile and clean energy carrier within a low-carbon economy. Its diverse production methods afford adaptability to changing resources and evolving technologies, making it a promising contender for addressing environmental challenges. However, the prospect of hydrogen leakage necessitates vigilant attention due to its multifaceted implications encompassing safety, environmental integrity, and economic viability.

This paper has introduced a novel modification to the EDAS method, complemented by the FMEA technique, to scrutinize and manage the risks associated with hydrogen leakage in steam-reforming hydrogen generation units. The thorough analysis of failure modes and their underlying causes has emerged as a pivotal tool for upholding the safety and efficiency of hydrogen production systems. Furthermore, recognizing the limitations of conventional methods, our incorporation of a novel similarity measure within the EDAS method addresses the challenges posed by outliers, enhancing the accuracy and robustness of decision-making processes.

The practical case study presented in this research underscores the tangible benefits of our proposed approach in managing hydrogen leakage risks. The systematic exploration of failure modes, coupled with the innovative modifications, contributes not only to informed decision-making but also to bolstering safety and reliability in hydrogen production systems. By bridging the gap between risk assessment and mitigation strategies, our framework aims to foster sustainable practices and align with the broader goals of energy transition and climate mitigation.

In essence, as hydrogen plays an increasingly pivotal role in the pursuit of sustainable energy solutions, our proposed modification to the EDAS method, combined with the insights gained from FMEA analysis, forms a robust platform for effectively managing hydrogen leakage risks. This research serves as a stepping stone towards a safer and more resilient hydrogen economy, highlighting the significance of interdisciplinary approaches in ensuring the responsible development and deployment of clean energy technologies.

References

1. Yazdi, M., Moradi, R., Pirbalouti, R.G., Zarei, E., Li, H.: Enabling safe and sustainable hydrogen mobility: circular economy-driven management of hydrogen vehicle safety. Processes **11** (2023). https://doi.org/10.3390/pr11092730
2. Chau, K., Djire, A., Khan, F.: Review and analysis of the hydrogen production technologies from a safety perspective. Int. J. Hydrogen Energy **47**, 13990–14007 (2022)
3. Faye, O., Szpunar, J., Eduok, U.: A critical review on the current technologies for the generation, storage, and transportation of hydrogen. Int. J. Hydrogen Energy **47**, 13771–13802 (2022)
4. Moradi, R., Groth, K.M.: Hydrogen storage and delivery: review of the state of the art technologies and risk and reliability analysis. Int. J. Hydrogen Energy **44**, 12254–12269 (2019)
5. Abohamzeh, E., Salehi, F., Sheikholeslami, M., Abbassi, R., Khan, F.: Review of hydrogen safety during storage, transmission, and applications processes. J. Loss Prev. Process Ind. **72**, 104569 (2021)
6. Groth, K.M., Hecht, E.S.: HyRAM: a methodology and toolkit for quantitative risk assessment of hydrogen systems. Int. J. Hydrogen Energy **42**, 7485–7493 (2017)
7. Baykara, S.Z.: Hydrogen: a brief overview on its sources, production and environmental impact. Int. J. Hydrogen Energy **43**, 10605–10614 (2018)
8. Wang, X., Zhang, C., Gao, W.: Risk assessment of hydrogen leakage in diesel hydrogenation process. Int. J. Hydrogen Energy **47**, 6955–6964 (2022)
9. ISO.: Basic considerations for the safety of hydrogen systems (2004)
10. Seiti, H., Fathi, M., Hafezalkotob, A., Herrera-Viedma, E., Hameed, I.A.: Developing the modified R-numbers for risk-based fuzzy information fusion and its application to failure modes, effects, and system resilience analysis (FMESRA). ISA Trans. **113**, 9–27 (2021)
11. Li, H., Yazdi, M.: An Advanced TOPSIS-PFS Method to Improve Human System Reliability BT - Advanced Decision-Making Methods and Applications in System Safety and Reliability Problems: Approaches, Case Studies, Multi-criteria Decision-Making, Multi-objective Decision-Making,. In: Li H, Yazdi M, editors., Cham: Springer International Publishing, pp. 109–125 (2022). https://doi.org/10.1007/978-3-031-07430-1_7
12. Chakraborty, S., Zavadskas, E.K., Antucheviciene, J.: Applications of WASPAS method as a multi-criteria decision-making tool. Econ Comput Econ Cybern Stud Res **49**, 5–22 (2015)
13. Yoon KP, Hwang C-L. Multiple attribute decision making: an introduction. Sage publications; 1995.

14. Zavadskas, E.K., Turskis, Z.: A new additive ratio assessment (ARAS) method in multicriteria decision-making. Technol. Econ. Dev. Econ. **16**, 159–172 (2010)
15. Stević, Ž., Pamučar, D., Puška, A., Chatterjee, P.: Sustainable supplier selection in healthcare industries using a new MCDM method: Measurement of alternatives and ranking according to COmpromise solution (MARCOS). Comput & Ind Eng 2020;140:106231

Multi-Criteria Decision-Making for Optimal Selection of Hydrogen Fuel Cell Types: Evaluating Sustainability and Performance Criteria

Hamidreza Seiti, Amir Mohammad Larni-Fooeik, and Ali Elkamel

Abstract Hydrogen fuel cells offer promising benefits for energy production, operating through chemical reactions between hydrogen and oxygen to generate electricity and water. Their advantages span from high efficiency and low emissions to potential as a renewable energy source. However, challenges such as economic costs and safety concerns, owing to hydrogen's flammability and explosiveness, must be addressed for successful integration. The paper reviews prominent fuel cell types including Proton Exchange Membrane Fuel Cells (PEMFC), Solid Oxide Fuel Cells (SOFC), Alkaline Fuel Cells (AFC), and Molten Carbonate Fuel Cells (MCFC), employing MCDA techniques to rank these cells in alignment with organizational criteria.

Keywords Hydrogen fuel cell · Sustainability · MCDM

1 Introduction

Hydrogen fuel cells represent a revolutionizing technology in the world of energy production and usage, providing a multitude of benefits and opportunities across various sectors [1]. They operate by harnessing the chemical reactions transpiring between hydrogen and oxygen and converting these reactions into electricity and pure water [2]. The advantages of hydrogen fuel cells are numerous and diverse, extending from their high efficiency and reduced emissions to their potential as a source of renewable energy [3]. Uniquely, unlike conventional batteries which require

H. Seiti (✉) · A. M. Larni-Fooeik
Department of Industrial Engineering, Iran University of Science and Technology, Tehran, Iran
e-mail: h_seiti@yahoo.com

A. Elkamel
Department of Chemical Engineering, Khalifa University, Abu Dhabi, UAE

Department of Chemical Engineering, University of Waterloo, ON, Canada

M. Yazdi (ed.), *Progressive Decision-Making Tools and Applications in Project and Operation Management*, Studies in Systems, Decision and Control 518,
https://doi.org/10.1007/978-3-031-51719-8_9

intermittent recharging, hydrogen fuel cells continually generate electricity as long as a steady supply of hydrogen and oxygen is maintained [4].

The fundamental structure of a hydrogen fuel cell comprises three key components: the anode, cathode, and an electrolyte membrane [5]. At the anode, hydrogen molecules undergo a process of electrolysis, splitting into protons and electrons; These protons traverse the electrolyte membrane to reach the cathode, while the electrons travel through an external circuit, thus generating electricity [6]. At the cathode, oxygen molecules interact with protons and electrons to create water, which is the sole byproduct of the reaction [7].

Applications for hydrogen fuel cells span a wide spectrum of industries that include power generation, transportation, industrial processes, and heating [8]. Their adaptability and potential to revolutionize the energy landscape make them an enticing solution for sustainable development [9]. Nevertheless, alongside these prospects exist prominent challenges and hazards that necessitate careful consideration and management for the technology's successful integration into our energy systems [10].

A primary challenge revolves around the economic aspects of hydrogen fuel cells, as they currently involve high costs and limited durability [11]. The materials used in their construction, such as platinum-based catalysts, are expensive and can degrade over time. Moreover, hydrogen fuel cells face risks related to their inherent properties. Hydrogen gas, although a promising energy carrier, is highly flammable and explosive over a broad range of concentrations [12]. Its invisibility and high combustion temperature further complicate safety considerations. Additionally, hydrogen gas can embrittle certain metals and pose asphyxiation hazards when confined in closed spaces [13].

One of the best approaches to consider and evaluate various aspects of different hydrogen cells is multi-criteria decision-making methods (MCDM). This paper aims to utilize well-known MCDM methods to select the optimal fuel cell based on an organization's critical criteria. The exploration begins with an overview of various fuel cell types. To facilitate the decision-making process, several MCDA techniques are employed to rank these fuel cells in accordance with the organization's established criteria. Section 2 delves into a detailed case study, while Sect. 3 presents the obtained results and conclusions.

2 Case Study

In this case study, the objective is to assess the diverse attributes of various fuel cell types and prioritize their alignment with the organization's critical criteria, which encompass factors such as efficiency, environmental impact, cost-effectiveness, and overall performance. Through a comprehensive analysis, we thoroughly examine four prominent fuel cell types: Proton Exchange Membrane Fuel Cells (PEMFC), Solid Oxide Fuel Cells (SOFC), Alkaline Fuel Cells (AFC), and Molten Carbonate

Fuel Cells (MCFC). These fuel cell technologies are meticulously evaluated and their comparative attributes are presented in Table 1.

In the context of this case study focused on the selection of hydrogen fuel cells, a comprehensive set of selection criteria and subcriteria has been meticulously curated. These criteria have been thoughtfully categorized into distinct aspects, including Economic, Safety, Environmental, Technical, Social and Ethical, as well as Regulatory and Legal considerations. Detailed explanations for each of these criteria and their corresponding subcriteria can be found in the informative Table 2. To ensure a holistic evaluation, a qualitative assessment of various hydrogen fuel cell options has been undertaken, measuring their performance against these predetermined criteria. The relative importance of each criterion and subcriterion has been assigned specific weights, signifying their relative significance within the decision-making process (Table 3).

The transition of the qualitative decision matrix in Table 3 into a quantitative framework is essential to apply deterministic MCDM methods effectively. This conversion involves assigning numerical values, using a 1 to 5 scale, to the linguistic descriptions, where 1 signifies minimal importance and 5 indicates the highest significance. This process not only streamlines the application of deterministic MCDM techniques but also enhances the consistency and comparability of the evaluation. The quantified matrix is presented in Table 4.

In the conclusive stages of this study, the quantified matrix, as depicted in Table 4, is subjected to four well-established MCDM methodologies, namely, WASPAS [14], TOPSIS [15], ARAS [16] and EDAS [17]. The resulting rankings are meticulously synthesized and presented in both Table 5 and Fig. 1. Notably, the outcomes of all methodologies, with the exception of the EDAS method, converge to designate the Solid Oxide Fuel Cell (SOFC) as the optimal alternative in accordance with both economic and sustainability criteria, solidifying its prominence in the selection process. Following suit, the Proton Exchange Membrane Fuel Cell (PEMFC) garners recognition as the second-best alternative. Given the incongruence of EDAS results with the other methods—a divergence attributed to EDAS' susceptibility to outliers and inherent limitations—its outcomes are judiciously disregarded.

These discerning results furnish decision-makers with a confident course of action, providing clear guidance for strategic investments within their organizations. With the robust endorsement of the SOFC hydrogen fuel cell as the premier choice across various economical and sustainable criteria, this research underscores its viability as a transformative and impactful technology for advancing energy sustainability and driving forward-looking initiatives.

Table 1 Hydrogen fuel cells and their descriptions

<table>
<tr><th>Fuel Cell Type</th><th>Definition</th><th>Description</th></tr>
<tr><td>Proton Exchange Membrane Fuel Cell (PEMFC)</td><td>A hydrogen fuel cell that operates at relatively low temperatures and uses a proton-conductive polymer electrolyte membrane as the separator between the anode and cathode. It converts hydrogen and oxygen into electricity and water. It is commonly used in applications like automobiles and portable electronics</td><td>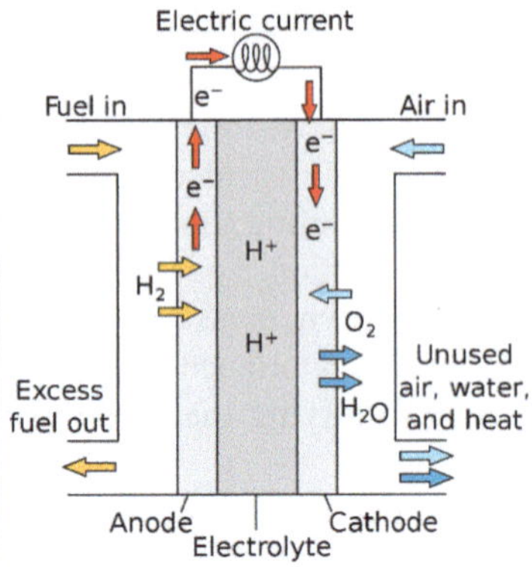
</td></tr>
<tr><td>Solid Oxide Fuel Cell (SOFC)</td><td>A high-temperature hydrogen fuel cell that utilizes a solid oxide electrolyte. It directly converts hydrogen and oxygen into electricity and water without the need for an external reformer. SOFCs are known for their high efficiency and adaptability to various fuels. They are suitable for stationary power generation and industrial applications</td><td>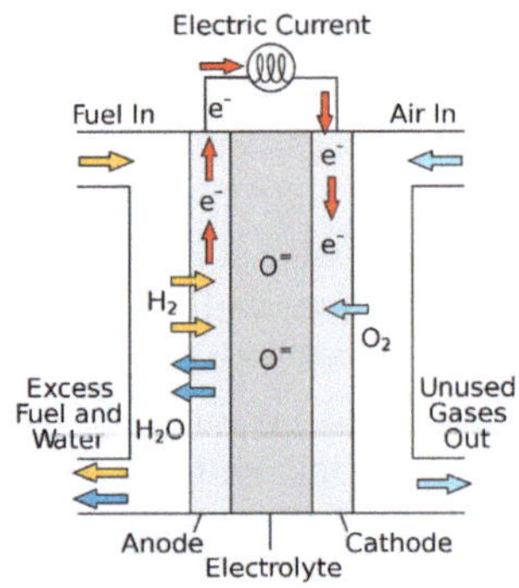
</td></tr>
<tr><td>Alkaline Fuel Cell (AFC)</td><td>An early type of hydrogen fuel cell that employs an alkaline electrolyte, typically potassium hydroxide solution. AFCs convert hydrogen and oxygen into electricity and water and are known for their high efficiency and rapid reaction rates. They have been used in space exploration and certain industrial applications</td><td>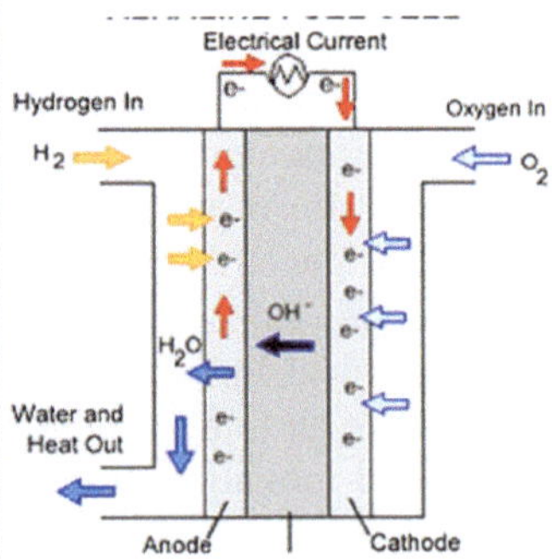
</td></tr>
</table>

(continued)

Table 1 (continued)

Fuel Cell Type	Definition	Description
Molten Carbonate Fuel Cell (MCFC)	A high-temperature hydrogen fuel cell that uses a molten carbonate electrolyte. It can convert a variety of fuels, including hydrogen and natural gas, into electricity and water. MCFCs are notable for their ability to capture carbon dioxide during operation, making them suitable for large-scale power generation and industrial applications	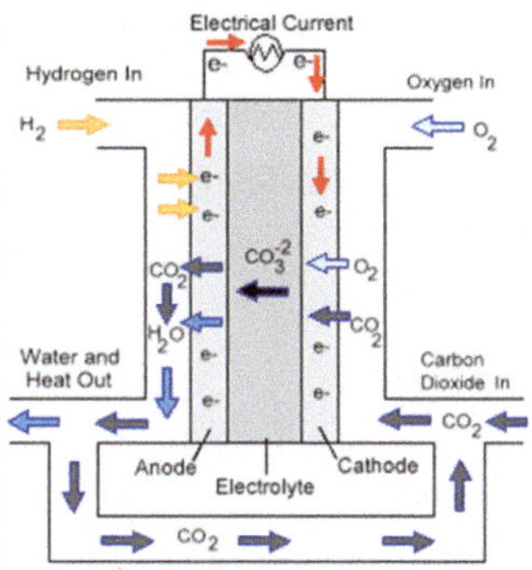
Phosphoric Acid Fuel Cell (PAFC)	A hydrogen fuel cell that operates at moderate temperatures and uses phosphoric acid as the electrolyte. PAFCs are commonly used for stationary power generation, especially in combined heat and power systems. They offer a longer lifespan and higher tolerance for impurities compared to some other fuel cell types	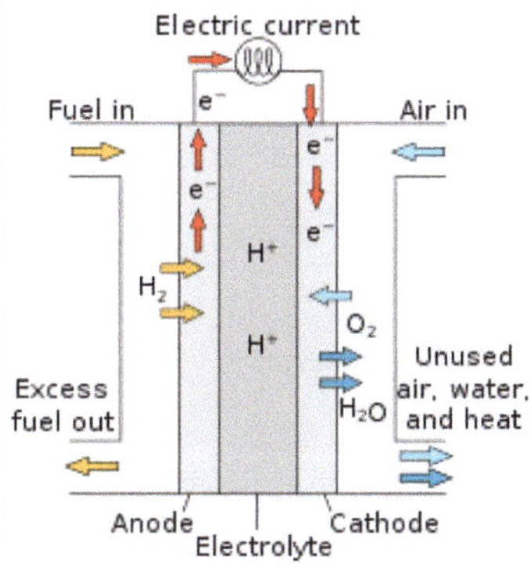

3 Conclusion

In conclusion, hydrogen fuel cells have emerged as a revolutionary technology with the potential to reshape energy landscapes across diverse sectors. Their ability to efficiently convert hydrogen and oxygen into electricity and water presents a clean and continuous power generation solution. However, their integration comes with challenges, including economic barriers and safety concerns due to hydrogen's flammability. This study employs Multi-Criteria Decision-Making methods to evaluate various fuel cell types and concludes that Solid Oxide Fuel Cells (SOFC) stand out as the optimal choice based on established criteria. SOFC's alignment with economic, safety, environmental, technical, and ethical considerations positions them as a leading contender in the quest for sustainable energy solutions. As hydrogen fuel cells advance, addressing these challenges while leveraging their transformative potential could catalyze a future powered by resilient and eco-friendly

Table 2 The selected criteria and their definitions

Criteria	Description
Economic Criteria	These criteria evaluate the economic aspects of the hydrogen fuel cell:
– Cost	Considers the overall cost of the fuel cell, including manufacturing, installation, and operational expenses
– Longevity and Durability	Examines the expected lifespan and durability of the fuel cell, impacting its long-term cost-effectiveness
– Fuel Availability	Assesses the availability and accessibility of the required hydrogen fuel for the fuel cell's operation
– Cost of Fuel	Evaluates the cost of the hydrogen fuel needed to power the fuel cell, influencing its economic feasibility
Safety Criteria	These criteria focus on the safety aspects of the hydrogen fuel cell:
– Flammability	Assesses the risk of flammability associated with hydrogen, considering its potential to ignite and cause fires
– Toxicity	Evaluates the level of toxicity of hydrogen, addressing potential health risks associated with exposure
– Asphyxiation Risk	Considers the risk of asphyxiation due to the displacement of oxygen by hydrogen gas, which can occur in confined spaces
– Operational Safety	Measures the overall safety of the fuel cell's operational processes, including safeguards against potential hazards
Environmental Criteria	These criteria gauge the environmental impact of the hydrogen fuel cell:
– Emissions	Evaluates the emissions produced during the fuel cell's operation and their potential impact on air quality and climate
– Resource Use	Assesses the consumption of resources, such as raw materials and energy, in the production and operation of the fuel cell
– Waste Management	Considers how waste generated during the fuel cell's lifecycle is managed, aiming for minimal environmental impact
– Sustainability	Examines the overall sustainability of the fuel cell, considering its environmental, social, and economic implications
Technical Criteria	These criteria pertain to the technical performance of the hydrogen fuel cell:
– Efficiency	Measures the efficiency of the fuel cell in converting hydrogen and oxygen into electricity, reflecting energy utilization

(continued)

Table 2 (continued)

Criteria	Description
– Power Output	Evaluates the electrical power output capacity of the fuel cell, influencing its applicability for various uses
– Operating Conditions	Considers the range of environmental conditions under which the fuel cell can operate effectively and reliably
– Start-up Time and Response	Analyzes the fuel cell's responsiveness and speed of starting up, influencing its usability in dynamic scenarios
Social and Ethical Criteria	These criteria address the societal and ethical implications of the fuel cell:
– Community Impact	Assesses the potential positive or negative effects the fuel cell may have on local communities and residents
– Job Creation	Evaluates the extent to which the fuel cell technology contributes to job creation and economic opportunities
– Ethical Considerations	Reflects on ethical aspects associated with the production, use, and disposal of the fuel cell and its components
Regulatory and Legal Criteria	These criteria consider the regulatory and legal aspects of the fuel cell:
– Regulatory Compliance	Examines how well the fuel cell complies with existing regulations and standards governing its operation and safety
– Permitting and Licensing	Considers the ease of obtaining necessary permits and licenses for the installation and operation of the fuel cell
– Liability and Legal Considerations	Assesses potential legal liabilities and responsibilities associated with the fuel cell's use

energy systems. Looking ahead, future studies will be focused on investigating various risks related to SOFC, ensuring a comprehensive understanding of their practical implementation and contributing to the ongoing refinement of this remarkable technology.

Table 3 The decision matrix and the weights of the criteria

Criteria	Benefit/Cost	Weight	PEMFC	SOFC	AFC	MCFC	PAFC
Economic Criteria	Benefit	**0.15**					
– Cost	Cost	0.40	High	Moderate	High	High	Moderate
– Longevity and Durability	Benefit	0.25	Low	High	Low	Moderate	Moderate
– Fuel Availability	Benefit	0.20	High	High	Low	High	High
– Cost of Fuel	Cost	0.15	Low	Low	High	Low	Low
Safety Criteria	Cost	**0.20**					
– Flammability	Cost	0.30	Low	Low	Low	Low	Low
– Toxicity	Cost	0.15	Low	Low	Low	Moderate	Moderate
– Asphyxiation Risk	Cost	0.20	High	High	High	High	High
– Operational Safety	Benefit	0.35	Moderate	Moderate	Moderate	Low	Low
Environmental Criteria		**0.15**					
– Emissions	Cost	0.30	None	None	None	CO2	CO2
– Resource Use	Cost	0.25	Moderate	Moderate	Moderate	High	High
– Waste Management	Benefit	0.15	Low	Low	Low	Moderate	Moderate
– Sustainability	Benefit	0.30	High	High	High	Moderate	Moderate
Technical Criteria		**0.20**					
– Efficiency	Benefit	0.40	60%	60%	50%	40%	
– Power Output	Benefit	0.25	< 1 kW – 100 kW	1 kW – 2 MW	1 – 100 kW	300 kW – 3 MW	5 – 400 kW
– Operating Conditions	Cost	0.20	< 120°C	500°–1,000°C	< 100°C	600°–700°C	150°–200°C
– Start-up Time and Response	Cost	0.15	Fast	Fast	Fast	Slow	Slow
Social and Ethical Criteria		**0.10**					
– Community Impact	Cost	0.30	Low	Low	Low	Moderate	Moderate

(continued)

Table 3 (continued)

Criteria	Benefit/Cost	Weight	PEMFC	SOFC	AFC	MCFC	PAFC
– Job Creation	Benefit	0.25	Moderate	Moderate	Moderate	Moderate	Moderate
– Ethical Considerations	Cost	0.45	Low	Low	Low	Low	Low
Regulatory and Legal Criteria		**0.10**					
– Regulatory Compliance	Benefit	0.60	Easy	Easy	Difficult	Difficult	Difficult
– Permitting and Licensing	Benefit	0.25	Easy	Easy	Difficult	Difficult	Difficult
– Liability and Legal Considerations	Cost	0.15	Low	Low	Moderate	Moderate	Moderate

Table 4 The quantified matrix

Criteria	Benefit/Cost	Weight	PEMFC	SOFC	AFC	MCFC	PAFC
Economic Criteria		0.15					
– Cost	Cost	0.40	5	3	5	5	3
– Longevity and Durability	Benefit	0.25	1	5	1	3	3
– Fuel Availability	Benefit	0.20	5	5	1	5	5
– Cost of Fuel	Cost	0.15	1	1	5	1	1
Safety Criteria		0.20					
– Flammability	Cost	0.30	1	1	1	1	1
– Toxicity	Cost	0.15	1	1	1	3	3
– Asphyxiation Risk	Cost	0.20	5	5	5	5	5
– Operational Safety	Benefit	0.35	3	3	3	1	1
Environmental Criteria		0.15					
– Emissions	Cost	0.30	1	1	1	4	4
– Resource Use	Cost	0.25	3	3	3	5	5
– Waste Management	Benefit	0.15	1	1	1	3	3
– Sustainability	Benefit	0.30	5	5	5	3	3
Technical Criteria		0.20					
– Efficiency	Benefit	0.40	4	4	3	2	4
– Power Output	Benefit	0.25	3	4	3	4	3
– Operating Conditions	Cost	0.20	1	5	3	5	3
– Start-up Time and Response	Cost	0.15	5	5	5	1	1
Social and Ethical Criteria		0.10					
– Community Impact	Cost	0.30	1	1	1	2	2
– Job Creation	Cost	0.25	3	3	3	3	3
– Ethical Considerations	Cost	0.45	1	1	1	1	1
Regulatory and Legal Criteria		0.10					
– Regulatory Compliance	Benefit	0.60	5	5	1	1	1
– Permitting and Licensing	Benefit	0.25	5	5	1	1	1
– Liability and Legal Considerations	Cost	0.15	2	2	3	3	3

Table 5 The ranking results

Hydrogen Cells	TOPSIS	Rank	EDAS	Rank	WASPAS	Rank	ARAS	Rank	COPRAS	Rank
PEMFC	0.706	2	0.551	5	0.868	2	0.962	2	0.243	2
SOFC	0.786	1	0.567	2	0.915	1	1	1	0.257	1
AFC	0.367	3	0.560	3	0.635	3	0.680	4	0.1705	4
MCFC	0.283	5	0.601	1	0.582	5	0.626	5	0.157	5
PAFC	0.341	4	0.554	4	0.633	1	0.683	3	0.1708	3

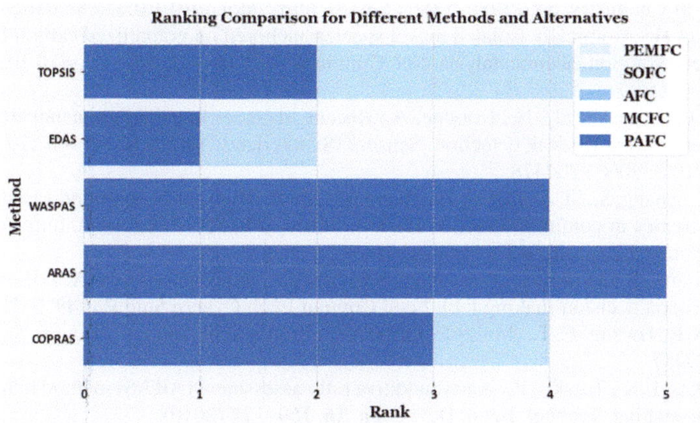

Fig. 1 Comparison results of different methods

References

1. Brandon, N.P., Kurban, Z.: Clean energy and the hydrogen economy. *Philosophical Transactions of the Royal Society A: Mathematical, Physical and Engineering Sciences*, vol. 375, no. 2098 (2017). https://doi.org/10.1098/rsta.2016.0400
2. Jin, Z., Li, P., Xiao, D.: A hydrogen-evolving hybrid-electrolyte battery with electrochemical/ photoelectrochemical charging from water oxidation. Chemsuschem **10**(3), 483–488 (2017). https://doi.org/10.1002/cssc.201601317
3. Adam, N., Schlicht, S., Han, Y., Bechelany, M., Bachmann, J., Perner, M.: Metagenomics meets electrochemistry: utilizing the huge catalytic potential from the uncultured microbial majority for energy-storage. *Front Bioeng Biotechnol*, vol. 8 (2020). https://doi.org/10.3389/fbioe.2020.00567
4. Gençer, E., Mallapragada, D.S., Maréchal, F., Tawarmalani, M., Agrawal, R.: Round-the-clock power supply and a sustainable economy via synergistic integration of solar thermal power and hydrogen processes. Proc Natl Acad Sci U S A **112**(52), 15821–15826 (2015). https://doi.org/10.1073/pnas.1513488112
5. Yazdi, M., Moradi, R., Pirbalouti, R.G., Zarei, E., Li, H.: Enabling safe and sustainable hydrogen mobility: circular economy-driven management of hydrogen vehicle safety. Processes 2023; 11 (2023). https://doi.org/10.3390/pr11092730
6. Abrevaya, X.C., Sacco, N.J., Bonetto, M.C., Hilding-Ohlsson, A., Cortón, E.: Analytical applications of microbial fuel cells. Part I: Biochemical oxygen demand. *Biosensors and*

Bioelectronics, vol. 63. Elsevier Ltd, pp. 580–590 (2015). https://doi.org/10.1016/j.bios.2014.04.034
7. Kim, J., Kwon, D., Kim, K., Hoffmann, M.R.: Electrochemical production of hydrogen coupled with the oxidation of arsenite. Environ. Sci. Technol. **48**(3), 2059–2066 (2014). https://doi.org/10.1021/es4046814
8. Ayers, K., Danilovic, N., Ouimet, R., Carmo, M., Pivovar, B., Bornstein, M.: Perspectives on Low-Temperature Electrolysis and Potential for Renewable Hydrogen at Scale (2019). https://doi.org/10.1146/annurev-chembioeng
9. Khaheshi, A., Rajabi, H.: Mechanical Intelligence (MI): a bioinspired concept for transforming engineering design. Adv. Sci. **9**(32) (2022). https://doi.org/10.1002/advs.202203783
10. Rajasekhar, K., Govindaraju, T.: Current progress, challenges and future prospects of diagnostic and therapeutic interventions in Alzheimer's disease. *RSC Advances*, vol. 8, no. 42. Royal Society of Chemistry, pp. 23780–23804 (2018). https://doi.org/10.1039/c8ra03620a
11. Fan, L. et al.: Atomically isolated nickel species anchored on graphitized carbon for efficient hydrogen evolution electrocatalysis. Nat. Commun. **7** (2016). https://doi.org/10.1038/ncomms10667
12. Ndaya, C.C., Javahiraly, N., Brioude, A.: Recent advances in palladium nanoparticles-based hydrogen sensors for leak detection. Sensors (Switzerland) **19**(20) MDPI AG (2019). https://doi.org/10.3390/s19204478
13. Tang, C., Zhang, S., Si, Z., Huang, Z., Zhang, K., Jin, Z.: High methane natural gas/air explosion characteristics in confined vessel. J. Hazard. Mater. **278**, 520–528 (2014). https://doi.org/10.1016/j.jhazmat.2014.06.047
14. Chakraborty, S., Zavadskas, E.K., Antucheviciene, J.: Applications of WASPAS method as a multi-criteria decision-making tool. Econ Comput Econ Cybern Stud Res **49**, 5–22 (2015)
15. Yoon, K.P., Hwang, C.-L.: Multiple attribute decision making: an introduction. Sage publications (1995)
16. Zavadskas, E.K., Turskis, Z.: A new additive ratio assessment (ARAS) method in multicriteria decision-making. Technol. Econ. Dev. Econ. **16**, 159–172 (2010)
17. Keshavarz Ghorabaee, M., Zavadskas, E.K., Olfat, L., Turskis, Z.: Multi-criteria inventory classification using a new method of evaluation based on distance from average solution (EDAS). Informatica **26**, 435–451 (2015)

Enhancing Supplier Selection Reliability: Integrated Variable Returns to Scale-Robust Parameter R Approach

Kehinde Adewale Adesina and Mohammad Yazdi

Abstract Effective decision-making requires a suitable supplier selection model that can consistently identify and recruit suppliers, especially in complex scenarios involving multiple variables. Recently, integrated-data envelopment analysis techniques, such as fuzzy-based methods, cross-efficiency, and cooperative games have demonstrated promising outcomes. However, these approaches have yet to address the challenge of accounting for variations caused by undetectable supplier indicators. To overcome this limitation, this study examines the relationships between input selection criteria and supplier indicators in a way that minimizes the impact of variations stemming from undetectable indicators on supplier performance responses. To achieve this, we propose an integrated and refined model that combines revamped variable returns to scale (VRS) with robust parameter estimation and multivariate multiple-dependent regressions. By incorporating the VRS methodology, our model ensures that the selection process accounts for variations in scale efficiency across suppliers, enabling a fair and accurate comparison. The robust parameter estimation technique also helps mitigate the influence of outliers and extreme observations, enhancing the model's reliability and robustness. Furthermore, the multivariate multiple-dependent regressions approach allows for considering multiple dependent variables simultaneously, enabling a comprehensive evaluation of supplier performance. This approach considers the interrelationships between performance indicators and ensures a more holistic assessment of supplier capabilities. In conclusion, our integrated refined model offers a comprehensive and effective solution for supplier selection in complex cases.

Keywords Multivariate multiple regression · Parameter optimization · Supplier selection multi-response problem · Data envelopment partitioning

K. A. Adesina
Independent Researcher, 32 Radburn Close, Harlow, Essex, UK

M. Yazdi (✉)
School of Computing, Engineering and Physical Sciences, University of the West of Scotland, London, UK
e-mail: adexkwale@gmail.com

M. Yazdi (ed.), *Progressive Decision-Making Tools and Applications in Project and Operation Management*, Studies in Systems, Decision and Control 518,
https://doi.org/10.1007/978-3-031-51719-8_10

1 Introduction

There is no doubt globalization is constantly bringing about new business strategies that are responsible for the competitiveness of the market especially in the aspect of supplier selection. The speed of transacting a business along service rate, communication and technological ideas for decision-making are always influenced in one way or the other. These issues have been complicated by the complex nature of the problems associated with the means of identifying, qualifying and quantifying supplier performance indicators for solving supplier selection problems in the supply chain management. The ultimate business assignment which is to ensure that demands are satisfied for profit-making is the sole responsibility of the concern organizations. The quest is to ensure that optimum supplier meet customers' requirements. Consequently, organizations now need to carefully take error-free strategies in appraising the array of suppliers with a view of selecting an optimum supplier that would guarantee success of the overall business objectives. Market competitiveness has conditioned the survival of businesses on the ability of the management to make quick and optimum decision toward selecting the right supplier.

To achieve this, factors such as limited product life cycle, pressure to satisfy the demand, and all those that determine the dynamics of the business need to be well examined in an effort toward manipulating various technologies in the supplier selection process. Beil [1] reported that supplier selection procedure requires huge financial obligations and most organizations tend to recoup these expenses from the expected return from the contracting suppliers. This study proposes a procedure through which decision makers can accurately appraise suppliers among vast potential suppliers. De Boer et al. [2], Ho et al. [3] and Wadhwa & Ravindran [4] have described various supplier criteria and indicators by implemented some selection models for solving supplier selection issues within the supply chain management domain. Habib [5] confirmed that diverse multi-performance exploits have been applied toward solving supplier selection menace especially those involving multi-performance indicators in profit-making supply chain management firms. According to Ma et al. [6] since selection parameters vary with the prevailing conditions, there can be no clear-cut or best procedure to assess and select supplier. Due to this assertion, different organizations have been adopting different techniques for solving the problems. Such procedures as reported by Mukherjee [7] have been broadly categorized into single and integrated models.

The single model uses the concept of mathematics, statistics and artificial intelligence while integrated model involves blending, incorporating or integrating two or more single model to solve the problem. Analytic hierarchy process (AHP), analytic network process (ANP) both in single and integrated models have been widely applied [8]. Classical models especially fuzzy integrated models have been used over time for trading-off between these supplier selections of different quantities [9]. Nazim et al. [10] and Nazim and Yaacob [11] proposed an integrated AHP-SCOR (Analytical Hierarchy Process-Supply Chain Operation Reference) model to improve the robustness of the supplier selection system and concluded that multi-parameters relevant for solving supplier selection problems are cost, quality, delivery and service. Elgarra et al. [12] described how the integrated SCOR a homogeneous model that is mainly used for business process redesigning, standardizing, and process dimensioning analysis could be utilized for best practices within the supply chain management. This model has been use as basis for organizations to fix problems within the supply chain management through filling the gaps in the chain efficiency. Agakishiyev [13] studied the application of Z-information technique for solving supplier selection problems where fuzzy and partially reliable information were construed as Z-number. Although Z-numbers [14–16] of the desired ideal and negative ideal solutions can be determined through different model, it can be then concluded that the use of partially reliable information was not a good indicator for the model.

Non-parametric data envelopment analysis techniques (DEA) have been proposed and widely applied as a single and integrated model. The reason for its wide adoption is in the ability of the model to handle multi-parameter cases. Thus, this study is basically interested in such previous studies involving the used data envelopment analysis-integrated concepts. A hybrid technique that incorporated multiple AHP, DEA and neural network (NN) considered by Sung & Krishnan [17] only achieved a combined supplier score which was used to rank the suppliers. Another integrated DEA-NN method of appraising supplier under incomplete information was proposed by Celebi and Bayrakar [18]. Desheng [19] added another hybrid technique with DEA-decision tree (DT-NN), hybrid DEA-AHP and activity-based costing (ABC). Similarly, Pandian et al. [20] applied a hybrid DEA-AHP where the weights of the variables were obtained through AHP, to analyses the supplier selection and purchasing decision making in the chemical industry. These hybrids produced enhanced overall efficiency and reduced indirect expenses. Integrated multiple multi-criteria decision making (MCDM) of Parthiban et al. [21] and the integrated fuzzy logic with strength-weakness-opportunity-threat (SWOT) into the DEA was proposed by Wadhwa & Ravindran [4]. According to Zeydan et al. [22], the combination of Fuzzy AHP, Fuzzy TOPSIS (Technique for Order Preference by Similarity to Ideal Solution) and DEA could be used. These exploits led to the use of a dummy input in the DEA models. A DEA cross efficiency design was conducted by Noorizadeh et al. [23] and Mahdiloo et al. [24] by eradicating unrealistic weighing plans amongst the DMUs to solve the problem thereby making it easier to completely rank them. For Saen [25], imprecise DEA can evaluate supplier selection from all available data both quantitatively and qualitatively failed in dimensioning the system legitimately. An augmented imprecise DEA methodology applied by Wu et al. [26] which was an improvement on Saen's exploit was able to properly handle imprecise data and also brought about an improved discriminatory tendency of the model. In another studies, the DEA integrated into risk assessment tools withing failure mode and effect analysis (FMEA) to improve the system safety performance aircraft landing over time [27, 28]. The results indicated that the proposed methodology has advantages to the common and conventional progressive FMEA tool like fuzzy developed FMEA (FDFMEA) [29].

Furthermore, a structured methodology proposed by Chen [9] integrated SWOT-Fuzzy-DEA-TOPSIS-MADA (multi-attribute decision making Analysis) with Delphi technique to successfully establish an evaluation framework for solving supplier selection in supply chain management. SWOT was used to identify germane criteria and indicators; fuzzy was used as a weighing scheme to determine the weight of the indicators. DEA was used to screen potential suppliers, TOPSIS and MADA was used for ranking, and Delphi technique utilized after the application of Chou [30] was used to monitor the performance of the suppliers. However, the technique was marred with lower discriminatory tendency of the decision making units (DMUs) due to the adoption of weak Constant Return to Scale (CRS) DEA model. This weakness is in the rigid relationships that exist between input criteria and indicators with the output. The relationship forces DMU to produce equal efficiency at both DEA model orientations. The aftermath is the tendency of the CRS model to produce misleading results capable of promoting DMU with inadequate weight vector as an efficient system. Similarly, Mukherjee [7] revealed that fuzzy based models are unsuitable because of the way they deal with uncertainties that are associated with the variations. Another issue with fuzzy is also the inability to assign suitable fuzzy number to variables. This manner of assigning fuzzy numbers could be prejudiced and circumstantial. The framework proposed with three phases. Phase 1 was the requirement and strategy analysis under which SWOT was used as a competitive strategy to identify and appraise criteria and indicators for supplier selection. In phase 2, a three-step candidate supplier assessment was introduced. Here, DEA was applied for weighing the decision indicators; fuzzy weighing technique estimated various weights and TOPSIS evaluated the suppliers. Delphi questionnaire technique in phase 3 was used to assess suppliers' performance. Ma et al. [6] proposed an adjustment to this framework of Chen [9] by adapting an integrated DEA Game cross efficiency. This approach considered all the DMUs as players within a team vying competitively for a common efficiency. By this, the relationships that exist between inputs and outputs are deemed nullified with a non-cooperative tendency. The two assumptions required for this approach was the use of an arbitrary efficiency and an assumed value for the free variable. The former is assigned to the desired DMU while the latter forms basis for estimating the efficiency score of the remaining DMUs.

Unfortunately, the likely interrelationships between criteria and indicators selected have not been considered by any of these previous methods. Those methods ignored that these criteria and indicators were selected randomly from among vast options. Even though their selection is based on standard concept of SWOT yet the effects of other undetectable indicators on the detectable indicators have not been examined. Our study view these previous frameworks as a kind of an on-line quality control based solely on inspecting suppliers as they are discharged from the SWOT and rejecting those that fall short of the specified acceptable supplier performance. In a clearer way, those previous decisions are based on achieving a balance between these desirable but incompatible supplier selection parameters. Without doubt, there exist variations within the system as a result of the interaction of undetectable parameters with other detectable parameters. A study of Cristian [31] attempted to solve a multi-objective problem of a sawmill with an optimization model, however, the characterization of a robust solutions was obtained. This was adduced to the presence of uncertainty in the weights of the variables which the studies said was difficult to determine accurately. Obviously, undetectable parameters are within the system influencing the decision making process indirectly. No model would optimally improve the supplier selection without fixing the issues of variations orchestrated by the undetectable indicators [32]. Little efforts have been dissipated on how to smoothen these effects of the variations. This study seeks to consider the relationships between input criteria and supplier performance through robust signal-to-noise ratio so that supplier performance responses are insensitive to the effects of variations due to the undetectable noise indicators.

This effort is aimed at contributing the followings (i) the weakness of the conventional data envelopment models that tends to allocate misleading scores to inefficient units due to its inherent problem of equivalent efficiency frontier at both orientations is reduced, (ii) the supplier performance responses are rendered insensitive to the effects of variations due to the uncontrollable (undetectable) indicators which could influence the accuracy and the sensitivity of the DEA-integrated models for solving supplier selection problems. Lastly, the optimum supplier selection is achieved; and other potential suppliers are ranked. This would assist the corporation or the decision-makers to quickly switch on to the next potential supplier without having to re-assess the suppliers in case the optimally selected supplier is suddenly incapacitated due to some unforeseen circumstances.

2 Data and Methods

A four procedural framework is presented. SWOT analysis of Chen [9] is adopted for discerning and evaluating supplier's criteria and performance indicators. The criteria and performance indicators of the possible suppliers are evaluated to select plausible suppliers. These plausible suppliers are analysed in the robust parameter design procedure where the signal-to-noise ratio is used to render the selection indicators insensitive to the effects of variations. The third procedure is the evaluation with DEA model. The revamped variable return to scale (VRS) is employed to determine the weight of the performance indicator of the plausible suppliers and probable suppliers are determined. In the fourth procedure, VRS penalization coefficient is used to select the optimum supplier from the list of probable suppliers. This is achieved by determining a probable supplier with the highest penalization coefficient. These procedures are explained as follows and have been depicted in Fig. 1.

Procedure A—Supplier selection requirement and strategy

SWOT analysis is adapted to determine the competitive strategy of suppliers and evaluates various criteria and indicators. The prevalent criteria and indicators for the evaluation and selection of the optimum supplier are selected.

Procedure B—Robust parameter procedures (robust design)

Step 1: selecting an adequate orthogonal array.
Step 2: conducting the experiment and data extraction.

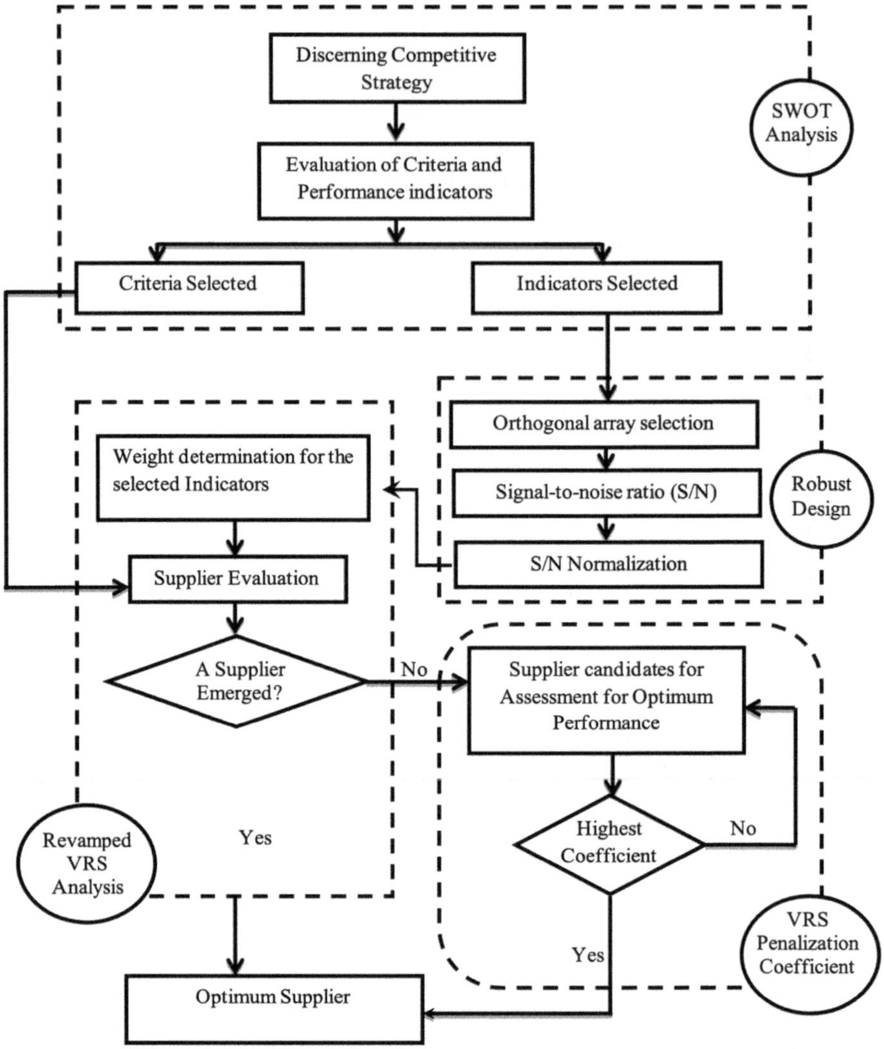

Fig. 1 Proposed revamped VRS signal-to-noise robust parameter framework for supplier selection optimization

Step 3: Estimation of signal-to-noise ratios (SN).
Step 4: Normalized signal-to-noise-ratio estimation NSNs.

Procedure C—Revamped facet VRS analysis

Step 1: Conducting VRS partitioning.
Step 2: Determining EP, WEP and SP.
Step 3: Determining the input and output weights using Eqs. (8–9)
Step 4: Estimating the upper bound restriction.

Step 5: Evaluating the efficiency score of the *SSDMU* using the revamped VRS model.
For steps 1–4, the multiplier model of MaxDEA 6.13 is used. Similarly, LINDO 13.0 is used to solve the linear programming formulations of the proposed models in step 5.

Procedure D—Optimization to select optimum SSDMU

To optimize and select optimum *SSDMU*, penalization coefficients of the efficient *SSDMUs* are estimated. The linear programming (LP) formulations of this model are solved using LINDO 13.0.

3 Theory/Calculation

According to the reported of Li and Zhu [33] and Taguchi et al. [32], the interrelationships within the selection criteria can be rendered insensitive to the effects of variations by robust parameter's signal-to-noise ratio (SN). The SN for each $SSDMU_{ij}$ is selected according to the desired quality status anticipated. These could be any of higher-the-better (HTB), lower-the-better (LTB) and nominal-the-better (NTB). Suppose n is number of time a particular indicator was examined against the y suppliers, then $SSDMU_{ij}$ is the number of suppliers examined, s_{ij}^2 is the variance of the suppliers' indicators determined for all input criteria i and indicators j, then Eqs. (1, 2 and 3) respectively states the quality expression of HTB, LTB and NTB.

$$\mathrm{SN} = -10\log\left(\frac{1}{\mathrm{n}}\sum_{\mathrm{i}=1}^{\mathrm{n}}\frac{1}{\mathrm{DMU_{ij}}}\right) \tag{1}$$

$$\mathrm{SN} = -10\log\left(\frac{1}{\mathrm{n}}\sum_{\mathrm{i}=1}^{\mathrm{n}}\mathrm{DMU_{ij}^2}\right) \tag{2}$$

$$\mathrm{SN} = -10\log\left(\frac{\overline{\mathrm{DMU}}_{\mathrm{ij}}^2}{\mathrm{s_{ij}^2}}\right) \tag{3}$$

Equations (4, 5 and 6) are used to obtain the normalized signal-to-noise-ratio ($\mathrm{NSN_{ij}}$) accordingly.

$$\mathrm{NSN_{ij}} = \frac{SSDMU_{ij} - min(DMU_{ij})}{max(SSDMU_{ij}) - min(DMU_{ij})} \tag{4}$$

$$\mathrm{NSN_{ij}} = \frac{\mathrm{min(DMU_{ij})} - SSDMU_{ij}}{\mathrm{max(DMU_{ij}) - min(DMU_{ij})}} \tag{5}$$

$$\mathrm{NSN_{ij}} = \frac{|SSDMU_{ij} - \mathrm{Target}| - \min(|\mathrm{DMU_{ij}} - \mathrm{Target}|)}{\max(|\mathrm{DMU_{ij}} - \mathrm{Target} - \min(|\mathrm{DMU_{ij}} - \mathrm{Target}|)} \tag{6}$$

The revamped VRS analysis emanated from the modified VRS concept of Daneshvar et al. [34]. This concept has been further studied and applied in other applications of Daneshvar and Adesina [35]. Therefore, finding the upper bound for standard VRS model denoted as ε from the u_o^- and u_o^+ determined through the weight vector of the self-evaluated DEA model.

Therefore:

$$\varepsilon = \max\{\mathrm{u_o^-}/\mathrm{u_0^+} \neq 1 \text{ for efficient DMUs}\} \tag{7}$$

The revamping is done by imposing partitioning within the modified VRS model to further enhance discrimination among the *SSDMUs*. With the insight given into the efficient point (EP) or strong efficient point (SEP) when $\theta = 1\text{and}\eta = 1$ (or vice versa), weak efficient point (WEP) when $\theta = 1\text{and}\eta < 1$ (or vice versa) and inefficient point when $\theta < 1\text{and}\eta < 1$, the optimal solution for each orientation of the standard VRS model u_o^* adequately belong to the inequality that $u_o^- \leq u_o^* \leq u_o^+$. The line of symmetry that exist between WEP and EP could be deciphered provided the efficiency score of the global optimal at either orientation can be a unity. Complete partitioning of the frontier into EPs and WEPs is possible provided the free variable u_o^* in the standard VRS model could be restricted to be strictly less than 1. Therefore, the input and output orientations of the VRS DEA models are determined according to [6].

For $SSDMU_{ij}$ under evaluation, y_{ro} is the output, x_{io} is the input; x_{ij} is the input, y_{rj} is the output to other SSDMU_{ij}; v_i, u_r are the input and output weight vectors respectively; u_o is the free variable, m is the total number of input and s is the total number of output. Therefore, the efficiency score θ at input orientation and score η at output orientation can be expressed (8–9).

$$\begin{aligned}
&\text{Max } \theta \\
&\text{S.t.} \sum_{i=1}^{m} \mathrm{v_i x_{io}} = 1 \\
&\sum_{r=1}^{s} \mathrm{u_r y_{ri}} - \sum_{i=1}^{m} \mathrm{v_i x_{ij}} + \mathrm{u_o} \leq 0 \\
&\mathrm{u_r} \geq 0 \quad \mathrm{r} = 1, \ldots \mathrm{s} \\
&\mathrm{v_i} \geq 0 \quad \mathrm{i} = 1, \ldots \mathrm{m} \\
&\mathrm{u_o} \text{ free}
\end{aligned} \tag{8}$$

$$
\begin{aligned}
&\text{Min } \eta \\
&\text{S.t.} \sum_{r=1}^{s} u_r y_{ro} = 1 \\
&\sum_{r=1}^{s} u_r y_{rj} - \sum_{i=1}^{m} v_i x_{ij} + u_o \leq 0 \\
&u_r \geq 0 \quad r = 1, \ldots s \\
&v_i \geq 0 \quad i = 1, \ldots m \\
&u_o \text{ free}
\end{aligned} \tag{9}
$$

The VRS penalization coefficient DEA model proposed by Gutierrez and Lozano [36] is employed. For $SSDMU_J$, J is the index of the supplier selection criteria to be evaluated with weight u_i, the penalization coefficient W_J is evaluated as.

$$
\begin{aligned}
&\text{Max } W_J \\
&\text{S.t.} \sum_{i=1}^{m} u_i NSD_{ij} \geq \ \nabla j \neq J \\
&\sum_{i=1}^{m} u_i NSD_{ij} = 1 \\
&u_i \geq W_J \qquad \nabla i \\
&W_J \geq 0
\end{aligned} \tag{10}
$$

4 Illustration of the Proposed Framework

The performance indicators of 12 textile suppliers of Taiwanese textile industry as described by Chen [9] is shown in Table 1. However, this study adapted the version that has been analyzed with DEA game cross-efficiency technique of Ma et al. [6] as the case study. Table 2 consists of three input variables and five quality responses. The responses are R&D rate, productivity, gross profit rate, quality discount (%), and inventory turnover ratio. Input variables are return rate, discount rate and operating expenses. For the robust parameter design, the quality attributes of LTB is set for all the responses. These input variables form the inner array of the orthogonal array (OA) for the robust parameter design for each (possible) supplier; thus L_{12} OA is adopted for the robust parameter design.

5 Results and Discussion

Table 3 presents the estimated SN and NSN used for the revamped VRS. The determination of the restriction for the upper bound variable ε is obtained to be 0.9656 (Table 4).

Table 1 Description of performance indicators for Taiwanian firms in selecting a textile supplier

Evaluation criteria	Performance indicators	Description
Quality	Return rate	Return rate = sales return/gross sale; (smaller return rate means sales of better quality products and higher customer acceptance)
	Discount rate	Discount rate = sales discount/gross sales; (smaller discount rate means sales of better-quality products and higher customer acceptance)
Cost	Gross profit rate	Gross profit rate = (net sales—cost of goods sold)/ sales; (larger supplier gross profit indicates stronger cost control ability)
	Quality discount	Suppliers offer the discount based on purchase quantity; (bigger quantity discount indicate higher cost ability)
Technology and production	R&D rate	R&D rate = R&D expense/sales; (higher supplier R&D rate denotes a stronger technology ability)
	Productivity	Higher supplier productivity means greater supply ability
Organizational management	Inventory turnover ratio	Inventory turnover ratio = cost of goods sold/ average inventory; (larger supplier inventory turnover ratio indicates stronger production/ marketing control ability)
	Operating expense rate	Operating expense rate = operating expense/net sales; (smaller supplier operating expenses rate expresses higher operating management efficiency)

$$\varepsilon = \mathrm{MaxU}_o^+/\mathrm{U}_o^- \neq 1$$

$$\varepsilon = \mathrm{Maxu}_o^+/\mathrm{u}_o^- \neq 1 = 0.96561$$

SSDMUs 2, 4,5,6,7,8,9,10,11 and 12 are all efficient at both orientations (Table 5); SSDMUs 1 and 3 turned out to be inefficient. Consequently, by partitioning, suppliers 1 and 3 are WEP while suppliers 2, 4, 5, 6, 7, 8, 9, 10, 11, 12 are either EPs or SEPs. On the application of the revamped VRS model, SSDMUs 2, 5, 7, 8 and 10 appeared on the frontier as efficient (plausible) suppliers. SSDMUs 1, 3, 4, 6, 9, 11 and 12 are inefficient. From this evaluation, these inefficient suppliers cannot be employed. Suppliers 1 and 3 are strictly inefficient and are thus the worst suppliers among the possible suppliers.

Five suppliers 2, 5, 7, 8 and 10 are efficient (probable) suppliers. The penalization coefficient, *SSDMU 2* gave the highest score of 0.323; therefore, *SSDMU* 2 is the optimal supplier on the OA. However, from the response graph of the average response values given in Fig. 2, *SSDMU 8* is returned as the optimal supplier. It will

Table 2 Parameter for the criteria and performance indicators for the 12 suppliers to a Taiwanian textile industry

Supplier	Input supplier selection criteria			Output supplier performance indicator				
(DMU)	Return rate	Discount rate	Operating expenses rate	R&D rate	Productivity	Gross profit rate	Quantity discount (%)	Inventory turnover ratio
1	0.06	0.66	5.73	1.11	201	0.01	7	0.67
2	0.54	0.22	2.92	1.13	267	9.69	7	6.02
3	1.11	0.5	8.38	2.12	311	6.36	5	5.8
4	0.15	0.48	5.68	1.57	361	6.42	5	6.17
5	0.19	0.41	4.16	1.5	300	9.51	10	6.76
6	1.28	0.5	7.01	3.08	310	13.81	7	7.48
7	0.01	0.01	5	2	250	5.41	8	7.04
8	0.42	0.13	2.82	1.04	398	6.82	7	11.16
9	0.65	1.05	3.83	1.66	375	7.51	5	5.17
10	0.25	0.07	2.64	2.62	103	1.43	8	5.16
11	0.72	0.18	4.25	2.09	164	6.71	6	12.45
12	0.13	1.37	5.55	2.52	200	2.98	6	6.36

Table 3 The signal-to-noise ratio for the supplier selection for the Taiwanian textile industry

Signal-To-Noise ratio (SN)				
0.91	26.02	-40.00	16.90	-3.48
1.06	48.53	19.73	16.90	15.59
6.53	49.86	16.07	13.98	15.27
3.92	51.15	16.15	13.98	15.81
3.52	49.54	19.56	20.00	16.60
9.77	49.83	22.80	16.90	17.48
6.02	47.96	14.66	18.06	16.95
0.34	52.00	16.68	16.90	20.95
4.40	51.48	17.51	13.98	14.27
8.37	40.26	3.11	18.06	14.25
6.40	44.30	16.53	15.56	21.90
8.03	46.02	9.48	15.56	16.07

be recalled that the application of the fuzzy base-DEA technique of Chen [9] obtained suppliers 2, 5, 7, 8 and 10 as efficient (probable) suppliers and selected supplier 7 as the optimal. The method of DEA game cross-efficiency of Ma et al. [6] ranked supplier 10 as the highest supplier but concluded that supplier 12 is the optimum supplier after a game cross efficiency convergence test. It is therefore, imperative to determine the global optimum value optimal value gotten from these previous

Table 4 The values of u_o^-, u_o^+ for efficient DMUs for the Hard disc case study

DMUs	u_o^-	u_o^+
1	0.71865	0.71865
2	0.2022	0.2022
3	0.5420	0.5420
4	0.3676	0.3676
5	0.4709	0.4709
6	0.0683	0.0683
7	1.1049	1.1049
8	0.6339	0.6339
9	1.0000	0.9656
10	0.6378	0.6378
11	0.0486	0.0486
12	0.1557	0.1557

approaches with the propose approach of this study. Multivariate multiple dependent general linear model (MMR) is employed to determine the global optimal value. The MMR is conducted on RStudio 1.1 version.

The multivariate multiple dependent regression is performed in R with responses captured in a cbind() function at 95% confidence interval as follows.

cbind(R&D rate, Productivity, Gross profit rate, Quantity discount (%), Inventory turnover rate) ~ return rate + Discount rate + Operating expense rate

Mlm1 < -- m(cbind(R&D rate, Productivity, Gross profit rate, Quantity discount (%), Inventory turnover rate) ~ Return rate + Discount rate + Operating expense rate, data = my_data)

lh.out < -- inearHypothesis(mlm1, hypothesis.matrix = c("Return rate = 0", "Discount rate = 0", "Operating expense rate = 0"))lh.out

Therefore, MMR model for the response is:

Response value < lm(cbind(R&D rate, Productivity, Gross profit rate, Quantity discount (%), Inventory turnover rate) ~ Return rate + Discount rate + Operating expense rate, data = my_data).

Table 6 shows that all the independent variables are positively related with R&D rate. Productivity is negatively related to operating expense rate and discount rate. Operating expense rate is negatively related with gross profit rate and inventory turnover rate. Quantity discount (%) is negatively related to all the independent variables. Check on how the independent (predictors) variable affect the model through multivariate test statistics MANOVA function is conducted as described in [37]. The MANOVA analysis, Pillai test statistic reveals an insignificant result that suggests that predictors as tested exist in the regression model.

Further multivariate analysis through similar tests of Wilks, Hotelling-Lawley and Roy (Table 7) produced the same results with Pillai test. Therefore, MMR analysis affirms that independent variables are necessary for the estimation of the dependent (supplier selection performance responses). For the prediction, using the input values

Table 5 Revamped VRS supplier selection procedure output

DMU	Input variables			Response					Score (θ)	Score (η)	Revamped VRS	Penalization coefficient
	Return rate	Discount rate	Operating expenses rate	R&D rate	Productivity	Gross profit rate	Quantity discount (%)	Inventory turnover ratio				
1	0.06	0.66	5.73	0.06	0.00	0.00	0.49	0.00	0.81	0.63	0.81	Inefficient
2	0.54	0.22	2.92	0.08	0.87	0.95	0.49	0.72	1.00	1.00	1.00	0.32
3	1.11	0.50	8.38	0.66	0.92	0.89	0.00	0.70	0.65	0.97	0.41	Inefficient
4	0.15	0.48	5.68	0.38	0.97	0.89	0.00	0.72	1.00	1.00	0.84	Inefficient
5	0.19	0.41	4.16	0.34	0.91	0.95	1.00	0.76	1.00	1.00	1.00	0.25
6	1.28	0.50	7.01	1.00	0.92	1.00	0.49	0.80	1.00	1.00	0.55	Inefficient
7	0.01	0.01	5.00	0.60	0.84	0.87	0.68	0.78	1.00	1.00	1.00	0.27
8	0.42	0.13	2.82	0.00	1.00	0.90	0.49	0.96	1.00	1.00	1.00	0.30
9	0.65	1.05	3.83	0.43	0.98	0.92	0.00	0.66	1.00	1.00	0.87	Inefficient
10	0.25	0.07	2.64	0.85	0.55	0.69	0.68	0.65	1.00	1.00	1.00	0.29
11	0.72	0.18	4.25	0.64	0.70	0.90	0.26	1.00	1.00	1.00	0.82	Inefficient
12	0.13	1.37	5.55	0.82	0.77	0.79	0.26	0.74	1.00	1.00	0.88	Inefficient

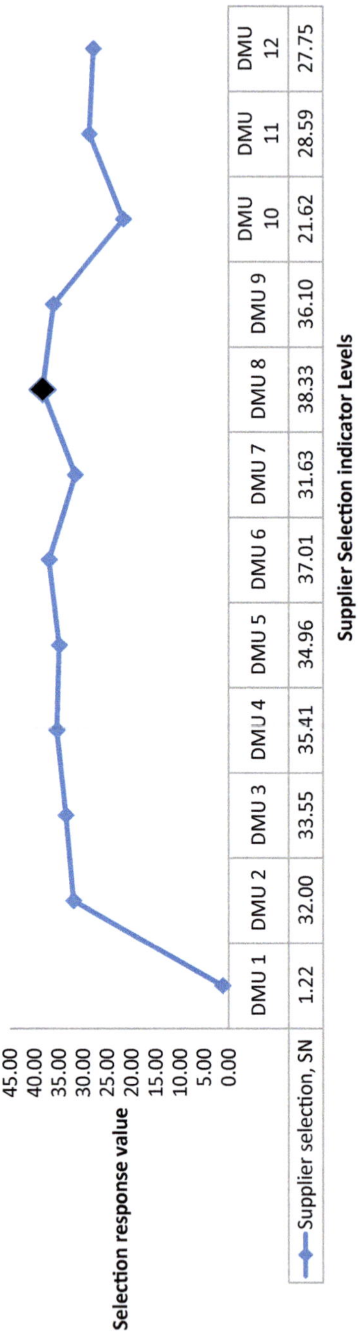

Fig. 2 Response graph showing the optimal Supplier selection factors setting using the proposed model (shaded points)

Table 6 Multivariate multiple dependent GLM and MANOVA analysis

```
Response R&D rate
Coefficients:
Estimate Std. Error t value Pr(>|t|)
(Intercept)  1.17774    0.59350   1.984    0.0825
a            0.48598    0.54630   0.890    0.3996
b            0.10970    0.53406   0.205    0.8424
c            0.08655    0.13709   0.631    0.5454

Response Productivity
Coefficients:
Estimate Std. Error t value Pr(>|t|)
(Intercept)   243.91      110.32   2.211     0.058
a             112.75      101.55   1.110     0.299
b              23.17       99.27   0.233     0.821
c             -10.67       25.48  -0.419     0.686

Response Gross profit rate
Coefficients:
Estimate Std. Error t value Pr(>|t|)
(Intercept)   5.4737      2.9056   1.884    0.0963
a             6.3950      2.6745   2.391    0.0438 *
b            -0.7417      2.6146  -0.284    0.7839
c            -0.3471      0.6711  -0.517    0.6190

Response Quantity discount (%)
Coefficients:
Estimate Std. Error t value Pr(>|t|)
(Intercept)   8.4294      1.2568   6.707 0.000152 ***
a            -1.1166      1.1569  -0.965 0.362742
b            -1.4476      1.1310  -1.280 0.236432
c            -0.1022      0.2903  -0.352 0.733972

Response inventory turnover ratio
Coefficients:
Estimate Std. Error t value Pr(>|t|)
(Intercept)   8.8911      2.5232   3.524    0.0078 **
a             2.9754      2.3226   1.281    0.2360
b            -1.8825      2.2705  -0.829    0.4311
c            -0.5528      0.5828  -0.948    0.3707

Type II MANOVA Tests: Pillai test statistic
  Df test stat approx F num Df den Df Pr(>F)
a  1   0.67143  1.63480      5      4 0.3272
b  1   0.33419  0.40155      5      4 0.8280
c  1   0.20704  0.20888      5      4 0.9418
```

Footnote: a = return rate, b = discount rate and c = operating expense rate

of the OA, optimal is *SSDMU 2*, for response value graph optimal is *SSDMU 8*, fuzzy based-DEA model gave *SSDMU 7* is the optimal, DEA game cross-efficiency model gave *SSDMU 10* as the optimal and convergence procedure of game cross-efficiency further selected SSDMU *12* as the optimal (Table 5).

Total anticipated improvement is determined as the sum of the predicted responses for the SSDMU. As given in Table 8, the OA optimal of the proposed model is a better predictor of the responses. However, DEA game cross efficiency model gave highest value of R&D rate. The OA optimal of the proposed model gave the highest predicted total anticipated response value of 303.89. Therefore, the optimal value of the OA *SSDMU 2* (supplier 2) is selected as the global optimal value for the

Table 7 Multivariate test

	Df	test stat	approx F	num Df	den Df	Pr(>F)
Pillai	3	1.227891	0.831478	15	18.00000	0.637305
Wilks	3	0.133972	0.817278	15	11.44364	0.649181
Hotelling-Lawley	3	3.848452	0.684169	15	8.00000	0.749470
Roy	3	3.019796	3.623755	5	6.00000	0.074216

Table 8 MMR response prediction output of the optimal model

Response Prediction model	Supplier selection performance indicators (Response)					Total anticipated improvement
	R&D rate	Productivity	Gross profit rate	Quantity discount (%)	Inventory turnover ratio	
DMU 2: (OA optimal)	1.7170	278.7411	7.7503	7.2096	8.4696	303.8877
DMU 7: (Fuzzy base DEA model)	1.1164	191.9265	3.7946	7.8929	6.1383	210.8687
DMU 8: (Response graph optimal)	1.6402	264.1927	7.0843	7.4841	8.3373	288.7386
DMU 10: Cross efficiency optimal	1.5354	245.5554	6.1042	7.7792	8.0439	269.0181
DMU 12: Convergence procedure of DEA game cross efficiency)	1.8716	231.0971	3.3625	5.7340	3.6311	245.6962

supplier selection problem of the Taiwanian textile industry under the condition of the given supplier criteria and indicators. Although our proposed model is not a raking based method, however, based on the results obtained, the efficient suppliers can be categorized according to order of performance as;

$$\text{Supplier } 2_{\text{DMU } 2} > \text{Supplier } 8_{\text{DMU } 8} > \text{Supplier } 10_{\text{DMU } 10} > \text{Supplier } 12_{\text{DMU } 12} > \text{Supplier } 7_{\text{DMU } 7}$$

This ranking becomes necessary in the situation when the optimal supplier become redundant or incapacitated in any way beyond what have been envisaged. The Corporation could rely on the ranking to engage the next efficient supplier; provided the prevalent conditions of the criteria and indicators remain unchanged.

6 Conclusions

Revamped VRS signal-to-noise robust parameter framework for the optimum supplier selection is proposed. SWOT was retained to determine various supplier selection criteria and indicators that will be used to appraise the suppliers. Robust signal-to-noise parameter design is applied for smoothing the effects of variations. By utilizing a partitioning-imposed modified VRS model, a revamped VRS procedure is engaged to estimate the weights of the input and output selection variables. VRS penalization is used to determine the optimal on the orthogonal array (OA). The optimal solution obtained from the response graph of the robust parameter design was different from the optimal value obtained from OA, thereby necessitating a confirmatory test. Multivariate multiple dependent general linear model (MMR) is used to check for the independent variables that are germane for the prediction of the responses. The followings can be concluded about the study.

- Adequate selection of the optimum supplier selection indicators setting is achieved through the enhanced discriminatory tendency within the efficient suppliers as a result of the imposed partitioning.
- The proposed model self-evaluated to estimate the input and output weight vectors. Therefore, there is no need for prior information and or assumptions like the previously applied fuzzy base DEA, DEA game cross efficiency and other DEA based methodologies.
- Multivariate multiple dependent regression model (MMR) could be used to determine the independent factor that is contributing to the estimation of the response. It could equally be deployed to determine the global optimal value.
- The framework is simple to utilize and implement for solving supplier selection problems.
- Generally, the proposed supplier selection approach could identify those inefficient suppliers promoted as efficient suppliers by the previously integrated standard DEA models used is the DEA based models such as the fuzzy based DEA and DEA game models. The benefits of the proposed framework include simplicity, enhanced discriminatory tendency, specificity and unambiguous outputs.
- Top logistic managers, operational managers, supply chain stakeholders, contracting and consulting firms and other decision makers would find this framework appropriate and useful.

The authors aim to expand the application of this concept to address the challenges of supplier selection in situations characterized by vague and incomplete data. This study has the potential to revolutionize the evaluation and selection of suppliers within the field of supply chain and logistics management. Moreover, the use of DEA in these tools can be further enhanced by incorporating meta-data analysis techniques such as bibliometric analysis [38, 39], which can be applied to large volumes of published works in this field. Such analysis would yield valuable insights into the current state-of-the-art and emerging trends in research related to supplier selection.

References

1. Beil, D.R.: Supplier selection, in: Wiley Encycl. Oper. Res. Manag. Sci., (2011). https://doi.org/10.1002/9780470400531.eorms0852
2. de Boer, L., Labro, E., Morlacchi, P.: A review of methods supporting supplier selection, Eur. J. Purch. Supply Manag. **7** 75–89 (2001). https://doi.org/10.1016/S0969-7012(00)00028-9
3. Govindan, K., Rajendran, S., Sarkis, J., Murugesan, P.: Multi criteria decision making approaches for green supplier evaluation and selection: a literature review, J. Clean. Prod. **98** 66–83 (2015). https://doi.org/10.1016/j.jclepro.2013.06.046
4. Wadhwa, V., Ravindran, A.R.: Vendor selection in outsourcing, Comput. Oper. Res. 34 (2007) 3725–3737. https://doi.org/10.1016/j.cor.2006.01.009
5. Habib, M.: Supply Chain Management (SCM): Its Future Implications, Open. J. Soc. Sci. **02**, 238–246 (2014). https://doi.org/10.4236/jss.2014.29040
6. Ma R., Yao, L., Jin, M., Ren, P.: The DEA game cross-efficiency model for supplier selection problem under competition, Appl. Math. Inf. Sci. **8** 811–818 (2014). https://doi.org/10.12785/amis/080242
7. Mukherjee, K.: Supplier selection criteria and methods: past, present and future. Int. J. Oper. Res. **27**, 356–373 (2016). https://doi.org/10.1504/IJOR.2016.078470
8. Hou, J., Su, D.: EJB-MVC oriented supplier selection system for mass customization. J. Manuf. Technol. Manag. **18**, 54–71 (2007). https://doi.org/10.1108/17410380710717643
9. Chen C.-T., Lin, C.-T., Huang, S.-F.: A fuzzy approach for supplier evaluation and selection in supply chain management, Int. J. Prod. Econ. **102** 289–301 (2006). https://doi.org/10.1016/j.ijpe.2005.03.009
10. Nazim, R.: A new approach to supplier selection problem: An introduction of AHP-SCOR integrated model, Int. J. Recent Innov. Trends Comput. Commun. **3** 338–346 (2015). https://doi.org/10.17762/ijritcc2321-8169.150169
11. Nazim, R., Yaacob, R.A.I.R.: Criteria for supplier selection: An application of AHP-SCOR integrated model (ASIM). Int. J. Supply Chain Manag. **6**, 284–290 (2017)
12. Elgazzar, S., Nicoleta, S., Nick, J.: An application of fuzzy AHP to SCOR performance measures : a case study of an Egyptian natural Original Citation, (2010)
13. Agakishiyev, E.: Supplier Selection Problem under Z-information, Procedia Comput. Sci. **102**, 418–425 (2016). https://doi.org/10.1016/j.procs.2016.09.421
14. Li, H., Guo, J.-Y., Yazdi, M., Nedjati, A., Adesina, K.A.: Supportive emergency decision-making model towards sustainable development with fuzzy expert system. Neural Comput. Appl. **33**, 15619–15637 (2021). https://doi.org/10.1007/s00521-021-06183-4
15. Yazdi, M., Hafezi, P., Abbassi, R.: A methodology for enhancing the reliability of expert system applications in probabilistic risk assessment, J. Loss Prev. Process Ind. 51–59 (2019). https://doi.org/10.1016/j.jlp.2019.02.001
16. Adesina, K.A., Yazdi, M., Omidvar, M.: Emergency decision making Fuzzy-Expert aided disaster management system BT—Linguistic methods under fuzzy information in system safety and reliability analysis, In: Yazdi, M. (Ed.), Springer International Publishing, Cham, pp. 139–150 (2022). https://doi.org/10.1007/978-3-030-93352-4_6
17. Ha, S.H., Krishnan, R.: A hybrid approach to supplier selection for the maintenance of a competitive supply chain, Expert Syst. Appl. **34** 1303–1311 (2008). https://doi.org/10.1016/j.eswa.2006.12.008
18. Çelebi, D., Bayraktar, D.: An integrated neural network and data envelopment analysis for supplier evaluation under incomplete information, Expert Syst. Appl. **35** 1698–1710 (2008). https://doi.org/10.1016/j.eswa.2007.08.107
19. Wu, D.: Supplier selection: A hybrid model using DEA, decision tree and neural network, Expert Syst. Appl. **36** 9105–9112 (2009). https://doi.org/10.1016/j.eswa.2008.12.039
20. Pitchipoo, P., Venkumar, P., Rajakarunakaran, S., Ragavan, R.: Decision model for supplier evaluation and selection in process industry: A hybrid DEA approach. Int. J. Ind. Eng. Theory Appl. Pract. **25**, 186–199 (2018)

21. Parthiban, P., Zubar, H.A., Katakar, P.: Vendor selection problem: a multi-criteria approach based on strategic decisions. Int. J. Prod. Res. **51**, 1535–1548 (2013). https://doi.org/10.1080/00207543.2012.709644
22. Zeydan, M., Çolpan, C., Çobanoğlu, C.: A combined methodology for supplier selection and performance evaluation, Expert Syst. Appl. **38** 2741–2751 (2011). https://doi.org/10.1016/j.eswa.2010.08.064.
23. Noorizadeh, A., Mahdiloo, M., Saen, R.F.: Suppliers ranking in the presence of undesirable outputs. Int. J. Logist. Syst. Manag. **11**, 354–374 (2012). https://doi.org/10.1504/IJLSM.2012.045918
24. Mahdiloo, M., Farzipoor Saen, R., Tavana, M.: A novel Data Envelopment Analysis model for solving supplier selection problems with undesirable outputs and lack of inputs, Int. J. Logist. Syst. Manag. **11** 285–305 (2012). https://doi.org/10.1504/IJLSM.2012.045915
25. Saen, R.F.: Suppliers selection in the presence of both cardinal and ordinal data, Eur. J. Oper. Res. **183** 741–747 (2007). https://doi.org/10.1016/j.ejor.2006.10.022
26. Wu, T., Shunk, D., Blackhurst, J., Appalla, R.: AIDEA: a methodology for supplier evaluation and selection in a supplier-based manufacturing environment. Int. J. Manuf. Technol. Manag. **11**, 174–192 (2007). https://doi.org/10.1504/IJMTM.2007.013190
27. Adesina, K.A., Yazdi, M., Zarei, E., Pouyakian, M.: Smart decision fuzzy-based data envelopment model for failure modes and effects analysis BT—Linguistic methods under fuzzy information in system safety and reliability analysis, In: Yazdi, M. (Ed.), Springer International Publishing, Cham, pp. 151–170 (2022). https://doi.org/10.1007/978-3-030-93352-4_7
28. Daneshvar, S., Yazdi, M., Adesina, K.A.: Fuzzy smart failure modes and effects analysis to improve safety performance of system : Case study of an aircraft landing system, Qual. Reliab. Eng. Int. (2020) 1–20. https://doi.org/10.1002/qre.2607
29. Yazdi, M., Daneshvar, S., Setareh, H.: An extension to Fuzzy Developed Failure Mode and Effects Analysis (FDFMEA) application for aircraft landing system. Saf. Sci. **98**, 113–123 (2017). https://doi.org/10.1016/j.ssci.2017.06.009
30. Chou, C.: Developing the e-Delphi system: a web-based forecasting tool for educational research, Br. J. Educ. Technol. 33 (2002) 233–236. https://doi.org/10.1111/1467-8535.00257
31. Christian, D.P.: Characterization of robust solutions of multi-objective optimization models with uncertain weights: Application in a Sawmil. Int. J. Ind. Eng. **25**, 29–39 (2018)
32. The second industrial revolution and information technology, In: Taguchi's Qual. Eng. Handb., 2004: pp. 1–24. https://doi.org/10.1002/9780470258354.ch1
33. Li, Y., Zhu, L.: Optimisation of product form design using fuzzy integral-based Taguchi method. J. Eng. Des. **28**, 480–504 (2017). https://doi.org/10.1080/09544828.2017.1346239
34. Daneshvar, S., Izbirak, G., Javadi, A.: Sensitivity analysis on modified variable returns to scale model in Data Envelopment Analysis using facet analysis. Comput. Ind. Eng. **76**, 32–39 (2014). https://doi.org/10.1016/j.cie.2014.07.016
35. Adesina, K.A., Daneshvar, S.: Integrated data envelopment-thermoexergetic optimization framework for multicomponent distillation system with multiexergetic response in the robust parameter design procedures, Energy Sources. Part A Recover. Util. Environ. Eff. **40**, 1491–1507 (2018). https://doi.org/10.1080/15567036.2018.1477876
36. Gutiérrez, E., Lozano, S.: Data envelopment Analysis of multiple response experiments, Appl. Math. Model. **34** 1139–1148 (2010). https://doi.org/10.1016/j.apm.2009.08.001
37. Hayden, R.W.: A review of: "An R Companion to Applied Regression , Second Edition, by Fox, J., Weisberg, S." J. Biopharm. Stat. 22 (2012) 418–419. https://doi.org/10.1080/10543406.2012.635980
38. Li, H., Peng, W., Adumene, S., Yazdi, M.: Cutting edge research topics on system safety, reliability, maintainability, and resilience of Energy-Critical Infrastructures BT—Intelligent reliability and maintainability of energy infrastructure assets, in: Li, H., Peng, W., Adumene, S., Yazdi, M., (Eds.), Springer Nature Switzerland, Cham, pp 25–38 (2023) https://doi.org/10.1007/978-3-031-29962-9_2
39. Yazdi, M., Mohammadpour, J., Li, H., Huang, H.-Z., Zarei, E., Pirbalouti, R.G., Adumene, S.: Fault tree analysis improvements: A bibliometric analysis and literature review, Qual. Reliab. Eng. Int. n/a (2023). https://doi.org/10.1002/qre.3271

Operation Management of Sky Magic: Consolidating Perspective of Risk and Safety

Tuong Quyen Vu, Rosita Moradi, and Mohammad Yazdi

Abstract This paper explores the crucial aspects of operation management in the context of Sky Magic, a cutting-edge technology that revolutionises the field of aerial entertainment. The objective is to consolidate perspectives on risk and safety to ensure Sky Magic performs efficiently and securely deployment. The study begins by comprehensively understanding the risks associated with Sky Magic operations. A risk assessment framework is developed by analysing historical data, identifying potential hazards, and considering various environmental factors. This framework enables identifying and prioritising potential risks, allowing for implementing proactive mitigation strategies. A robust safety management system is proposed to guarantee safety during Sky Magic performances. This system encompasses establishing standard operating procedures, rigorous training protocols, and effective communication channels. It also includes implementing advanced technological solutions such as remote monitoring systems, real-time data analysis, and predictive maintenance to enhance safety measures.

Keywords Operation management · Aerial entertainment · Risk assessment · Safety management system

---

T. Q. Vu
School of Media Design, College of Technology and Design, University of Economics, Ho Chi Minh City, Vietnam

R. Moradi
University of Wollongong, Northfields Avenue, Wollongong, NSW 2522, Australia

M. Yazdi (✉)
School of Computing, Engineering and Physical Sciences, University of the West of Scotland (UWS), London, UK
e-mail: mohammad.yazdi@mq.edu.au

Faculty of Science and Engineering, Macquarie University, Sydney, Australia

M. Yazdi (ed.), *Progressive Decision-Making Tools and Applications in Project and Operation Management*, Studies in Systems, Decision and Control 518,
https://doi.org/10.1007/978-3-031-51719-8_11

1 Introduction

The Unmanned Aerial Vehicle (UAV), also known as a drone, has gained significant popularity due to its practical applications and research appeal, particularly in areas such as rescue operations, surveillance, and environmental monitoring [1–3]. One of the cutting-edge applications of UAV technology is in visual performance shows, where multiple UAVs are employed. An impressive example of this is Intel's successful performance featuring 100 quadrotors working in coordination with an orchestra [3].

A UAV can be defined as an aircraft equipped with onboard computers and sensors, enabling it to fly autonomously. This autonomous flight capability is made possible through advanced technology that allows the UAV to navigate and operate without direct human control. In the following, by overviewing an extensive literature review [4–8], help of AI tool [9], and interview of subject matter experts (SMEs), we summarised a couple of reasons that why risk assessment in UAV shows as sky magic are vital.

Risk assessment of UAV shows is crucial due to several reasons:

- Safety: UAV shows involve multiple uncrewed aerial vehicles flying near each other and the audience. Assessing the potential risks helps identify and mitigate hazards, ensuring the safety of both participants and spectators. This includes evaluating the risk of collisions, equipment failure, or unpredictable weather conditions that may threaten people or property.
- Regulatory Compliance: Many countries have regulations and guidelines for UAV operation, mainly when used in public spaces or near crowds. Conducting a risk assessment helps ensure compliance with these regulations, avoiding legal issues and penalties.
- Property Protection: UAV shows often occur in areas with valuable infrastructure or sensitive equipment. By assessing the risks, organizers can identify potential hazards that may damage property, such as UAVs veering off course or crashing. Preventive measures can then be implemented to minimize the risk of property damage.
- Reputation Management: A successful and safe UAV show enhances the reputation of the organizers and the technology itself. Conversely, accidents or incidents during a show can harm public perception. Conducting a risk assessment helps mitigate potential risks and ensures a smooth and well-executed event, enhancing the overall reputation of the organizers and the UAV technology industry.
- Insurance Considerations: Insurance providers often require risk assessments to determine coverage options and premiums for UAV shows. By conducting a thorough risk assessment, organizers can provide insurers with accurate information about the event's safety measures, reducing potential liabilities and securing appropriate insurance coverage.

Therefore, a comprehensive risk assessment of UAV shows is essential to ensure safety, regulatory compliance, property protection, reputation management, and insurance considerations. It helps identify potential risks and implement appropriate mitigation measures, ensuring a successful and secure event. In the following section, we conducted a brief and general risk assessment with the format of (I) Hazards, (II) Risks, and (III) Safety interventions actions. The idea behind of the UAV risk assessment is to reduce the amount of risk to an acceptable and/or enhance the total safety of the programs [10–15].

2 Risk and Safety Assessment of Sky Magic: UAV

- (I) UAV Show involving a choreographed flight of multiple UAVs, (II) UAV loss of control and crash landing due to loss of signal (power outage) resulting in third-party or worker injuries, (III) UAVs to be programmed with an emergency 'back-to-base' style auto-return function, UAV Shows Pilots to use batteries that are within their optimal recharge cycles life and reliable, To arrange for permits and approvals so the Pilots can have a test flight before the show and check all systems and signal strength, UAV Shows Pilots ensure that batteries are fully charged in ideal conditions before each show and apply a 'buddy-check' pre-flight check system to prevent any errors, omissions or bias.
- (I) UAV Show involving a choreographed flight of multiple UAVs, (II) UAV loss of control and crash landing due to adverse weather conditions—wind, dense low-altitude clouds, fog, rain—resulting in third party or worker injuries, (III) UAV Shows Pilots confirm environmental conditions in which the UAVs may be safely operated—referring to original manufacturer specifications, UAV Shows Pilots establish a regime for daily checks of weather forecasts data and referring to Observatory Hill wind data when available, UAV Shows Pilots firmly agree in writing that the UAV Show will not take place on conditions that are 'no fly' conditions under the manufacturer specifications or are at 80% of these conditions, To develop a pre-flight checklist for the UAV Show to be filled and cross-signed by risk representatives before and after each show. Checklist to include environmental data, pre-operational checks, and records of occurrences.
- (I) UAV Show involving a choreographed flight of multiple UAVs, (II) Pilot distractions, errors or omissions resulting in an accident, (III) To allocate a specific work area for UAV pilots, providing them with the following: A clear line of sight to the UAVs at all times, Delineation/isolation from other workers, to communicate to other event crew to avoid distracting the pilots at all times unless it is an emergency, UAV Shows Pilots implement a 'buddy-check' system for pre-flight and after-flight checks for error capture, to carry out compliance checks before the flight and avoid distracting pilots during UAV Show flight (unless it is a critical safety matter or an emergency).
- (I) UAV Show involving a choreographed flight of multiple UAVs, (II) External electronic and radio interference disrupting the UAV operation and potential failure, (III) UAV Shows Pilots use internal 'know-how' to establish controls and protect the UAV operation from external radio-electronic interferences, Test

UAV controls and choreography in a controlled environment to confirm the reliability, Test and confirm 'back-to-base' mode/regime of every UAV so there is assurance that should external accidental or malicious interference occurs—the UAVs will automatically return and safely land at the landing area/trays.

- (I) UAV Show involving a choreographed flight of multiple UAVs, (II) Mechanical failure of a UAV due to inadequate maintenance, repair and inspection regime causing crash landing and associated injuries to third parties, (III) Risk SMEs engage a reputable UAV Shows Pilots and insist on project planning and delivery with a high degree of quality and reliability, UAV Shows Pilots ensure that pre-flight checks are carried out before each show, with a 'buddy check' check system to prevent any errors, omissions or bias.
- (I) UAV Show involving a choreographed flight of multiple UAVs, (II) Electrical or electronic failure of a UAV due to poor maintenance, repair and inspection regime causing crash landing and associated injuries to third parties, (III) UAV Shows Pilots to use batteries that are within their optimal recharge cycles life and reliable, and UAV Shows Pilots to demonstrate to Risk SMEs records of inspections and maintenance (and any fault events and rectification) of the UAVs showing continual commitment to quality and reliability, UAV Shows Pilots ensure that pre-flight checks are carried out before each show, with a 'buddy check' check system to prevent any errors, omissions, or bias, UAV Shows Pilots carry out visual and operational checks of each UAV after each show.
- (I) UAV Show involving a choreographed flight of multiple UAVs, (II) UAV loss of control and crash-landing due to loss of battery power resulting in third-party or worker injuries, (III) UAV Shows Pilots select and use UAVs that provide reliable live battery levels always read and never operate UAVs with less than 25% power remaining, UAV Shows Pilots to use batteries that are within their optimal recharge cycles life and reliable, UAV Shows Pilots ensure that pre-flight checks are carried out before each show, with a 'buddy check' check system to prevent any errors, omissions, or bias, UAV Shows Pilots carry out visual and operational checks of each UAV after each show.
- (I) UAV Show involving a choreographed flight of multiple UAVs, (II) UAV loss of control and crash landing due to loss of pilot signal (distance/radius exceeded) resulting in third party or worker injuries, (III) UAV Shows Pilots design the UAV Show flight plan so that UAVs are controlled at not more than 0.5 of their signal radius, UAV Shows Pilots, and RAS develop a site plan taking into account spatial projection of the flight path and coordinate this with all concerned, Risk SME arrange for permits and approvals so the Pilots can have a test flight before the show and check all systems and signal strength, Risk SME set a process in place for monitoring UAV flight plans compliance and ensuring that pilots do not fly beyond 50% of the signal radius.
- (I) UAV take-off and landing zone in a public space, (II) Unauthorised access to the area resulting in an injury, (III) Isolate the UAV take-off and landing zone using flat foot crowd control barrier positioned following UAV Shows safety requirements, Signpost the perimeter with access prohibition signage that includes text and

symbolic messaging, Roster security as a deterrent during UAV Show set-up, operation and pack-down on each operation day.

3 Conclusion

The conclusion summarizes the paper's key findings, highlighting the importance of a comprehensive approach to risk assessment and safety management in Sky Magic operations. It emphasizes the need for ongoing monitoring, evaluation, and improvement of the safety management system. The collusion concludes with a forward-looking perspective, acknowledging the potential for further advancements in aerial entertainment and the importance of adapting the operation management strategies accordingly.

- Understanding Risks in Sky Magic Operations: This section dives into the comprehensive understanding of risks associated with Sky Magic operations. It discusses the analysis of historical data to identify potential hazards and examines various environmental factors that can affect operational safety. The aim is to build a solid foundation for the subsequent risk assessment framework.
- Risk Assessment Framework: This section develops a risk assessment framework for Sky Magic operations. The framework incorporates the findings from the previous quarter and enables the identification and prioritization of potential risks. The assessment includes quantitative and qualitative analysis, allowing for a thorough evaluation of risks. The framework acts as a proactive tool for implementing effective mitigation strategies.
- Robust Safety Management System: This section proposes a robust safety management system to ensure safety during Sky Magic performances. It encompasses establishing standard operating procedures, rigorous training protocols, and effective communication channels. The emphasis is on developing a culture of safety and accountability among the operational staff. Integrating advanced technological solutions, such as remote monitoring systems, real-time data analysis, and predictive maintenance, is explored to enhance safety measures.

References

1. Kendoul, F.: Survey of advances in guidance, navigation, and control of unmanned rotorcraft systems. J. Field Robot. **29**, 315–378 (2012). https://doi.org/10.1002/rob.20414
2. Casbeer, D.W., Kingston, D.B., Beard, R.W., McLain, T.W.: Cooperative forest fire surveillance using a team of small unmanned air vehicles. Int. J. Syst. Sci. **37**, 351–360 (2006). https://doi.org/10.1080/00207720500438480
3. Kim, H.-J., Ahn, H.-S.: Realization of swarm formation flying and optimal trajectory generation for multi-drone performance show. In: 2016 IEEE/SICE International Symposium on System Integration, pp. 850–855 (2016). https://doi.org/10.1109/SII.2016.7844106

4. Yazdi, M.: A review paper to examine the validity of Bayesian network to build rational consensus in subjective probabilistic failure analysis. Int. J. Syst. Assur. Eng. Manag. **10**, 1–18 (2019). https://doi.org/10.1007/s13198-018-00757-7
5. Dargahi, M.D., Naderi, S., Hashemi, S.A., Aghaiepour, M., Nouri, Z., Sahneh, S.K.: Use FMEA method for environmental risk assessment in ore complex on wildlife habitats. Hum. Ecol. Risk Assess. **22**, 1123–1132 (2016). https://doi.org/10.1080/10807039.2015.1106912
6. Li, H., Yazdi, M., Huang, H.-Z., Huang, C.-G., Peng, W., Nedjati, A., Adesina, K.A.: A fuzzy rough copula Bayesian network model for solving complex hospital service quality assessment. Complex Intell. Syst. (2023). https://doi.org/10.1007/s40747-023-01002-w
7. Yazdi, M., Ding, Y., Adumene, S., Shafie, P.: Progressive system safety and reliability analysis: a sustainable game theory approach, Qual. Reliab. Eng. Int. n/a (2022). https://doi.org/10.1002/qre.3247
8. Li, H., Yazdi, M.: What are the critical well-drilling blowouts barriers? A progressive DEMATEL-game theory BT. In: Li, H., Yazdi, M. (eds.) Advanced Decision-Making Methods and Applications in System Safety and Reliability Problems: Approaches, Case Studies, Multi-criteria Decision-Making, Multi-objective Decision-Making, Fuzzy Risk-Based Models, pp. 29–46. Springer International Publishing, Cham (2022). https://doi.org/10.1007/978-3-031-07430-1_3
9. OpenAI, ChatGPT [Software]. https://openai.com (2021)
10. Li, H., Peng, W., Adumene, S., Yazdi, M.: Cutting edge research topics on system safety, reliability, maintainability, and resilience of energy-critical infrastructures BT. In: Li, H., Peng, W., Adumene, S., Yazdi, M. (eds.) Intelligent Reliability and Maintainability of Energy Infrastructure Assets, pp. 25–38. Springer Nature Switzerland, Cham (2023). https://doi.org/10.1007/978-3-031-29962-9_2
11. Yazdi, M., Kabir, S., Kumar, M., Ghafir, I., Islam, F.: Reliability analysis of process systems using intuitionistic fuzzy set theory BT. In: Garg, H. (ed.) Advances in Reliability, Failure and Risk Analysis, pp. 215–250. Springer Nature Singapore, Singapore (2023). https://doi.org/10.1007/978-981-19-9909-3_10
12. Yazdi, M.: Acquiring and sharing tacit knowledge in failure diagnosis analysis using intuitionistic and Pythagorean assessments. J. Fail. Anal. Prev. **19**, 369–386 (2019). https://doi.org/10.1007/s11668-019-00599-w
13. Li, H., Yazdi, M.: An advanced TOPSIS-PFS method to improve human system reliability BT. In: Li, H., Yazdi, M. (eds.) Advanced Decision-Making Methods and Applications in System Safety and Reliability Problems: Approaches, Case Studies, Multi-criteria Decision-Making, Multi-objective Decision-Making, pp. 109–125. Springer International Publishing, Cham (2022). https://doi.org/10.1007/978-3-031-07430-1_7
14. Li, H., Yazdi, M.: Developing failure modes and effect analysis on offshore wind turbines using two-stage optimization probabilistic linguistic preference relations BT. In: Li, H., Yazdi, M. (eds.) Advanced Decision-Making Methods and Applications in System Safety and Reliability Problems: Approaches, Case Studies, Multi-criteria Decision-Making, Multi-objective Decision-Making, Fuzzy Risk-Based Models, pp. 47–68. Springer International Publishing, Cham (2022). https://doi.org/10.1007/978-3-031-07430-1_4
15. Li, H., Yazdi, M.: Stochastic game theory approach to solve system safety and reliability decision-making problem under uncertainty BT. In: Li, H., Yazdi, M. (eds.) Advanced Decision-Making Methods and Applications in System Safety and Reliability Problems: Approaches, Case Studies, Multi-criteria Decision-Making, Multi-objective Decision-Making, Fuzzy Risk-Based Models, pp. 127–151. Springer International Publishing, Cham (2022). https://doi.org/10.1007/978-3-031-07430-1_8

Research on Reliability Design Analysis Application to Gas Turbines

Meiyu Yang, Wei Wang, Yihao Zhang, Jitao Cai, and Wenwen Liu

Abstract The complex mechanical structure and harsh working environment of gas turbines have posed significant challenges to their reliability. Reliability design analysis is an essential technical activity that can identify and eliminate potential defects and weaknesses in products, which contributes to providing crucial theoretical support and optimization strategies for gas turbine design and reliability improvement. However, current gas turbine reliability design analysis is plagued by problems such as a lack of available data, incomplete reference standards, and insufficient systematic application. This paper compares the previous application studies on gas turbine reliability design with the requirements of national military standards and product reliability design analysis. The development trends and shortcomings regarding existing application research are also summarized and analyzed. Based on these findings, the paper suggests future directions and core contents for gas turbine reliability design analysis, which could provide a useful reference for this approach's systematic application and development.

Keywords Gas turbine · Reliability design analysis · Application research · Operation research

M. Yang · W. Wang (✉) · Y. Zhang
School of Reliability and System Engineering, Beihang University, Beijing, China
e-mail: w.wei@buaa.edu.cn

J. Cai
School of Emergency Management and Safety Engineering, China University of Mining and Technology, Beijing, China

W. Liu
China Shipbuilding Industry Corporation, Seventy-Three Research Institute, Harbin, China

M. Yazdi (ed.), *Progressive Decision-Making Tools and Applications in Project and Operation Management*, Studies in Systems, Decision and Control 518,
https://doi.org/10.1007/978-3-031-51719-8_12

1 Introduction

The gas turbine stands as the indisputable "crown jewel" within the manufacturing industry, revered for its exceptional attributes of high power output, unwavering operational stability, and extended operational lifespan. Its prominence is exemplified by its status as the epitome of advanced power generation globally, wielding profound significance in realms as diverse as national defence security, energy sustainability, and industrial competitiveness. This recognition underscores gas turbines' pivotal role in shaping modern technological landscapes.

Across the spectrum of developed nations, a profound emphasis has been placed on the continual evolution of gas turbine technology. This resolute focus has propelled the rapid growth of gas turbine capabilities, culminating in technological feats that continually push boundaries and defy expectations.

Notably, gas turbines have found their versatile applications woven into the fabric of various industrial domains, spanning electric power generation, aviation propulsion, and maritime propulsion systems. Beyond their industrial prowess, gas turbines assume a mantle of paramount importance within the confines of military apparatus. Often standing as the primary power source, they serve as vital components in military equipment, diligently ensuring the machinery's operational readiness.

Within this context, the reliability of gas turbines emerges as an unequivocal imperative. These turbines' seamless and dependable operation becomes an elemental factor, directly influencing the safety and efficacy of military operations. In essence, the reliability of gas turbines is the linchpin safeguarding the unhindered and secure functioning of military equipment, thereby underpinning the broader realms of national security and strategic defence.

The contemporary landscape presents intricate challenges confronting modern gas turbines, necessitating adept solutions for their conquest. Foremost among these challenges is the integration of cutting-edge technologies during the developmental phase, aimed at augmenting the technical prowess of gas turbines. This pursuit, while commendable, has led to an intricate and multifaceted turbine structure, thereby diversifying the spectrum of potential failure triggers. This complexity underscores the need for a comprehensive approach to address these vulnerabilities.

Compounding these intricacies is the diverse range of environments gas turbines operate in. Their deployment spans from high-altitude locales with thin air and low temperatures to the unrelenting rigours of demanding settings, such as corrosive marine environments characterized by elevated temperatures, pressures, and velocities. Adding to the challenge is the dynamic nature of their operational milieu, characterized by rapidly fluctuating conditions driven by many environmental variables. This dynamic interplay necessitates that naval gas turbines be imbued with an elevated level of reliability, ensuring their sustained and effective functionality across mission profiles.

Recently, China has demonstrated a heightened focus on the reliability of naval gas turbines. In response, the nation has introduced a comprehensive framework of reliability indices, thoughtfully aligned with the necessities of significant equipment.

This strategic approach serves as the bedrock upon which further enhancements are built. Guided by the tenets outlined in "GJB 450A-2004 General Requirement for Materiel Reliability Program," the development unit tasked with product realization leverages this influential reference to meticulously refine the very fabric of reliability design. By meticulously scrutinizing analysis methodologies and technical protocols, the developmental stage becomes crucial for heightened reliability, underpinning the overarching pursuit of achieving stellar performance, enduring reliability, and a prolonged product lifespan for naval gas turbines.

The process of reliability design analysis stands as a pivotal cornerstone, wielding power to unearth latent flaws and vulnerabilities within products. While this methodology has found resounding success in electronic products, its seamless application to conventional mechanical counterparts, particularly exemplified by gas turbines, remains a formidable undertaking. The intricate web of challenges stems from the complicated architectural makeup, non-uniform components, and an array of potential failure modes and mechanisms inherent to gas turbines' mechanical nature.

Consequently, the endeavour to orchestrate a reliability design analysis for gas turbines emerges as a focal point that reverberates domestically and globally. The complexity inherent to these mechanical marvels underscores the need for a specialized and concerted engineering effort to conquer the intricacies. Pursuing reliability in this context is not merely an auxiliary endeavour; instead, it stands as a quintessential imperative, pivotal to enhancing gas turbines' operational integrity and longevity.

In both domestic and international engineering spheres, the relentless pursuit of practical reliability design analysis for gas turbines is a testament to the engineering community's unwavering commitment. Recognizing the challenge, the quest to unravel the labyrinthine complexities of these mechanical behemoths is a rallying point that transcends geographical boundaries, representing a collective aspiration to harness the full potential of gas turbines while safeguarding their operational reliability.

In this study, our foremost objective is to delineate a comprehensive framework for gas turbine reliability design and analysis. To this end, we embark upon a systematic exploration, meticulously charting the essential work components mandated by the exacting criteria stipulated in the "GJB 450A-2004 General Requirement for Materiel Reliability Program." Our approach rests upon an astute recognition of the intricate nuances inherent to gas turbines, encompassing their specific structural configuration and the peculiarities of their operational utilization. This dual-pronged perspective ensures a methodologically sound foundation and an operational contextualization that resonates with real-world scenarios.

At the crux of our investigative journey lies a comprehensive landscape survey, where existing applications concerning gas turbine reliability design and analysis are meticulously examined. This undertaking assumes a multifaceted dimension, delving into critical aspects such as reliability analysis methodologies, applicability across diverse scenarios, and the focal points that govern their functional realm. By meticulously unravelling these threads, we aspire to unearth insights that transcend theoretical abstraction, offering a coherent narrative bridging the gap between theory and practicality.

In the culmination of our explorative endeavour, we cast a discerning eye on the present landscape, critically examining its limitations and deficiencies. Within this analytical framework, we endeavour to synthesize potent suggestions, laying the groundwork for future strides in the realm of gas turbine reliability design and analysis. The culmination of these insights not only forms a strategic compass for upcoming undertakings but also functions as an invaluable point of reference, steering the course for effective reliability design and analysis endeavours about gas turbines. By unveiling these intricate layers, we aspire to enhance the operational fabric of gas turbines and illuminate the trajectory that governs their future reliability-driven evolution.

2 Overview of Gas Turbine and Its Reliability Index

2.1 Overview of Gas Turbine

In this study, our foremost objective is to delineate a comprehensive framework for gas turbine reliability design and analysis. To this end, we embark upon a systematic exploration, meticulously charting the essential work components mandated by the exacting criteria stipulated in the "GJB 450A-2004 General Requirement for Materiel Reliability Program." Our approach rests upon an astute recognition of the intricate nuances inherent to gas turbines, encompassing their specific structural configuration and the peculiarities of their operational utilization. This dual-pronged perspective ensures a methodologically sound foundation and an operational contextualization that resonates with real-world scenarios.

At the crux of our investigative journey lies a comprehensive landscape survey, where existing applications concerning gas turbine reliability design and analysis are meticulously examined. This undertaking assumes a multifaceted dimension, delving into critical aspects such as reliability analysis methodologies, their applicability across diverse scenarios, and the focal points that govern their functional realm. By meticulously unravelling these threads, we aspire to unearth insights that transcend theoretical abstraction, offering a coherent narrative bridging the gap between theory and practicality.

In the culmination of our explorative endeavour, we cast a discerning eye on the present landscape, critically examining its limitations and deficiencies. Within this analytical framework, we endeavour to synthesize potent suggestions, laying the groundwork for future strides in the realm of gas turbine reliability design and analysis. The culmination of these insights not only forms a strategic compass for upcoming undertakings but also functions as an invaluable point of reference, steering the course for effective reliability design and analysis endeavours about gas turbines. By unveiling these intricate layers, we aspire to enhance the operational fabric of gas turbines and illuminate the trajectory that governs their future reliability-driven evolution.

The pursuit of elevating gas turbine technical prowess and efficiency to align with exacting operational demands necessitates an unwavering commitment to infuse advanced technology throughout the developmental continuum. This strategic endeavour rests upon integrating cutting-edge material technologies and state-of-the-art manufacturing methodologies. These bedrock principles work harmoniously, engendering meticulous precision and uncompromising quality of gas turbine components and materials. This holistic approach fosters a transformative leap, wherein the amalgamation of innovation becomes the hallmark of gas turbine production.

In parallel, the infusion of advanced control technologies assumes paramount significance. This infusion is orchestrated with a twofold objective: the calibration of gas turbines to achieve a seamless degree of automation and the orchestration of control mechanisms with unparalleled precision. This concerted orchestration amplifies operational efficiency and begets a level of control granularity hitherto unattainable.

Of equal importance is the evolution of combustion dynamics. This journey navigates the terrain of advanced combustion chamber designs and innovative fuel injection technologies. The intertwined synergy of these elements is elemental, for its ushers forth an era of optimal combustion efficiency coupled with exceptional emission performance. This synergy of function invariably steers gas turbines toward a trajectory of heightened operational excellence and environmental stewardship.

The ascent toward enhanced efficiency and longevity further necessitates the infusion of avant-garde turbine design paradigms. Integrating multi-stage turbines and cutting-edge blade cooling technologies unfolds as a transformative endeavour. These innovations collectively steer gas turbines toward an apex where efficiency and endurance intersect. However, it's imperative to recognize that this integration heralds a paradigm shift; while innovations enrich the tapestry of gas turbine advancements, they concurrently introduce the prospect of complexity. Introducing these pioneering technologies and methodologies can engender intricate processes, manufacturing intricacies, or designs, subsequently conferring a heightened difficulty upon the overall gas turbine structure. This complicated landscape, in turn, ushers forth a diverse panorama of potential failure triggers.

In this intricate dance between innovation and complexity, the pathway to augmenting gas turbine performance emerges as a symbiotic pursuit. A strategic balance must be forged between innovation's impetus and the diligence required to mitigate emergent complexities. As such, the seamless marriage of pioneering advancements with a steadfast commitment to ensuring reliability must remain at the forefront of this transformative journey (Fig. 1).

The operational theatre of most gas turbines is a crucible of formidable environmental challenges, characterized by a maelstrom of hostile factors including soaring temperatures, elevated pressures, breakneck speeds, pervasive humidity, and the corrosive ravages of nature. This multifaceted onslaught leaves an indelible imprint upon their performance and longevity, casting the formidable shadow of reliability as a pervasive challenge that demands meticulous attention.

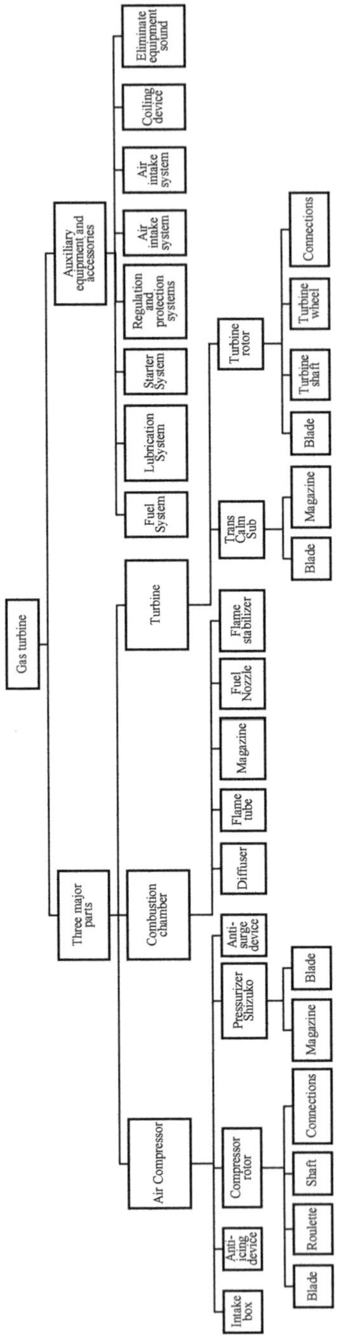

Fig. 1 Basic components of a gas turbine

Classifying gas turbines through the prism of their operational domains, a trichotomy emerges: aviation, marine, and industrial power generation. The aviation iteration of these engineering marvels stands at the threshold of extremity, contending with temperatures that often exceed the 1000 °C marks. At this precipice of heat, high temperature asserts itself as the paramount adversary. Endurance becomes the hallmark, as these turbines navigate an intricately choreographed ballet of high loads, rapid revolutions, cyclical thermal stresses, erratic heating gradients, and the uneven strains that accompany lofty flights in altitudinous realms.

Conversely, the marine counterpart of gas turbines embarks on a tumultuous maritime odyssey. Here, the backdrop is one of tempestuous seas and relentless conditions—high winds, surging waves, and torrential downpours—all conspire to render the marine gas turbine's existence a ceaseless struggle. The theatre of marine operations begets another adversary: the insidious and tenacious grip of salt spray corrosion. Alongside this corrosion, severe vibrations emerge as an equally potent nemesis, relentlessly chipping away at performance and resilience.

A distinct narrative unravels in the realm of industrial power generation gas turbines, where relative stability becomes a guiding beacon. However, beneath this veneer of stability lurk subtler assailants. Grid peaking-induced thermal fatigue and the ceaseless cycle of start-stop operations wield their toll, prodding the turbine into a delicate dance with operational longevity.

In this intricate tale of challenges and domains, a common thread emerges: the paramount significance of meticulous engineering and reliability measures. As these gas turbines thread the labyrinthine passages of their operational domains, their success hinges upon a symphony of innovation and robust design, harmonizing the harmonious interplay between engineering brilliance and the unrelenting demands of their operational landscapes.

The complexity of the structural composition of gas turbines, the technology used, and the harshness of the working environment will directly lead to a wide variety of failure modes, high frequency, and great hazards, so it is especially important to design and analyze their reliability and improve the reliability level of gas turbines.

2.2 Gas Turbine Reliability Index

The reliability of a gas turbine refers to its ability or likelihood to function without failure during its design time and under the specified conditions [1]. Reliability is a crucial aspect of gas turbine design analysis and verification, which is commonly measured by various indicators. Typically, for the entire gas turbine system, which is repairable, its reliability is described using indicators such as reliability, average time between failures, and average repair time. In contrast, for the power transmission shaft in a gas turbine, which is a non-repairable component, reliability and failure rate are the common indicators used to describe its reliability. The specific meanings are:

– Reliability

The reliability of a gas turbine refers to the likelihood of the turbine performing a specific function without any failures or issues within the predetermined time frame and under specific conditions. This probability is typically represented by the letter R and is dependent on time, making it a function of time. The reliability function, also known as R(t), expresses this probability over time.

– Failure rate:

Also known as the failure of efficiency, it refers to the probability that a gas turbine will fail per unit of time after that moment, with a premise of that this gas turbine has not failed up to a certain moment in operation. It is generally expressed as the ratio between the number of failures and the duration of operation within a specified time.

– Mean time between failures (MTBF):

Refers to the average trouble-free operation time of a gas turbine within a certain period. It is usually expressed as the ratio of the total operating time to the number of failures that occur during that period.

– Mean time to repair (MTTR):

The average time required to repair a gas turbine from a faulty condition to a condition in which it can operate normally is known as the Mean Time to Repair (MTTR). It is measured as the ratio of the total time for restorative maintenance of a gas turbine to the total number of failures of the repaired product at any given level of maintenance, under specified conditions and within a specified period.

3 Gas Turbine Reliability Design Analysis Work Determination

To study the current status of gas turbine reliability design analysis applications, as well as to summarize the shortcomings of existing research and future development trends, we first screened the reliability design analysis items that need to be considered for gas turbines in accordance with the relevant requirements of "GJB 450A-2004 General Requirement For Materiel Reliability Program" Then, starting from the keywords "reliability design analysis of gas turbine" and "failure analysis of gas turbine", the literature on the reliability work of gas turbines was sorted out. The selected literature was further analyzed and summarized to determine the current state of application of gas turbine reliability design analysis work projects. Finally, ideas for future research on the application of gas turbine reliability design analysis were proposed.

3.1 Gas Turbine Reliability Design Analysis Work Determination

National military standard serves as the fundamental basis for reducing development risks, ensuring, and enhancing the quality of military equipment, as well as providing a vital framework for the scientific management of military equipment development, production, and use. In order to achieve the required level of reliability, military products are required to undergo reliability work throughout their life cycle in strict accordance with the provisions outlined in "GJB 450A-2004 General Requirement for Materiel Reliability Program". This helps to standardize military products and reliability work, establish a systematic and rule-based approach to reliability work, ensure the successful completion of reliability work for military products, and significantly improve their overall reliability level.

The "*GJB 450A-2004 General Requirement for Materiel Reliability Program*" is a comprehensive compass for directing reliability endeavours across a spectrum of equipment, including gas turbines. This directive encapsulates an array of work items and core tenets about equipment reliability design and analysis, conveniently presented in Table 1.

Since gas turbines belong to the domain of mechanical products, the crux of their reliability design and analysis is intricately interwoven with structural and mechanism reliability considerations. This intricate web entails a meticulous assessment of various facets, such as the mechanical integrity of the structure, susceptibility to fatigue, wear and tear dynamics, propensity for fractures, and other related influences stemming from external loads. Furthermore, the scope extends to encompassing the foreseen functional breakdowns and failures due to the complex interplay of forces during operation.

The gas turbine reliability design and analysis framework is meticulously constructed around pivotal work facets in light of these complexities. These encompass the formulation of an exhaustive reliability model, reasonable allocation of reliability measures, prognostication of reliability levels, comprehensive analysis of failure modes, intricate exploration of potential impacts and hazards, rigorous fault tree analysis, utilization of finite element analysis techniques, and an exhaustive evaluation of durability thresholds.

The conventional application of these reliability design and analysis methodologies offers a dual-pronged advantage. Firstly, it aids in attaining predefined reliability benchmarks by orchestrating a well-defined reliability modelling, allocation, and projection process. Secondly, it is a proficient tool for pinpointing latent vulnerabilities within gas turbines. This is facilitated through the systematic adoption of Failure Mode and Effects Analysis (FMEA) and its contextual Hazard Analysis (CA) iteration (FME(C)A), alongside the analytical prowess of Fault Tree Analysis (FTA), Finite Element Analysis, and the comprehensive insights afforded by Durability Analysis.

In pursuing enhancing the reliability quotient of gas turbines, this holistic approach lends itself to the proposal of constructive enhancements. By adopting these proposed

Table 1 Equipment reliability design and analysis work items and main content

Work projects	Main content
Building reliability models	Establish a function-based reliability block diagram and corresponding mathematical model based on the product structure, usage requirements and usage constraints
Reliability allocation	Assigning quantitative reliability requirements to defined product levels according to the reliability model
Reliability projections	Using reliability models, anticipate the basic and mission reliability of the product and evaluate whether the proposed design solution meets the specified quantitative reliability requirements
Failure mode, impact, and hazard analysis	Determine all possible failure modes of components, parts, equipment, and software in the design and manufacturing process through systematic analysis, as well as the causes and effects of each failure mode, in order to identify potential weak points and propose improvement measures
Fault tree analysis	Use deductive level-by-level analysis to find the various possible causes of a certain failure event (top event), know the most basic causes, and determine potential hardware and software design defects through the analysis of logical relationships in order to take improvement measures
Potential analysis	Under the assumption that all components and devices are normal, analyze and confirm potential states that can cause undesired functions or inhibit the desired functions
Circuit tolerance analysis	Analyze the impact of the circuit components' parameter deviations and parasitic parameters on the circuit performance tolerances within the specified operating temperature range, and propose corresponding improvement measures based on the analysis results
Development of reliability design guidelines	Develop and implement reliability design guidelines to guide designers in the reliability design of products
Component, part and raw material selection and control	Select and use components, parts, and raw materials according to the characteristics of the developed products
Identify reliability critical products	Reliability critical products through FMECA, FTA or other analysis methods, list and focus on their control, and propose their control methods and test requirements

(continued)

Table 1 (continued)

Work projects	Main content
Determine the impact of functional testing, packaging, storage, handling, transportation, and maintenance on product reliability	Testing and analysis to determine the impact of functional testing, packaging, storage, handling, transportation, and maintenance team product reliability. Can be used to predict product failure rates and design trade-offs
Finite element analysis	Analyze and evaluate the mechanical strength and thermal characteristics of the product during the design process to identify the weaknesses of the load-bearing structure and materials and the overheating part of the product as early as possible so that design improvement measures can be taken in time
Durability analysis	Identify parts that may have premature wear and tear failures, determine the root cause of the safeguard and possible corrective actions to be taken

measures, the reliability landscape of gas turbines can be elevated to new heights, reinforcing their operational dependability, and bolstering their stature within the realm of mechanical engineering.

3.2 *Literature Screening for Gas Turbine Reliability Design Analysis*

Based on the research objectives of this paper, the focus of the literature search is on "gas turbine reliability design" and "gas turbine reliability analysis". Initially, over 120 relevant papers were retrieved from literature indexing databases, including China Knowledge Network, Web of Science, and SpringerLink. However, some of these papers were focused on other mechanical products, such as aero-engines, or used reliability design and analysis methods that were not specified in "GJB 450A-2004 General Requirement for Materiel Reliability Program", which are beyond the scope of this study. Thus, to ensure the relevance and quality of the literature selected for this paper, specific criteria were used to further screen and sort the retrieved literature.

- Be written in Chinese or English, with a complete structure;
- Peer-reviewed journal articles or conference articles;
- The object of study is a gas turbine system, equipment, unit, or component;
- The subject matter is closely aligned with one or more of the main work items considered in the reliability design and analysis of gas turbines in the "GJB 450A-2004 General Requirement for Materiel Reliability Program";

- To propose the implementation, improvement or optimization of reliability design and analysis methods in gas turbines

3.3 Gas Turbine Reliability Design Analysis Research Profiles

Overview of the Current Situation and Summary Analysis: Based on the research conducted on the application of statistical reliability design analysis in the field of gas turbines, it was found that the current application of gas turbine reliability design analysis aligns with the work items that are mainly considered for gas turbines in the "GJB 450A-2004 General Requirement for Materiel Reliability Program". However, there is a lack of significant referenceable literature in the area of reliability allocation and expectation, and thus there is a need to increase research on the reliability index design to ensure the basic realization of the inherent reliability of gas turbines. The existing research on gas turbine reliability design analysis mainly includes qualitative and semi-quantitative analysis methods, reliability design analysis improvement methods, and reliability design analysis by means of simulation.

3.3.1 Qualitative and Semi-quantitative Methods Analysis Methods

In the realm of gas turbine reliability design analysis, select scholars have opted for a nuanced approach by harnessing qualitative and semi-quantitative analysis methodologies. These methods, which encompass techniques such as reliability modelling, Failure Mode and Effects Analysis (FME(C)A), and Fault Tree Analysis (FTA), exhibit an astute capability to address the intricate challenges inherent in this domain.

For instance, the work of Zhu [2] exemplifies the application of these methods. He deftly constructed a reliability logic block diagram for an aero-reform gas turbine combustion chamber, rooted in the granularity of its compositional mechanism. Subsequently, a meticulous FMECA analysis was conducted on key components, including the combustion chamber magazine, flame barrel, nozzle, fuel main tube, and front seal tight ring. This analytical endeavour culminated in the identification of specific failure modes, which, in turn, paved the way for surgically tailored improvement strategies aimed at preempting each distinct type of failure.

Chen [3] charted a unique path in the arena of combustion turbine generator set reliability design. His approach encompassed an intricate determination of the logical interplay amongst components bolstered by functional relationships. This intricate tapestry was further woven through the analysis of both unit systems and component intricacies within the fuel system. Consequent to this, a corresponding reliability model was established, and reliability indices were judiciously allocated utilizing methodologies such as equal distribution.

Shang et al. [4] carved a meticulous pathway by creating an elaborate fault tree dedicated to gas turbine failures. The central premise involved employing five common failure modes as intermediary events, subjecting them to a step-by-step analysis to elucidate their fundamental components and events.

In a distinct vein, Wei [5] embarked on the establishment of a fault tree for an industrial gas turbine grappling with anomalous slip oil consumption during operation. This endeavour entailed an intricate dissection of potential units and components that could induce oil leakage within the unit. Ultimately, the root cause was pinpointed to the failure of the power turbine bearing chamber seal.

Fang et al. [6] traversed the intricacies of a combined solid oxide fuel cell-gas turbine (SOFC-GT) cycle system utilizing FTA as their compass. This methodological journey involved a layer-by-layer deducing process, ceaselessly probing until the bedrock event at the bottom was illuminated. This analytical odyssey was further empowered through the utilization of "FreeFtr," a fault tree analysis software, which facilitated the derivation of probabilities related to top event occurrence, bottom event probability importance, and critical importance.

In sum, these scholars have illustrated the potency of qualitative and semi-quantitative methodologies, specifically reliability modelling, FME(C)A, and FTA, in navigating the complex terrain of gas turbine reliability design analysis. These methods not only dissect intricate failure pathways but also illuminate targeted strategies for fortification, thereby fostering a more dependable landscape for gas turbine applications.

4 Reliability Design Analysis Improvement Method

Researchers have also made efforts to improve gas turbine reliability design and analysis methods to make them more relevant to actual engineering situations. Conventional reliability theory often represents the wear and tear of gas turbine components as either complete failure or complete normal, which is a significant deviation from reality. Additionally, due to the large base of the average failure interval of gas turbines, it is challenging to obtain sufficient failure statistics from actual cases. Therefore, reliability design analysis mostly relies on a combination of a small amount of collected data and the empirical values of engineers and technicians.

In order to address the limitations of traditional reliability theory in gas turbine design and analysis, many researchers have explored the use of fuzzy theory and techniques. For instance, Yan et al. [7] developed a fuzzy fault tree to analyze the starting failure of marine gas turbines, overcoming the limitations of traditional fault trees by incorporating fuzzy information from actual engineering feedback to calculate the probability interval of starting failure under different confidence levels. Li et al. [8] conducted a fuzzy fault tree analysis of a gas turbine fuel system and calculated the range of failure rates for intermediate and top events with different confidence levels based on fuzzy data provided by users' feedback and staff's experience. Yang and Hong [9] introduced fuzzy triangular numbers in a modified FMEA to conduct reliability analysis of industrial gas turbine systems and ranked the risks of seven typical failure modes common to gas turbines in order of priority, with wear identified as the weakest link among them. Wen and Duan [10] proposed a new gas turbine reliability allocation method based on fuzzy theory, which applies fuzzy decision theory and single-factor fuzzy evaluation to allocate reliability indexes based on reliability data of similar models and expert experience. The application of fuzzy theory in gas turbine reliability design and analysis is effective in addressing the challenges of small sample events and limited failure data, thus making research results more relevant to engineering reality.

There are also scholars who have combined traditional reliability design analysis methods with other analysis techniques to address the challenging problems of complex gas turbine reliability analysis issues and limited reliability data. For instance, Bai [11] employed a combination of hierarchical analysis (AHP) and fault tree analysis to analyze gas turbine start-up failures. Bai established a two-by-two comparison matrix between intermediate events and between intermediate and bottom events. By normalizing the matrix, the risk rate of each failure event can be ranked, which has been shown to be feasible in actual engineering situations.

5 Simulation-Based Reliability Design Analysis

Currently, some scholars are combining mathematical and mechanical analysis methods in the process of gas turbine reliability design analysis to achieve accurate analysis through optimization algorithms. For example, Liu et al. [12] used a parametric finite element model to obtain the stress distribution under creep relaxation conditions for a heavy gas turbine combined rotor turbine end. They used a combination of Monte Carlo Latin hypercube sampling and response surfaces on the ANSYS finite element platform to analyze the effects of material parameters, rotational speed, and structural dimensions on the reliability and sensitivity of the rotor system. Shetty et al. [13] focused on turbine blades in gas turbines, performing finite element analysis of the load applied to the thermal engine and using optical imaging technology to model and simulate the blades to obtain solid models. They analyzed that under the maximum stress condition, the annular gap and tip clearance between the dynamic blade tip and the outer cassette changed irregularly, resulting in friction between the blade and the cassette wall, leading to its failure. In addition, He and Sun [14] used a neural network to map the highly nonlinear performance of gas turbines based on actual operation data to optimize the design of both BP and RBF networks. Through the implementation of the algorithm and the comparative analysis of the results, the RBF network showed a better approximation ability and was more suitable to achieve the reliability expectation. Zhang and Li [15] used the symmetrical support plate seat of a gas turbine as the research object and used a stochastic finite element method combining the finite element method and Monte Carlo method to analyze and obtain its tangential ultimate thermal stress probability density distribution. They then calculated the reliability of the support seat under the action of the ultimate thermal stress by integrating the MATLAB software. By adopting simulation methods for gas turbine reliability design analysis and using mathematical models instead of the actual system for analysis, researchers can identify reliability weak points, discover possible potential failures as early as possible, and improve them. This approach has the advantages of low cost, high speed, and high accuracy, ultimately leading to an improved product reliability design level.

Existing Shortcomings and Future Prospect: Based on the above summary analysis, the current reliability design and analysis work for gas turbines essentially addresses the key items specified in national military standards. However, according to the publication dates of the collected literature, it can be observed that industries have only recently begun to pay more attention to gas turbine reliability design and analysis. Despite this, the following deficiencies still exist:

- At present, gas turbine reliability work in China is not systematic and standardized, as well as cannot meet the needs of effective reliability design analysis;
- The reliability design analysis anticipates a lack of targeted research on the assignment of reliability indicators to gas turbines;
- Improved reliability design analysis methods have not been well applied to engineering practice.

In response to the above three shortcomings, the future of reliability design analysis work for gas turbines has the following three perspectives:

- In the future, it is necessary to strengthen the gas turbine reliability design analysis according to the relevant requirements and to conduct extensive research on gas turbine reliability design analysis methods;
- In the future, regarding gas turbine reliability design analysis, it is necessary to focus on reliability allocation projections to fundamentally ensure the achievement of gas turbine reliability targets.
- In the future, we should combine the high efficiency and accuracy pursued in research with the ease of operation and high practicality pursued in engineering practice to find a more suitable reliability design analysis method for gas turbines.

6 Discussion and Conclusion

This paper aims to compare the existing gas turbine reliability design analysis methods with the work items specified in the "GJB 450A-2004 General Requirement for Materiel Reliability Program" and to evaluate whether the current reliability design analysis methods fully cover the work items specific to gas turbines. Additionally, this paper investigates whether these methods have undergone any improvements during their application process.

It has been observed that the existing gas turbine reliability design and analysis applications align with the requirements of the "GJB 450A-2004 General Requirement for Materiel Reliability Program". However, there is a lack of targeted research in reliability allocation projections. Many researchers have incorporated fuzzy mathematics into gas turbine reliability design and analysis work, combining traditional reliability design analysis with other analytical and mathematical methods to obtain an improved reliability design and analysis method, which is beneficial for carrying out practical and accurate gas turbine reliability design and analysis work. In the future, it is essential to focus on systematic and in-depth research on reliability allocation and anticipation in gas turbine reliability design analysis to find a method that meets the needs of scientific research and practical engineering applications.

Furthermore, to gain a more comprehensive understanding of the alignment between existing gas turbine reliability design analysis methods and the stipulated work items in the "GJB 450A-2004 General Requirement for Materiel Reliability Program," additional investigation is warranted. Exploring how these methods address the unique challenges and intricacies of gas turbine systems will shed light on potential gaps that must be addressed.

As the field of gas turbine technology evolves, it becomes increasingly important to assess whether the reliability design analysis methods have evolved as well. This paper has illuminated the current state of these methods, but it also raises the question of their adaptability over time. Have any improvements or refinements been made to these methods as they have been put into practice? Evaluating the practical applications of these methods in real-world scenarios could uncover valuable insights into their effectiveness and identify areas for enhancement.

Moreover, introducing fuzzy mathematics into the gas turbine reliability design and analysis realm presents an intriguing avenue for advancement. The fusion of traditional approaches with innovative analytical techniques offers the potential for more accurate and robust results. This integrated approach acknowledges the complexities inherent in gas turbine systems and seeks to provide a more holistic understanding of their reliability performance. As such, future research endeavours should continue to explore the synergies between these mathematical methodologies and their application in gas turbine reliability design and analysis.

A notable area that requires focused attention is reliability allocation projections. The identified need for targeted research indicates an opportunity for development. A more nuanced and comprehensive exploration of reliability allocation methods within the context of gas turbines is essential for both scientific rigour and practical application. By delving into this aspect, researchers and practitioners can devise strategies to optimally allocate resources and efforts to enhance the overall reliability of gas turbine systems, in some cases can refer to the following references [16–18].

In conclusion, this paper has taken significant strides in assessing the alignment between gas turbine reliability design analysis methods and established standards. However, the journey towards robust and comprehensive reliability design analysis for gas turbines is ongoing. Continued exploration of evolving methodologies, adaptability over time, and integration of innovative mathematical techniques will be pivotal in meeting the evolving demands of scientific research and practical engineering applications in gas turbine reliability design analysis.

References

1. Qu, S., Mou, Y.: Gas turbine reliability technology. China Sci. Technol. Inf. (16), 169–171 (2014)
2. Zhu, T.: Failure mode, effects and criticality analysis of aero-derivative gas turbine combustor. Aeroengine **48**(01), 33–39 (2022)
3. Chen, X.: A preliminary study on the reliability design of gas turbine generator sets. Heilongjiang Sci. Technol. Inf. **05**, 21 (2017)
4. Shang, W., Wang, W., Qi, P., et al.: Gas turbine engine fault diagnosis based on conditions of ruled-based ang fault tree. J. Mech. Electr. Eng. **30**(07), 798–801+14 (2013)
5. Wei, C., Peng, T., Fang, Y.: Fault analysis of abnormal oil consumption in an industrial gas turbine. Therm. Turbine **50**(04), 296–299 (2021)
6. Fang, K., Liang, Q., Luo, J., et al.: Reliability study of solid oxide fuel cell-gas turbine combined cycle system based on fault tree. J. Nav. Univ. Eng. **32**(04), 99–105 (2020)
7. Yan, D., Liang, Q., Shao, M., et al.: Analysis of reasons for start failure of the gas turbine with fuzzy fault tree. China Shiprep. **28**(01), 40–42 (2015)
8. Li, L., Liang, Q., Liang, Y.: Fuzzy fault tree analysis of a gas turbine fuel system. In: IOP Conference Series: Earth and Environmental Science, vol. 237(2) (2019)
9. Yang, H., Hong, X.: Reliability analysis of gas turbine based on the failure mode and effect analysis. In: Proceedings of the Asia-Pacific Power and Energy Engineering Conference (APPEEC), Wuhan, Peoples R China, pp. 25–28 (2011)
10. Zhao, D., Wen, W., Duan, C.: Reliability allocation of GT Based on fuzzy theory. Gas Turbine Technol. **04**, 44–47 (2003)
11. Bai, Z.: Reliability analysis of gas turbine based on analytic hierarchy process and FTA. In: Proceedings of 2018 International Conference on Mathematics, Modeling, Simulation and Statistics Application (MMSSA2018), Shanghai, China, pp. 190–193 (2018)
12. Liu, Q., Shen, R., He, J., et al.: Simulation and analysis of reliability of combined rotor turbine of heavy gas. Comput. Emul. **36**(02), 183–186+200 (2019)
13. Shetty, P., Mishra, R., Prithvi, S., et al.: Finite element approach for failure analysis of a gas turbine blade. J. Fail. Anal. Prev. **18**, 1210–1215 (2018)
14. He, X., Sun, F.: Research on reliability prediction of gas turbine based on neural network. Gas Turbine Technol. **22**(02), 1–4+8 (2009)
15. Zhang, Y., Li, B.: The strength reliability analysis of symmetric struts supports of gas turbine based on stochastic finite element. Mach. Des. Res. **28**(01), 46–48 (2012)
16. Yazdi, M., Mohammadpour, J., Li, H., et al.: Fault tree analysis improvements: a bibliometric analysis and literature review. Qual. Reliab. Eng. Int. **39**, 1639–1659 (2023). https://doi.org/10.1002/qre.3271
17. Li, H., Yazdi, M.: Advanced decision-making methods and applications in system safety and reliability problems. In: Studies in Systems, Decision and Control, p. 211. Springer Cham (2022)
18. Yazdi, M., Moradi, R., Pirbalouti, R.G., Zarei, E., Li, H.: Enabling safe and sustainable hydrogen mobility: circular economy-driven management of hydrogen vehicle safety. Processes **11** (2023). https://doi.org/10.3390/pr11092730

Prioritization of HSE Management Risks in the Management of Construction Projects (Non-Mechanized Tunnel) by Monte Carlo Modeling Method

Adel Gholami, Mahdi Mozaffari, and Reza Ghasemi pirbalouti

Abstract Given the extensive scope of construction operations, it's evident that construction activities rank among the leading causes of accidents in many countries. This underscores the critical need for in-depth research on health, safety, and environmental (HSE) issues within the construction industry. In this context, identifying and prioritising key indicators influencing non-mechanised tunnel construction accidents, particularly within high-risk construction projects, is a pivotal strategy to mitigate accidents and reduce the associated human and financial toll, thereby enhancing overall HSE conditions within the project. One noteworthy example of this imperative is the Tehran Shomal Freeway, a national mega project in Iran integral to the North–South Corridor initiative. Its initial phase comprises 39 tunnels spanning a total length of 31.5 kms, and it has encountered a multitude of HSE challenges, including tunnel collapses, landslides, and occupational accidents leading to fatalities. To gain insights into this complex scenario, a comprehensive approach was adopted. Initially, HSE experts' opinions on tunnel construction within the Tehran Shomal Freeway project were gathered through a structured questionnaire, employing the Delphi method to categorise these insights into 11 primary indicators and 74 sub-indices. Subsequently, a Monte Carlo modelling technique was applied, generating 2000 random data sets under the Skew Normal distribution. These data were then evaluated against three predefined success criteria and synthesised using the Copeland method to prioritise the leading indicators. Among the findings, it became apparent that "management" and "design and implementation" risks exerted the most substantial influence, while "Human Resources" risks had a relatively lower impact on HSE management within non-mechanised tunnel construction projects.

A. Gholami (✉) · M. Mozaffari
Civil and Environmental Engineering Department, Amirkabir University of Technology (AUT), Tehran, Iran
e-mail: adel.gh@aut.ac.ir

R. G. pirbalouti
Department of Mathematical and Industrial Engineering, Polytechnique Montréal, Montreal, Canada

M. Yazdi (ed.), *Progressive Decision-Making Tools and Applications in Project and Operation Management*, Studies in Systems, Decision and Control 518,
https://doi.org/10.1007/978-3-031-51719-8_13

Keywords Health · Safety · Environment · Tunneling · Freeway · Management · Monte carlo · Bootstrap · Copeland

1 Introduction

The construction sector consistently ranks as a top contributor to accidents worldwide, primarily due to its expansive reach. Within this industry, the presence of diverse risks and hazardous factors in construction sites has transformed it into one of the most perilous sectors globally. Despite representing only about 7% of the world's workforce, the construction industry bears a disproportionate burden, accounting for 30–40% of global injuries. Notably, among various construction endeavors, tunnel construction projects carry the highest risk quotient, with an incidence rate ranging from 40–50%. This heightened risk profile can be attributed to the unique structural characteristics of tunnel projects, often marred by inadequate knowledge of ground conditions, surrounding environments, and the inherent uncertainties, frequently culminating in severe damages and incidents [1, 2].

The risk management process stands out as a paramount component within project management, and a paradigm shift in contemporary perspectives now regards it as the very foundation and core essence of effective project management [3, 4]. Effective and timely risk management is essential for both preventing the occurrence of negative consequences and mitigating their impact on project objectives. Identifying and proactively addressing risks not only reduces the likelihood of accidents but also enhances the prospects of achieving the project's goals [5, 6]. The efficacy of a systematic risk management process hinges significantly on the accurate and timely identification of risks. Within this framework, the identification of health, safety, and environmental (HSE) risks assumes paramount significance. Specifically, when it comes to pinpointing HSE risks in the context of tunnel construction, extensive research in this domain has been ongoing since the 1980s. Back in 1987, Dudek et al. laid the foundation for categorizing risks associated with underground space projects. They delineated these risks into three distinct categories: structural, contractual, and functional. This categorization was primarily driven by considerations related to the temporal aspects and longevity of underground structure [7]. The efficacy of a systematic risk management process hinges significantly on approximately a decade later, in 1996, Yogaranpan and colleagues conducted a comprehensive study aimed at discerning the risks associated with underground space projects. Their research categorized these risks into four distinct types: natural hazards (such as floods and earthquakes), external factors (including economic and political variables), internal considerations (pertaining to strategic and planning aspects), and risks stemming from human errors (such as accidents resulting in damage) [8, 9]. In 2004, as understanding in the domain of underground rail projects advanced, Brown et al. delineated the risks associated with tunneling projects into four overarching categories. These categories encompassed: Risk of Damage: This pertains to potential harm or damage to both personnel and equipment involved in the project. Risk of Non-Compliance with Standards and Criteria: This category focuses on the potential for not meeting established industry standards and criteria, encompassing safety, quality, and regulatory requirements. Risks of Delays: These encompass the various factors and uncertainties that could lead to project schedule disruptions, potentially resulting in missed deadlines and increased project duration. Risks of Cost Escalation: This category addresses the factors and contingencies linked to cost overruns, which could result in budgetary concerns and financial strain on the project [10, 11]. In 2005, Wagner et al. delineated this division differently, classifying it into four distinct categories: Construction and design risks Income and demand risks Operation and maintenance risks Other risks, encompassing factors such as changes in laws, tax regulations, and other miscellaneous considerations [12]. In 2006, Einstein et al. conducted a comprehensive review that addressed the diverse array of natural indicators associated with risks in underground projects. These encompassed factors such as the structural composition and attributes of geological formations, geological heterogeneity, underground water conditions, geological processes, as well as events tied to seismic activity, erosion, karst formations, and geothermal heat [13]. Upon reviewing previous research on the identification and categorization of risks associated with tunnel construction, it becomes apparent that the approach to risk classification has varied depending on the specific requirements of each project. Consequently, the design and evaluation of tunnel risks have been subjects of discussion, with a particular focus on risk acceptance and safety management. In the context of risk mitigation and sound decision-making, there is a need to explore the utilization of multiple criteria for risk reduction and to determine which criteria are most apt for assessing risk acceptance [14, 15]. In such a way that even seasonal changes (climate) are effective in civil accidents [16] and on the other hand, by examining civil accidents, it can be concluded that unsafe actions also constitute approximately 80 to 90% of the causes of civil accidents [17]. In this regard, the higher the percentage of preventive operations implemented in a strategy, the lower the loss rate of civil accidents [18]. The primary factors contributing to the rise in construction accidents can be categorized into three distinct groups: Environmental Factors: These encompass elements in the construction surroundings that significantly impact safety, such as weather conditions, terrain, and site-specific hazards. Interactions Among Identified Indicators: This group explores the complex interplay between various safety-related indicators, shedding light on how they influence and exacerbate each other within the construction context. Correlation Between Construction Value and Assembly Production: This group investigates the intrinsic link between the scale and scope of construction projects and the corresponding increase in assembly production, as it pertains to safety and accident risk [19, 20]. Also, the use of online warning systems in tunnel construction can increase the level of safety [21]. Safety in construction matters requires care and planning throughout the life of the project from design to maintenance [22]. Despite the evident significance of effectively managing HSE risks in tunnel construction, as underscored by recent studies, a notable gap persists in the field. This gap pertains to the systematic identification, categorisation, and prioritisation of HSE risks observed over the years in tunnel construction projects.

Addressing this gap represents a crucial aspect of recent research endeavours in construction safety and risk management.

The primary objective of this research is to formulate a comprehensive model that not only identifies but also prioritises the many HSE risks that have surfaced in tunnel construction projects. This endeavour holds paramount importance within the broader context of project risk management. The rationale behind this pursuit is that it becomes possible to proactively plan and mitigate potential HSE incidents through the prioritisation and systematic categorisation of these identified risks. This research seeks to empower construction projects with the tools necessary for preemptive risk management, thereby enhancing overall safety and environmental standards.

However, realising this research's objectives has its challenges. One of the most significant limitations is the relatively small pool of experts available, particularly in a developing country like Iran, where this research is being conducted. Furthermore, the variability in expertise within the HSE field of non-mechanised tunnel construction projects adds complexity to the task.

To overcome these limitations, this research introduces innovative approaches and methodologies. These strategies aim to bridge the expertise gap by facilitating collaboration and knowledge-sharing among experts, both locally and globally. This research strives to ensure a more inclusive and diverse perspective on HSE risks in tunnel construction by harnessing technology and creating platforms for expert input and consensus-building. Ultimately, it seeks to offer practical solutions that can be applied in Iran and similar contexts worldwide, thereby contributing to safer and more environmentally responsible construction practices.

1.1 Research Structure

This research unfolds five distinct sections, each contributing to a comprehensive understanding of the subject matter. Section 1: Introduction and Research Context The initial segment serves as the gateway to our study. It provides an overview of the research's general context and introduces the fundamental background of the proposed research design. This section delves into the research's approach, objectives, underlying necessity, and inherent limitations. These aspects are elucidated through an extensive review of existing literature and prior studies, focusing on risk assessment and Health, Safety, and Environment (HSE) management within the domain of tunnel construction. This step is pivotal in establishing the research's framework and showcasing its innovative contributions to the field. Section 2. Research Methodology and Data Collection The second segment of the research framework unveils the research methodology. Here, we delineate the chosen research method, elaborate on the study's target population, and outline the statistical sample selected for analysis.

Additionally, we expound upon the methods and tools employed for data collection, shedding light on the meticulous techniques used in data analysis. These elements collectively steer us toward the attainment of the research objectives.

Section 3. Data Analysis and Findings In the third part, we delve into the core of our research—the results of data analysis. This section unveils the empirical findings from our methodical examination of the collected data. It offers insights into the intricate web of information we have gathered, providing a lucid depiction of our discoveries. Section 4. Discussion, Conclusions, and Implications The penultimate segment is dedicated to an in-depth discussion and synthesis of the results obtained for each research question. We critically analyze the implications of these findings and draw meaningful conclusions.

Additionally, we offer a concise summary of the practical implications of our research, illuminating avenues for future investigations. This section encapsulates our research journey's essence, offering closure and a springboard for further exploration. Section 5: Practical Implications and Future Research The concluding part of our research extends beyond mere closure. It encapsulates the practical implications drawn from our findings, emphasizing their relevance to real-world scenarios. Furthermore, we outline potential directions for future research endeavours, paving the way for continued growth and development. This section bridges the insights gleaned from our study and their application in the broader context, ensuring that our research serves as a catalyst for positive change.

2 Methodology

To enhance comprehension and provide a more detailed depiction of the method, we have illustrated an overview in Fig. 1. This visual representation is a valuable aid in elucidating the key steps and components of the technique, offering a clear and concise roadmap for a deeper understanding. In Fig. 1, each element has been strategically arranged to elucidate the sequential flow of the method, facilitating a visual grasp of its intricacies and relationships between different components. This graphical representation not only simplifies complex processes but also serves as a quick reference point, allowing stakeholders to easily navigate the method's intricacies.

First step: Gathering information from experts

(a) **Collecting data and information**
This research employed a comprehensive approach, combining library and field methods to gather diverse data and information. In the library section, we extensively reviewed the literature, drawing upon theoretical foundations and research background primarily sourced from Persian and Latin articles, books, and publicly accessible theses—this rigorous examination of existing knowledge provided a solid theoretical framework for building our research. We actively engaged with experts in the field section to acquire real-world insights. We sought the perspectives of two distinguished university professors with expertise in tunnel construction projects. We enlisted the input of three

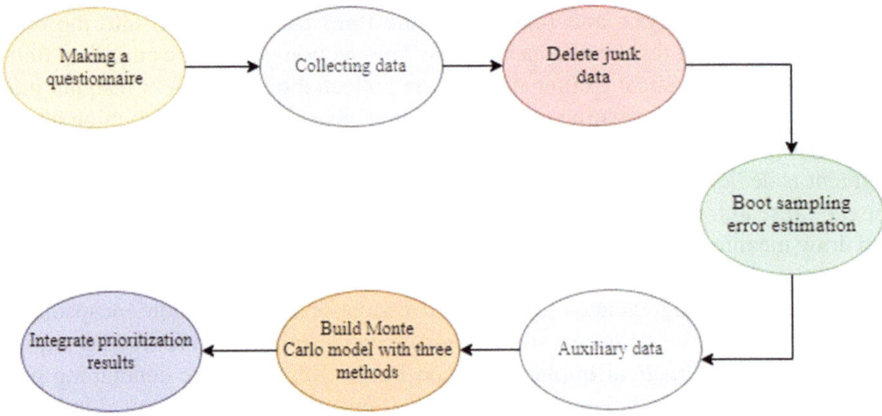

Fig. 1 Overview of the method

experienced executive specialists specializing in Health, Safety, and Environment (HSE) within the context of tunnel construction. This approach ensured that our research was grounded in both theoretical principles and practical wisdom, offering a well-rounded perspective on the subject matter. Through this meticulous data collection process, we identified and categorized the various risks impacting the HSE management of tunneling projects. These risks were organized into 11 primary indicators, resulting in a comprehensive inventory of 74 specific risk indicators, as detailed in Table 1. This categorization and enumeration of risks serve as a valuable resource for stakeholders in the field, enabling a deeper understanding of the multifaceted challenges associated with HSE management in tunnel construction projects and providing a foundation for effective risk mitigation strategies.

- Risks of natural disasters [23–27]
- Environmental and biological risks [23, 24, 26, 28, 29]
- Legal risks [25, 28]
- Economic risks [23–25, 30, 31]
- Management risks [23–26, 32]
- Contractual risks [23, 26, 30]
- Project damage and loss risks [23–26, 30, 33–35]
- Human Resources risks [23–26, 30, 36]
- Risks of equipment resources and materials [25, 26, 30, 33, 37–40]
- Technical risks of design and implementation [23, 30, 31, 37]
- Planning risks [23–25, 33, 39]

Table 1 Project risks related to HSE management at the macro level

Risks of the Project Related to HSE Management at the Macro level

Environmental and Biological Risks

1. Negative Impacts of Noise, Visual, Traffic, and Air on Residents and Local Businesses During Construction (Due to Heavy Explosions and Devices, etc.)
2. Loss of agricultural production, impacts on tourism and regional economy
3. Risk of environmental pollution (such as water, soil, and air) from the impact of road operations on local people
4. Impacts of traffic on the local road network and communications
5. Site Security (such as Theft of Equipment)
6. Potential landslides due to construction activity
7. Risk of impacts on the drinking water basin
8. Risk of impacts on surface water flow, including floods
9. Risk of impacts on groundwater flow
10. Risk of potential impacts on endangered species and endangered ecological communities, native plants, animal habitats, etc.
11. Risk of potential impacts on aquatic ecosystems and groundwater-dependent communities
12. Potential Impact on Native and Non-Native Heritage
13. Potential for ground pollution
14. Potential hazards related to the management, storage, and transportation of hazardous materials during construction
15. Potential effects of waste during construction (such as waste production, disposal, and collection)
16. The comfort/quality of life of local people

Technical risks of design and implementation

1. Contractor and Consultant Competence: Experience, Capacity and Ability
2. Startup problems, inaccuracies in implementation, superficiality, rework
3. Technological difficulties in performing the task manually or manually
4. Strategic Issues: Ability to maintain, operate and train
5. Dig tunnels (such as NATM or explosions)
6. Unreliable and insufficient data and information and inadequate documentation
7. Experience and skills of specialist personnel in working with HSE requirements
8. Continuous testing, inspection and monitoring (HSE requirements audit)
9. Consolidation of important positions on-site and workshop equipment (such as including offices, technical equipment, transport, and labor camp)

Legal Risks

1. Local and international regulations, licenses and approvals, environmental standards, forestry, national pastures, natural habitats, etc.
2. Compliance with environmental standards (contenders and owners along the way)
3. Changes in laws and regulations (local and international)
4. Unions and workers' associations, and insurance

Project planning Risks

1. Changes made by the client during project implementation
2. Planning risk, project definition weakness, project completion time
3. Emergency response guidelines (such as fire, flood, accident, and collision inside a tunnel)
4. Lack or weakness in the economic and environmental technical justification
5. Lack of strategic vision in the project
6. Drainage paths from tunnels and underground stations
7. Traffic Management / Control Systems
8. Effective communication systems

Equipment and materials resource Risks

1. Classification and prioritization, hazard identification, and information systems for hazardous products (such as raw materials)
2. Lack of available equipment and availability / timely availability
3. Storage, transportation, and use of hazardous substances (such as explosives and fuel)
4. Quality of equipment and warranty of sellers and insurance
5. Change the type and specifications of materials and equipment during construction

Contractual Project Risks

1. Risk of ambiguity in the description of duties and obligations in contracts
2. Contractual risk of unrealistic bidding and competitive bidding
3. Cost control risk during design and implementation (employer-related)
4. Contractors Policy for Tender (HSE Risk Index)
5. Nationality of contractors
6. Penalty for delays
7. Synchronize tasks and payments, etc.

Disaster Risks

1. Earthquake disasters and landslides
2. Hurricane and flood disasters
3. Natural disasters of fire

Economic Risks

1. Market conditions (such as price fluctuations, inflation, interest rates,exchange rates, and trades)
2. Actual cost coverage based on adjustment rates
3. Stock rate changes, financing conditions,government economic policies
4. Taxes, duties and customs duties

Project management Risks

1. Organizational risk management, poor control, unrealistic goals, buildability
2. Risk of disproportionate expertise and experience of managers/professionals with their position
3. Risk of inaccurate distribution of financial, human, and equipment resources
4. Failure to define the tasks of the teams involved in the project
5. Client Type: Private or Governmental
6. HSE Risk Management
7. Committee of HSE issues

Project Human Resources Risks

1. Damage and injuries of Human Resources related to work (such as altitude, electricity, chemicals, and mechanical equipment)
2. Wages and incentives such as rewards and penalties
3. Nationality of staff and workers and ethnicity
4. Workers' personal and family problems and being away from the family
5. Personal protective equipment and workwear
6. Workforce training appropriate to project needs

These indicators wield substantial influence over the comprehensive HSE risk management of the project, a topic that has been extensively examined and deliberated in various scholarly references. These well-documented insights underscore the critical role these indicators play in shaping the risk management landscape of construction projects. When we delve into the realm of non-mechanized tunnel construction projects, we recognize that HSE management risks operate at a macro level, encompassing a multifaceted spectrum of challenges. To comprehensively capture and categorize these challenges, an intricate analysis has led to their classification into 11 distinct categories, as detailed in Table 1. This categorization serves as a roadmap, delineating the various facets of HSE risks inherent to non-mechanized tunnel construction projects. It enables project stakeholders to gain a nuanced understanding of the complexities involved and, more importantly, to strategize and implement effective risk mitigation measures across these diverse categories.

(b) **Construction and validity of the questionnaire**

(1) **Validity of the Questionnaire**

Various methods exist to assess the validity of a measurement instrument, with content validity emerging as one of the most widely employed approaches. In this study, content validity was pivotal in evaluating the questionnaire's effectiveness in capturing the relevant constructs. Utilizing the Delphi technique, a systematic and iterative process was employed to ensure the content validity of the questionnaire.

The initial phase of content validation involved soliciting the input of subject matter experts, notably professors and specialists in the field. These individuals were chosen for their deep knowledge and expertise in tunnel construction and health, safety, and environmental (HSE) matters. To commence this assessment, an open questionnaire was distributed to the experts, inviting them to offer their perspectives on the collected indicators and sub-indices.

Subsequently, a closed questionnaire, meticulously designed by the researcher, was administered to these experts. This questionnaire aimed to elicit their judgments regarding the appropriateness of the identified indicators and sub-indices. This step allowed for a more structured evaluation of the content validity, as experts were asked to provide explicit feedback on the relevance and accuracy of the questionnaire's components.

Finally, a standardized questionnaire for modelling was employed to synthesise and formalise this valuable input. The Monte Carlo, a robust statistical technique, was used to gather and process information. This quantitative approach helped distil the qualitative judgments of the experts into a comprehensive and data-driven assessment of content validity.

The research process drew upon the insights and expertise of two esteemed university professors from the Faculty of Civil and Environmental Engineering at Amir Kabir University of Technology in Tehran,

Iran. In addition to academic perspectives, the study benefited from the practical experience and insights of three seasoned specialists in the realm of HSE within tunnel construction projects, who were affiliated with the Tehran Shomal Freeway Company.

Incorporating the perspectives of these experts ensured that the content validity of the questionnaire was rigorously assessed and confirmed, enhancing the reliability and accuracy of the research instrument for the comprehensive evaluation of HSE risks in non-mechanized tunnel construction projects, with particular relevance to the Tehran Shomal Freeway initiative.

(2) **Reliability of the questionnaire**

Before performing any analysis on the collected data and statistical inference, the reliability of the measurement tool must be ensured [38]. Cronbach's alpha is a measure of the internal consistency of the observed variables. The closer the obtained percentage is to 100%, it indicates the greater reliability of the questionnaire, and basically, Cronbach's alpha coefficient above 70% is considered as good and acceptable reliability for the questionnaire [41].

(3) **Questionnaire and how to answer**

A well-structured questionnaire was developed to comprehensively address the research objectives, consisting of two main sections: demographic characteristics and specialized sections. The technical teams focused on 11 key indicators, encompassing 74 questions designed in a 5-point Likert scale format.

The questionnaire served as a vital tool in gathering valuable insights from experts. These experts were tasked with evaluating the significance of each indicator based on their extensive knowledge and experience. Using a range of scores from 1 to 5, experts assigned values to the arrows, with one denoting low importance and 5 signifying very high priority. This approach allowed for a nuanced assessment of the relative significance of each indicator as perceived by the experts.

Subsequently, the collected data, enriched with the experts' assessments, was meticulously entered into the widely used statistical software, SPSS. This data analysis platform was employed to conduct a rigorous analysis, providing a deeper understanding of the weighted importance of each sub-index within the 11 overarching indicators. This method, known as Monte Carlo modelling, enabled a robust and data-driven approach to prioritize and allocate importance scores to these sub-indices.

By employing this methodological framework, the research project was able to harness the collective expertise of these professionals and derive objective, data-driven insights. These insights contributed to a comprehensive understanding of the HSE dynamics within non-mechanized tunnel construction projects. They laid the foundation for informed decision-making and targeted risk management strategies, ultimately enhancing the overall safety and environmental performance of such endeavours.

(c) **Sampling error**

The resampling method based on the existing sample tries to better estimate the parameters of the statistical population or calculate the error of the estimator [42]. In this research, to estimate the sampling error of prioritizing 11 categories in the Monte Carlo model using the bootstrapping method using SPSS software and with 95% confidence and the number of N (the reason for N in the fourth stage of Monte Carlo model construction is explained) Sampling and confidence interval type BCa (more accurate and will reduce bias) and simple random sampling method have been done.

The closer the bootstrap number of each of the 11 main indicators is to zero from two perspectives:

1—the standard deviation of the average and 2- the 5% of the interrupted average; It will have less error.

The second stage: the influence of expert weight

Acknowledging the diversity of expertise within a group of professionals, it becomes evident that various factors come into play when assessing their qualifications. These factors encompass a wide range of attributes, including but not limited to experience, educational background, age, and their respective positions within their organizations. To effectively account for these differences and establish a more nuanced evaluation framework, a method was devised to assign relative weights to indicators. This method leveraged a combination of these diverse criteria and utilized the Likert scoring scale to assess specific actions. The resulting weighted indicators are thoughtfully presented in Table 2, offering a comprehensive and well-rounded perspective on the expertise of individuals within the group. This approach not only promotes fairness and equity in the assessment process but also ensures that a multidimensional view of expertise is considered, thereby enhancing the overall quality and depth of the evaluations conducted.

To gauge the influence of experts and their respective rankings on the evaluation process, a systematic approach is employed. Initially, the individual weights of all experts are aggregated, creating a unified metric that reflects the collective expertise involved in the assessment. This consolidation of expert weights is crucial for establishing a fair and comprehensive evaluation framework. To effectively harness the diverse perspectives offered by each expert, their respective weights are incorporated into the data using Eq. 1 [43].

This method ensures that the experts' opinions carry varying degrees of influence based on their unique levels of expertise. Consequently, it introduces a nuanced dimension to the data, reflecting the diversity in expert viewpoints without altering the raw data itself. Moreover, this process has a dual benefit: it not only preserves the original data but also transforms the responses on the 5-point Likert scale from discrete values into continuous variables. This transformation serves a twofold purpose. Firstly, it enables the application of rigorous statistical analyses on the dataset, enhancing the depth and precision of insights derived from it. Secondly, it accommodates the inherent variability in expert opinions, allowing for a more

Table 2 Weighted criteria of experts

No	Expert weighting criterion	Benchmark ranking	Score
1	Age	Less than 30 years	1
		31–40 years	2
		41–50 years	3
		50 years and up	4
2	Education	BSc	1
		MSc	2
		Ph.D	3
3	Experience	Less than 5 years	1
		5 - 10 years	2
		11–15 years	3
		16–20 years	4
		Over 20 years	5
4	Employment	Expert/Officer	1
		Supervisor/Consultant	2
		Manager	3

fine-grained examination of the subject matter. In essence, this approach marries the collective wisdom of experts with the nuances of their expertise, enriching the evaluation process. It not only maintains the integrity of the data but also empowers statistical analyses to uncover subtleties and trends that may remain hidden in a discrete dataset. This thoughtful integration of expert input ensures a more comprehensive and robust assessment, facilitating well-informed decision-making and yielding valuable insights for various applications.

$$\mathrm{y_i} = \frac{(x_{i-n} - \overline{\mathrm{X}})}{(x_{max} - \mathrm{X_{min}})} + 1 \tag{1}$$

Then the weight of each expert is multiplied in the answers to the questions so that the effect of the level of expertise is applied in the answers given.

The third step: find the distribution function of the data

Before embarking on any statistical analysis in research, the pivotal first step is the careful selection of the most suitable statistical method. This choice lays the foundation for the subsequent implementation of statistical techniques and the generation of meaningful test statistics that, in turn, lead to logical and informed conclusions regarding research hypotheses. Among all the preparatory measures, gaining a profound understanding of the underlying data distribution stands as a fundamental priority. The selection of the appropriate statistical method is akin to charting a course for your research journey. It dictates the tools, techniques, and strategies you'll employ to extract meaningful insights from your data. Different research questions and data types call for specific statistical approaches, and making an astute choice at this stage can greatly impact the rigor and validity of your findings. Central to this decision-making process is an in-depth grasp of the data distribution. Recognizing how your data is distributed allows you to make informed decisions about which statistical methods are most applicable. Whether your data follows a normal distribution, exhibits a skewed pattern, or adheres to some other distribution, this knowledge guides your choice of statistical tests and helps ensure the assumptions underlying those tests are met. Furthermore, an understanding of data distribution aids in assessing the appropriateness of parametric versus non-parametric statistical methods, helping you select the most accurate and powerful tools for your analysis. It also plays a pivotal role in determining sample size requirements, as different distributions may necessitate larger or smaller sample sizes to achieve meaningful results. In essence, the careful consideration of data distribution and the subsequent selection of the right statistical method constitute the bedrock of sound statistical research. This process ensures that your analysis is not only statistically valid but also relevant and capable of providing meaningful insights into the research questions at hand. Ultimately, by prioritizing this critical step, you set the stage for a successful and rigorous research endeavor [44].

The fourth step: adding data with r studio software

In this critical phase of the process, we embark on the task of generating N random data points based on the distribution function derived from the preceding step, utilizing the versatile R Studio software. To carry out this task effectively, it is imperative to precisely define and calculate the necessary parameters that will drive this data generation process. In essence, these parameters serve as the guiding elements that breathe life into our distribution function, enabling it to produce the desired dataset. The process of specifying and calculating these parameters involves a meticulous understanding of the chosen distribution, its parameters, and the desired characteristics of the random data we intend to generate. For instance, in the context of a Gaussian or normal distribution, two crucial parameters are mean (μ) and standard deviation (σ). These values are instrumental in determining the central tendency and dispersion of the generated data, thereby shaping the distribution's profile. The precision and accuracy with which these parameters are specified and calculated are pivotal to ensure that the generated data faithfully represents the underlying distribution. In cases where more complex distributions are involved, such as skewed or multimodal distributions, a broader set of parameters may be required to capture the nuances of the distribution accurately. Furthermore, the selection of an appropriate random number generation method or algorithm within R Studio is another facet of this step that warrants careful consideration. Different methods may be better suited to specific distributions or data characteristics, and the choice can significantly impact the quality and reliability of the generated dataset. Ultimately, the effectiveness of this parameter specification and calculation process is directly linked to the fidelity of the generated data to the real-world distribution under investigation. It is a crucial step in ensuring the validity and utility of subsequent analyses and insights drawn from the generated dataset, making it a pivotal aspect of the overall data generation and analysis workflow.

(a) Determining the range of data to build a monte carlo model

To perform this step, first remove the numbers that are outside the range of the highest and lowest data values so that the remote data are removed from the selection criteria.

The highest value = the highest possible expert weight × 5 (very high importance).

The lowest value = the lowest possible expert weight × 1 (very low importance).

Then Monte Carlo modeling is done under the following three conditions.

1. The probability of winning is 50%, in such a way that the sum of the sub-indices of each of the main indicators exceeds 50% of the maximum value that the sub-indices can have.
2. The probability of winning is greater than the average of the sub-indices of the main indicators of the generated data from the average of 5% of the raw data of each of the main indicators.
3. The probability of winning is the average of the sub-indices of the main indicators of the generated data in the 95% confidence interval for the average of the raw data of each of the main indicators (Fig. 2).

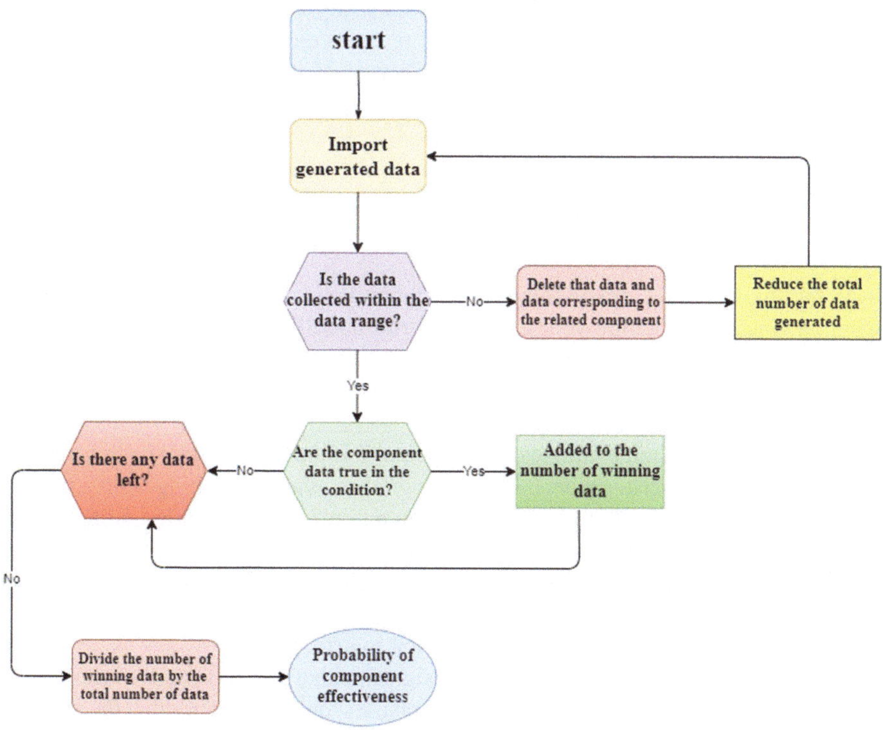

Fig. 2 General flowchart of Monte Carlo modeling

(b) **Finding the number of data additions required for the research**

When dealing with statistical modelling, a key concern arises when the available population data is insufficient for robust modelling. In such scenarios, a critical question emerges: How much data generation is necessary to prioritize the risks identified through the Monte Carlo modelling method? A systematic approach is employed to determine the optimal amount of data required to address this challenge.

A specific indicator, coupled with one of the more universally applicable winning conditions, is randomly selected as the starting point in this process. Subsequently, the data generation process is initiated, and additional data is continuously incorporated into the model. This iterative procedure continues until a point is reached where further data additions do not significantly alter the obtained results. This point is often referred to as "convergence."

Convergence, in this context, signifies the stage at which the model's outcomes stabilize or remain relatively consistent despite the continued injection of additional data. It serves as a crucial milestone in the modelling process, indicating that the model has gathered sufficient data to make reliable risk prioritization assessments.

Determining convergence is a pivotal aspect of data-driven modelling, as it not only informs the adequacy of the dataset but also provides insights into when diminishing returns are encountered in data generation. This methodology ensures that resources are optimally allocated towards data collection, balancing data quantity and modelling accuracy. Ultimately, achieving convergence in data generation is a key step in enhancing the reliability and utility of Monte Carlo modelling in risk assessment and decision-making processes.

2.1 Integration of Research Modeling Results

Given the diverse outcomes yielded by the three distinct prioritization methods, or "winning conditions," employed in Monte Carlo modelling, the Copeland method is vital to consolidate these findings and arrive at robust conclusions. This methodology hinges on creating a pairwise comparison matrix among the identified indicators, facilitating informed decision-making.

The Copeland method evaluates one indicator's relative preference or superiority over another in multi-criteria decision-making. This assessment hinges on a numerical score assigned to each comparison. When an indicator is deemed preferable to another option across various multi-criteria decision-making techniques, it receives a score of 1 in the paired comparison matrix. Conversely, in cases where no clear majority consensus exists or when the votes are evenly split between indicators, a score of 0 is assigned in the paired comparison matrix.

To interpret these scores, a value of 1 signifies that the indicator in the row is preferred over the hand in the column. In contrast, a score of 0 implies that the arrow in the queue is favoured over the one in the row. This process is repeated systematically for all the identified indicators, culminating in constructing a comprehensive pairwise comparison matrix.

Crucially, the sum of scores within each row of this matrix effectively quantifies the extent to which each indicator dominates its counterparts. This cumulative indicator dominance metric provides a clear and quantitative insight into the hierarchy of hands, shedding light on their respective importance within the study context. By employing the Copeland method, the research can derive actionable insights and prioritize key indicators effectively, ultimately enhancing the decision-making process within the project's health, safety, and environmental management. [45].

3 Results

To effectively conduct research and gain valuable insights into the subject matter, selecting an appropriate case study that aligns with the research indicators is imperative. In this context, the Tehran Shomal Freeway project was chosen as the ideal case study for several compelling reasons.

The Tehran Shomal Freeway project is relevant due to its historical and strategic significance within the Iranian infrastructure landscape. Initially proposed in 1974, this ambitious project was conceived to address pressing transportation needs. Its primary objective is to establish a swift and cost-effective connection between Iran's northern and central regions. Furthermore, it aims to facilitate enhanced communication and trade ties with neighbouring countries in the north of Iran.

A pivotal aspect that makes the Tehran Shomal Freeway project an exceptional case study is its role in regional connectivity. Once completed, this freeway will form an integral part of the broader "Tehran Shomal Freeway," representing the shortest and most efficient link between the Caspian Sea, specifically Nowshahr Port, and the Persian Gulf. This strategic corridor is poised to facilitate regional transit, trade, and economic development.

In summary, the Tehran Shomal Freeway project serves as an ideal case study due to its historical significance, strategic importance in addressing regional transportation needs, and potential to reshape trade and connectivity dynamics in Iran's northern and central regions. This selection ensures that the research will yield valuable insights that are not only academically relevant but also have practical implications for regional development and infrastructure planning [46] (Fig. 3).

3.1 The Statistical Population of the Study (First Phase of Tehran Shomal Freeway Project)

The National Mega Project, known as the Tehran Shomal Freeway, forming a vital component of the North–South Corridor in Iran, is not immune to the significant

Fig. 3 Map of Tehran Shomal Freeway as part of the North–South Corridor in Iran

challenges HSE risks pose. The inaugural phase of this ambitious endeavour spans an impressive 32 kms and encompasses a total of 39 tunnels, collectively measuring 31.5 kms in length. This stretch, extending from the confluence of Azadegan Highway and Shahid Hemmat Highway, traces a path through various regions, including Ken, Solqan, Imamzadeh Agil, Talon Tunnel, Laniz Valley, and culminating at the Shahristanak T-road.

Despite its grandeur and strategic importance, the Tehran Shomal Freeway project has encountered many HSE challenges within its initial section. These challenges are as diverse as they are formidable, encompassing dangerous occurrences such as tunnel collapses, landslides, and grievous occupational accidents leading to fatalities. Such incidents underscore HSE management's intricate and multifaceted nature within large-scale infrastructure projects, demanding rigorous attention to safety protocols, environmental preservation, and the workforce's well-being.

Navigating these challenges is essential to ensure the successful completion of the Tehran Shomal Freeway, safeguard the lives of those involved and protect the environment. The intricate web of tunnels, valleys, and terrain covered by this project necessitates a comprehensive and proactive approach to HSE risk mitigation, integrating cutting-edge technology, expert insights, and a steadfast commitment to the principles of health, safety, and environmental stewardship. In doing so, the Tehran Shomal Freeway project can not only fulfil its infrastructural objectives but also serve as a model for responsible and sustainable mega-projects, setting a precedent for safe, efficient, and environmentally conscious development (Fig. 4).

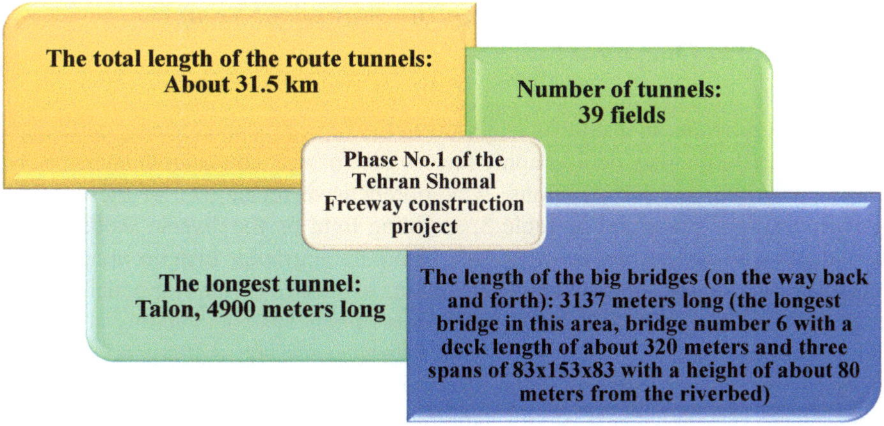

Fig. 4 Study case characteristics [47]

The Tehran Shomal Freeway, a prominent component of Iran's National Mega Project and an integral part of the North–South Corridor, stands as a striking example of a colossal infrastructure endeavor. However, like many other construction projects of its scale, it is not immune to the challenges posed by HSE risks. In fact, the initial phase of this ambitious project, which encompasses 39 tunnels spanning a total length of 31.5 km, has encountered a myriad of HSE challenges, including but not limited to tunnel collapses, landslides, and tragic occupational accidents resulting in loss of life. This pivotal phase of the Tehran Shomal Freeway project, linking Tehran to Doab and Shahristanak, extends over approximately 32 km. It commences at the confluence of Azadegan Highway and Shahid Hemmat Highway, serving as a vital artery of transportation. As it winds its way through the landscape, it traverses significant locales such as Ken, Solqan, Imamzadeh Agil, Talon tunnel, and Laniz valley, ultimately culminating at the Shahristanak T-road junction. The significance of this infrastructure endeavor cannot be overstated, not only for its role in connecting regions and facilitating commerce but also in the profound responsibilities it bears regarding the safety of its workers and the preservation of the environment. As Tehran Shomal Freeway navigates through challenging terrain and intricate engineering feats, it underscores the critical importance of comprehensive HSE management and meticulous risk mitigation strategies to ensure the well-being of its workforce and the sustainable development of the surrounding ecosystems. In doing so, it exemplifies the ongoing efforts to balance progress and safety in complex construction projects on a grand scale.

3.2 *The Demographic Results of the Statistical Sample (General Questions)*

This section provides an insightful overview of the survey respondents' general characteristics, including key demographic factors such as age, educational background, work experience, and job positions. A comprehensive breakdown of these demographic details is presented in Table 3, shedding light on the diverse profile of the individuals who participated in the survey. These demographic insights are essential in contextualizing the perspectives and opinions expressed by the respondents in the subsequent sections of this study, enabling a more nuanced understanding of the data and its implications.

Table 3 provides a comprehensive overview of the demographic characteristics of the survey respondents, shedding light on the composition of the study's sample group. The data reveals several key insights:

Table 3 Demographic results

Age	Abundance	Frequency	The cumulative percentage
Less than 30 years	9	31.2	31.2
Between 30 and 40 years	16	57.1	89.3
Between 41 and 50 years	2	7.1	96.4
More than 50 years	1	3.6	100
Total	28	100	—
Level of education	**Abundance**	**Frequency**	**The cumulative percentage**
Bachelor's degree	10	35.7	35.7
Master's degree	18	64.3	100
P.H.D	0	0	100
Total	28	100	—
Work experience	**Abundance**	**Frequency**	**The cumulative percentage**
Less than 5 years	10	35.7	35.7
Between 5 and 10 years	9	32.1	67.9
Between 11 and 15 years	4	14.3	82.1
Between 15 and 20 years	4	14.3	96.4
More than 20 years	1	3.6	100
Total	28	100	—
Side	**Abundance**	**Frequency**	**The cumulative percentage**
The expert	6	21.4	21.4
Supervisor/consultant	15	53.6	75
The manager	7	25	100
Total	28	100	—

1. **Work Experience**: A significant portion of respondents (35.7%) reported having less than five years of work experience in the field. On the other hand, a minority (3.6%) indicated having over 20 years of experience, reflecting a wide range of professional backgrounds and tenure within the industry.
2. **Age Distribution**: The age distribution of the respondents is notable, with a substantial majority (57.1%) falling within the age bracket of 30 to 40 years. In contrast, only a tiny fraction (3.6%) of respondents are over 50, illustrating a predominantly youthful workforce participating in the survey.
3. **Education Levels**: An impressive majority (64.3%) of respondents possess a master's degree, emphasizing the high educational attainment within the sample. Strikingly, none of the respondents holds doctorate degrees, indicating a specific academic profile prevalent among participants.
4. **Occupational Roles**: Regarding professional roles, the data indicates that the majority (53.6%) of respondents are in positions classified as supervisors or consultants. Conversely, a smaller proportion (21.4%) identifies themselves as experts within their respective fields, indicating a diverse range of job functions within the sample.

This insightful breakdown of respondent demographics underscores the diversity and varying degrees of experience within the surveyed population. It also highlights the prominence of master's degree qualifications and the prevalence of mid-career professionals within the study cohort, offering valuable context for the subsequent analysis and findings.

3.3 Validity and Reliability of the Questionnaire

Cronbach's alpha values, a measure of the internal consistency and reliability of the research indicators, were meticulously computed using SPSS version 26 software. The outcome of this rigorous analysis is thoughtfully showcased in Table 4, where the reported values exhibit an exceptionally high level of reliability.

This robust reliability assessment underscores the trustworthiness of the research indicators employed in the study. It signifies that the various elements comprising these indicators consistently align with one another, bolstering confidence in the validity and accuracy of the research findings. This high reliability further fortifies the foundation upon which the study's conclusions and subsequent implications are built, enhancing the overall rigour and credibility of the research.

Table 4 Cronbach's alpha

Reliability statistics	
Cronbach's alpha	N of items
0.959	74

3.4 Estimation of Sampling Error

Table 5 reveals insightful findings regarding sampling indicators impacting the HSE management within non-mechanized tunnel construction projects. These findings shed light on the variability and significance of different risk categories. Notably, environmental and biological risks emerge with the lowest recorded values. In contrast, management risks take the forefront, displaying the highest values across multiple statistical metrics, including mean, standard deviation, and 5% trimmed mean error.

This disparity in statistical measures underscores the critical role played by management factors in shaping the HSE landscape of such construction endeavours. Management-related risks exhibit higher mean values and more significant variability, as indicated by the standard deviation error. Furthermore, the 5% trimmed mean error corroborates the sustained prominence of management risks, even when outliers are factored out, emphasizing their consistent impact on HSE management within non-mechanized tunnel construction projects.

This observation closely examines the specific management practices and strategies employed in these projects, highlighting the need for comprehensive risk assessment and mitigation plans that prioritize the management dimension. Furthermore, it underscores the potential benefits of allocating resources and attention to bolstering management practices, as doing so can lead to enhanced safety, environmental responsibility, and overall project success in the complex and challenging realm of tunnel construction.

Table 5 Sampling error by bootstrapping method

No	Indicators	Mean standard deviation	5% Trimmed mean
1	Economic risks	0.1810	0.1803
2	Legal risks	0.2190	0.2314
3	**Environmental and biological risks**	**0.177**	**0.1789**
4	Natural disaster risks	0.2166	0.2475
5	Project damage risks	0.1856	0.1923
6	Contractual risks	0.2319	0.2418
7	**Management risks**	**0.2372**	**0.2529**
8	Planning risks	0.1985	0.2076
9	Technical risks of design and implementation	0.2118	0.2312
10	Equipment and materials resources risks	0.2045	0.2125
11	Human resources risks	0.2273	0.2409

3.5 Construction of Monte Carlo Research Model

As described in part 2, how to build the Monte Carlo model of the research, after collecting the required data (first step), the level of expertise of each expert is calculated (second step), which is shown in Table 6 of these values.

To find the distribution function of the data (third step), one of the most common tests is the data normality test, the Shapiro–Wilk test is more efficient for data less than 50 [48]. Given that experts may have assigned varying high or low scores to the questions, it's essential to acknowledge that the data distribution might exhibit skewness. Consequently, evaluating the skewness of the data becomes a crucial step in the analysis process. This assessment helps identify the nature of the distribution required for data augmentation. A portion of the results from the normality test is summarized in Table 6, shedding light on the data's underlying characteristics.

In data augmentation, understanding the skewness of the data is pivotal. It guides decisions regarding appropriate statistical techniques and transformations to address the data's non-normal distribution. This proactive approach allows for more accurate modelling and analysis, ensuring that conclusions drawn from the data are robust and reliable.

Table 7, which provides a glimpse into the normality test results, is a valuable reference point for researchers and analysts. It aids in assessing the extent of skewness and informs subsequent data manipulation strategies to enhance the overall quality and validity of the analysis.

All factors have a normal distribution with a probability of 95%, and a Sig above 0.05 indicates a normal data distribution.

Considering that the data distribution has a normal distribution with some skewness, the mean, standard deviation and delta (Azzalini function parameter - Skew normal distribution function) are needed for data augmentation, which is calculated in Eq. 2 [49].

Table 6 Expert weight after calculation

No	Expert weight calculated	No	Expert weight calculated	No	Expert weight calculated
1	1.211045	11	0.847409	21	1.029227
2	1.211045	12	1.392864	22	0.847409
3	1.120136	13	1.120136	23	0.847409
4	1.029227	14	1.211045	24	1.574682
5	1.029227	15	0.665591	25	0.847409
6	1.211045	16	0.938318	26	0.847409
7	1.301955	17	0.938318	27	0.847409
8	0.7565	18	0.847409	28	0.847409
9	0.665591	19	1.029227	Average	1.0000063
10	0.7565	20	1.029227		

Table 7 Shapiro–Wilk normality test

Sub-indices	Statistic	df	Sig.	Sub-indices	Statistic	df	Sig.
F1	0.934	28	0.053	F6	0.951	28	0.209
F2	0.935	28	0.080	F7	0.935	28	0.083
F3	0.966	28	0.471	F8	0.935	28	0.083
F4	0.935	28	0.083	F9	0.966	28	0.468
F5	0.977	28	0.769	F10	0.971	28	0.598

$$\delta = \frac{\alpha}{\sqrt{(1 + \alpha^2)}} \tag{2}$$

δ= Azzalini parameter delta coefficient
α= Skewness coefficient.

Table 8 shows some of the necessary parameters for adding data and running the Monte Carlo model (fourth step) using R STUDIO software.

As illustrated in Fig. 5, it becomes evident that as we progressively generate more random numbers for the economic risk indicator, the probability of winning one of the bets approaches a state of convergence. This convergence signifies that the solution to the problem remains relatively stable, exhibiting minimal fluctuations with the continued expansion of the dataset. Notably, the probability values associated with data generated from 2000 iterations onward show a consistent convergence pattern, especially when considering the fundamental principles of the Monte Carlo method, which entail a balanced 50% chance of winning.

The graph in Fig. 5 demonstrates that the likelihood of achieving a specific outcome stabilises as we perform more iterations of the Monte Carlo simulation. In this context, the product pertains to the probability of success in the economic domain. This stabilisation occurs so that, beyond a certain point (marked by data generated from 2000 iterations onward), the variation in the probability of winning becomes negligible, signifying a robust and dependable estimate.

Table 8 Part of R STUDIO software inputs

No	Average	Standard deviation	Alpha skewness coefficient	Delta Azzalini function parameter	No	Average	Standard deviation	Alpha skewness coefficient	Delta Azzalini function parameter
1	4.015	1.209	−0.484	−0.436	6	3.719	1.512	−0.735	−0.592
2	3.506	1.150	−1.319	−0.797	7	3.106	1.286	−0.886	−0.663
3	3.678	1.156	−0.385	−0.359	8	3.415	1.396	−1.276	−0.787
4	2.831	1.197	−1.211	−0.771	9	3.974	1.350	−0.632	−0.534
5	3.845	1.609	−0.375	−0.351	10	3.231	1.212	0.136	0.135

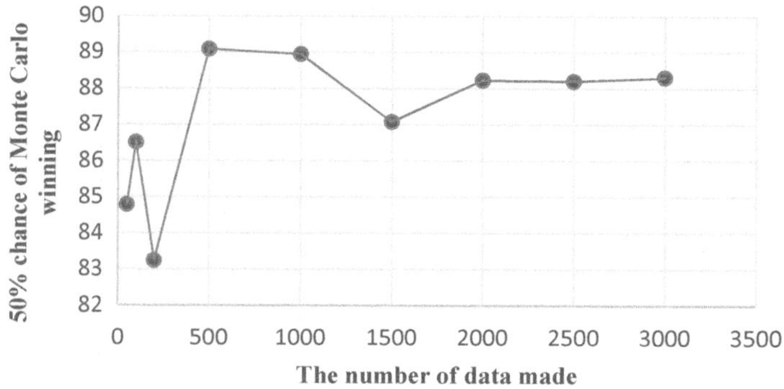

Fig. 5 Plot of calculated probability values for different constructed data counts

This observation has significant implications for decision-making processes, suggesting that further increasing the number of random trials may not yield substantially different results. It underscores the point at which a sufficient confidence level is reached in the Monte Carlo analysis, providing decision-makers with a reliable basis for evaluating and addressing economic risks.

In the context of our Monte Carlo modelling in the fourth stage of part 2, it's noteworthy that we have identified a range of values for the data generated by R STUDIO software. Specifically, the highest range value observed is 8.3275, while the lowest is 0.6655. These values serve as critical benchmarks for assessing the validity and relevance of the data generated.

It's imperative to recognize that the data generated by the R STUDIO software is intended to adhere to the characteristics of the normal distribution of Chula, which is fundamental to our primary research objectives. Therefore, any data points that fall outside this established range should be considered outliers and, as such, warrant removal from the dataset. This meticulous data curation process is essential to ensure that the dataset remains aligned with the primary research objectives and that the subsequent analysis is based on a representative and meaningful set of data points.

By excluding values that are outside the specified range, we maintain the integrity of our research data and enhance the accuracy and reliability of our findings. This rigorous approach not only upholds the standards of scientific research but also ensures that our conclusions are grounded in data that is directly relevant to the normal distribution of Chula, as intended for our investigation.

3.5.1 The Results of Monte Carlo Modeling with Three Conditions

The impact of each of the indicators of HSE management of construction projects (tunneling) in the range of data with three different conditions was determined by Monte Carlo modeling, which is shown in Table 9.

Table 9 Research indicators

No	Indicators	50% Chance of winning	Average disconnected 5%	95% Confidence interval for the mean
1	Economic risks	88.2	20.0136	42.474
2	Legal risks	82.56	26.5878	42.1544
3	Environmental and biological risks	96.91	15.1351	51.2162
4	Natural disaster risks	92.98	**59.0373**	46.2365
5	Project damage risks	88.23	52.9281	52.2099
6	Contractual risks	**80.64**	11.4824	36.7454
7	Management risks	99.02	35.817	**63.1038**
8	Planning risks	94.41	21.4285	48.843
9	Technical risks of design and implementation	**99.54**	30.6015	60.5754
10	Equipment and materials resources risks	96.11	29.7474	49.7979
11	Human resources risks	90.1	**7.8902**	**25.0428**

In the meantime, in the method of technical risks of design and implementation with 99.54%, the highest impact; and contractual risks with 80.64% the lowest impact; And in method two, the risks of natural disasters are the highest with 59.0373%; Human resources risks are the lowest 7.8902%; And in the third method, management risks are 63.1038% the most; And human resource risks have had the least impact of 25.0428% in HSE management of tunnel construction projects.

3.6 Integration of Research Modeling Results

The results of the Copeland method are shown in Table 10.

It is worth mentioning that the negative scores created in the final result are due to the difference between the rows (the score of two-by-two wins) and the columns (the score of two-by-two losses). This means that the indicators that have a negative score have more defeats than victories and are therefore placed in a lower priority.

Table 10 Copeland prioritization

Priority	Indicators	Copeland score
1	Management risks	7
	Technical risks of design and implementation	
2	Environmental and biological risks	5
3	Project damage risks	1
	Planning risks	
	Equipment and materials resources risks	
4	Natural disaster risks	-1
5	Legal risks	-5
6	Economic risks	-7
7	Contractual risks	-9
8	Human resources risks	-11

4 Discussion and Conclusion

Implementing a HSE management system is a strategic and operational decision for an organization. The success of the HSE management system depends on leadership, commitment and participation at all levels of the organization from top to bottom [50, 51].

Performance evaluation is a pivotal component within an organization's strategic management system, profoundly impacting the overall operational framework. In this context, a meticulous examination of key performance indicators (KPIs) within Health, Safety, and Environment (HSE) management for non-mechanized tunnel construction projects assumes a central role. Such an evaluation catalyzes identifying pertinent needs and critical issues, fostering motivation, and enhancing communication concerning HSE management within the construction industry. The significance of recognizing and understanding the relevance of each indicator cannot be overstated.

The research presents a robust and scientifically sound method for formulating an HSE management model tailored to non-mechanized tunnel construction projects. A notable advantage of the prioritization method is its capacity to yield meaningful insights even when working with limited data sets, provided that high-quality, expert-driven data, systems, and other relevant indicators are employed. Furthermore, the application of this method extends beyond the confines of HSE management.

The study employs the National Mega Project of the Tehran Shomal Freeway in Iran, a pivotal component of the North–South Corridor, as a compelling case study. This project, encompassing 39 tunnels spanning 31.5 kms, has encountered many HSE challenges, including tunnel collapses, landslides, and tragic occupational accidents. The prioritization results underscore the paramount importance of specific variables: "design and execution technical risks" and "management risks" are accorded the highest priority, followed by "Environmental and biological risks" with the second highest priority. The third tier of importance comprises "damage and loss risks," "planning risks," and "equipment and material resource risks." "Risks of natural disasters" take the fourth priority, with "Legal risks" and "Economic risks" following in the fifth and sixth positions, respectively. "Contractual risks" and "Human Resources risks" are ranked seventh and eighth, respectively, emphasizing their significance in HSE management.

Notably, Environmental and biological risks are of lesser concern, whereas management risks occupy the highest tier. Statistical measures, including mean standard deviation error and 5% trimmed mean error, demonstrate the accuracy and reliability of the indicator sampling process in assessing factors affecting HSE management within non-mechanized tunnel construction projects.

This model aligns seamlessly with established HSE management systems, including HSE-MS, ISO 45001:2018, ISO 14001:2015, and ISO 9001:2015. It underscores the pivotal role of leadership, unwavering commitment, and active participation at all organizational levels, from project inception and structural design to contractor selection, financing decisions, and the engagement of human resources. It challenges the conventional wisdom, suggesting that, contrary to earlier studies indicating human error as the primary cause of accidents, the system's responsibility is to proactively anticipate and implement control measures to mitigate the consequences of potential incidents. It advocates for a holistic and proactive approach to HSE management beyond simply attributing accidents to human error.

4.1 Practical Suggestions

Building upon the insightful findings and thorough analysis derived from this research, proposing a comprehensive plan for practically implementing these results is prudent. This strategic move becomes especially significant as we transition into the subsequent phases (Phases 2 to 4) of the Tehran Shomal Freeway project or analogous large-scale construction endeavours. This endeavour serves multiple purposes: it aims to validate and maximize the effectiveness of the research outcomes while also contributing to enhancing HSE practices in high-impact construction projects.

The roadmap for implementing these research outcomes should encompass several key components:

Integration into Project Planning: The HSE insights from this research should be seamlessly woven into the project planning process for Phases 2 to 4. This involves incorporating the identified indicators and prioritized risks into the project's strategy and decision-making framework.

Training and Capacity Building: An essential facet of the implementation plan should involve training and capacity-building programs for all project stakeholders. This ensures that the knowledge and best practices identified in the research are disseminated and put into practice effectively.

Monitoring and Evaluation: Establishing a robust system for continuous monitoring and evaluation is crucial. This involves real-time tracking of HSE performance indicators, prompt identification of deviations, and corrective action as necessary.

Regular Review and Adaptation: Given the dynamic nature of construction projects, it is vital to periodically review and adapt the HSE strategies in response to changing circumstances and emerging risks. This iterative approach ensures that safety measures remain current and effective.

Documentation and Reporting: A comprehensive documentation and reporting mechanism should be implemented to record HSE incidents, near-misses, and lessons learned. This data will be invaluable for future project phases and researchers interested in similar endeavours.

Furthermore, the findings from this research have the potential to serve as a launching point for further investigations in the field of construction safety and environmental management. Researchers interested in large-scale construction projects, such as the Tehran Shomal Freeway or comparable endeavours, may find fertile ground for exploration by delving deeper into the nuances of HSE management within such contexts. The knowledge generated from these future research endeavours can contribute to the continual improvement of construction industry practices and, ultimately, the safety and well-being of workers and the environment. Thus, this research holds immediate practical value and lays the groundwork for ongoing scholarly inquiry and advancement in the field.

References

1. Fany, F., Mardani, A., Alizadeh, A., Abbasian Jahromi, H.: Investigating the costs of damages caused by accidents in construction projects (a case study of excavation accidents), First Int. HSE Conf. Constr. Mining, Oil Gas Proj. (2015)
2. Li, H., Yazdi, M.: A Holistic question: is it correct that Decision-Makers neglect the probability in the risk assessment method? BT—Advanced Decision-Making methods and applications in system safety and reliability problems: Approaches, case studies, Multi-criteria Deci, in: Li, H., Yazdi, M. (Eds.), Springer International Publishing, Cham, pp. 185–189 (2022). https://doi.org/10.1007/978-3-031-07430-1_10
3. Turner, J.: The Handbook of project-based Management: Improving the Processes for achieving strategic objectives. McGraw-Hill, Landon (1999)
4. P.M.I. PMI, Project management body of knowledge (pmbok® guide), Proj. Manag. Institute, Inc. Newt. Square, PA, USA. (2017)
5. Sayyadi, M., Ahmad Reza. Rajab Zadeh, Ali.: Hossein poor, Mojtaba. Hayati, Risk ranking in tunneling projects using the ideal similarity method (TOPSIS), in: Eighth Iran. Tunn. Conf., Iranian Tunnel Association, Tehran, (2010)
6. Li, H., Yazdi, M.: Dynamic Decision-Making Trial and Evaluation Laboratory (DEMATEL): Improving safety management system BT—Advanced Decision-Making methods and applications in system safety and reliability problems: Approaches, case studies, Multi-criteria Decision-Maki, In: Li, H., Yazdi, M. (Eds.), Springer International Publishing, Cham, pp. 1–14 (2022). https://doi.org/10.1007/978-3-031-07430-1_1
7. Duddeck, H.: Risk Assessment and Risk Sharing in Tunneling. Tunneling Undergr. Sp. Technol. **2**, 315–317 (1987)

8. Yogaranpan, Y.M.: Risk management, the key to success in management of construction projects in general and underground projects in particular, in: Ninth Aust. Tunneling Conf. Break. New Gr., p. 499 (1996)
9. Gholamizadeh, K., Zarei, E., Omidvar, M., Yazdi, M.: Fuzzy sets theory and human reliability: Review, applications, and contributions BT—Linguistic methods under fuzzy information in system safety and reliability analysis, In: Yazdi M. (Ed.), Springer International Publishing, Cham, pp. 91–137 (2022). https://doi.org/10.1007/978-3-030-93352-4_5
10. Reilly, G.A.: J. J., & Arrigoni, Management and Control of Cost and Risk for Tunneling and Infrastructure Projects, (2005)
11. Li, H., Guo, J.-Y., Yazdi, M., Nedjati, A., Adesina, K.A.: Supportive emergency decision-making model towards sustainable development with fuzzy expert system. Neural Comput. Appl. **33**, 15619–15637 (2021). https://doi.org/10.1007/s00521-021-06183-4
12. Wagner, H.: Risk Evaluation and Control in Underground Construction, (2006)
13. Einstein, H.H.: Use of decision aids for tunneling, Geotechnical risk in rock tunnels, (2006)
14. Beard, A.N.: Tunnel safety, risk assessment and decision-making. Tunn. Undergr. Sp. Technol. **25**, 91–94 (2010)
15. Pirbalouti R.G., Dehkordi, M.K., Mohammadpour, J., Zarei, E., Yazdi, M.: An advanced framework for leakage risk assessment of hydrogen refueling stations using interval-valued spherical fuzzy sets (IV-SFS), Int. J. Hydrogen Energy. (2023). https://doi.org/10.1016/j.ijhydene.2023.03.028.
16. Liao, C.-W.: Pattern analysis of seasonal variation in occupational accidents in the construction industry. Procedia Eng. **29**, 3240–3244 (2012)
17. Oswald, D., Smith, S., Sherratt, F.: Accident investigation on a large construction project: an ethnographic case study. Procedia Manuf. **3**, 1788–1795 (2015)
18. Alarcón, L.F., Acu˜na, D., Diethelm, S., Pell, E.: Strategies for improving safety performance in construction firms Accident, Anal. Prev. 107–118 (2016)
19. Hoła, B., Nowobilski, T., Szer, I., Szer, J.: Identification of factors affecting the accident rate in the construction industry. Procedia Eng. **208**, 35–42 (2017)
20. Li, H., Yazdi, M.: Developing failure modes and effect analysis on offshore wind turbines using Two-Stage optimization probabilistic linguistic preference relations BT—Advanced Decision-Making methods and applications in system safety and reliability problems: approaches, In: Li, H., Yazdi, M. (Eds.), Springer International Publishing, Cham, pp. 47–68 (2022). https://doi.org/10.1007/978-3-031-07430-1_4
21. Zhou, C., Ding, L.Y.: Safety barrier warning system for underground construction sites using Internet-of-Things technologies. Autom. Constr. **83**, 372–389 (2017)
22. Martínez-Aires, M.D., López-Alonso, M., Martínez-Rojas, M.: Building information modeling and safety management: A systematic review. Saf. Sci. **101**, 11–18 (2018)
23. Ansari, M.: Risk management process in projects and proposing appropriate model for development plans in Iran, University of Tehran, (2006)
24. Baloi, A.D.: D., & Price, Modelling global risk factors affecting construction cost performance, Int. J. Proj. Manag. 21 261–269 (2003)
25. Touran, A.: Risk assessment in fixed guideway transit system construction, Washington, (1994)
26. Chapman, R.J.: The controlling influences on effective risk identification and assessment for construction design management. Int. J. Proj. Manag. **19**, 147–160 (2001)
27. Jiang, G.-J., Chen, H.-X., Sun, H.-H., Yazdi, M., Nedjati, A., Adesina, K.A.: An improved multi-criteria emergency decision-making method in environmental disasters. Soft. Comput. (2021). https://doi.org/10.1007/s00500-021-05826-x
28. Environmental risk analysis, Sydney, (2008)
29. Majlis, Labor Law of the Islamic Republic of Iran, Tehran, (1991)
30. McCabe, B.: Construction engineering and project management III: Monte Carlo simulation for schedule risks, in: Proc. 35th Conf. Winter Simul. Driv. Innov., pp. 1561–1565 (2003)
31. Pillay, A., Williams, A. E., El Mardi, J. R., Al-Lawati, M. O., Al-Hadabbi, S. M. H., Al-Hamdi, M. H.: Risk assessment of chromium and arsenic in date palm leaves used as livestock feed, Environ. Int. **29** 541–545 (2003)

32. Li, H., Yazdi, M.: What are the critical Well-Drilling blowouts barriers? a progressive DEMATEL-Game Theory BT—Advanced Decision-Making methods and applications in system safety and reliability problems: Approaches, case studies, Multi-criteria Decision-Making, Multi-obj, In: Li, H., Yazdi M. (Eds.), Springer International Publishing, Cham, pp 29–46 (2022). https://doi.org/10.1007/978-3-031-07430-1_3
33. Underground construction (Tunneling), Washington, D.C, (2003)
34. Li, H., Yazdi, M.: Stochastic game theory approach to solve system safety and reliability Decision-Making problem under uncertainty BT—Advanced Decision-Making methods and applications in system safety and reliability problems: approaches, case studies, Multi-criteria De, In: Li, H., Yazdi, M. (Eds.), Springer International Publishing, Cham, pp. 127–151 (2022). https://doi.org/10.1007/978-3-031-07430-1_8.
35. Yazdi M., Ding Y., Adumene, S., Shafie, P.: Progressive system safety and reliability analysis: A sustainable game theory approach, Qual. Reliab. Eng. Int. n/a (2022). https://doi.org/10.1002/qre.3247.
36. Li, H., Yazdi, M.: How to deal with toxic people using a fuzzy cognitive map: improving the health and wellbeing of the human system BT—Advanced Decision-Making methods and applications in system safety and reliability problems: approaches, case studies, Multi-criteria D, In: Li, H., Yazdi, M. (Eds.), Springer International Publishing, Cham, pp. 87–107 (2022). https://doi.org/10.1007/978-3-031-07430-1_6.
37. Nasir, L., McCabe, D., Hartono, B.: Evaluating risk in construction–schedule model (ERIC–S): construction schedule risk model, J. Constr. Eng. Manag. **129** 518–527 (2003)
38. Ahmadi S., Ali. Nahayi, Alireza. Nahayi: Comprehensive description of research methods (quantitative, qualitative and combinational paradigms, strategies, plans and approaches), Tolid Danesh, Tehran, (2008)
39. Thevendran, M.J., Mawdesley, V.: Perception of human risk factors in construction projects: an exploratory study, Int. J. Proj. Manag. **22** 131–137 (2004)
40. Li, H., Peng, W., Adumene, S., Yazdi, M.: Advances in failure prediction of subsea components considering complex dependencies BT—Intelligent reliability and maintainability of energy infrastructure assets, In: Li, H., Peng, W., Adumene, S., Yazdi, M. (Eds.), Springer Nature Switzerland, Cham, pp. 93–105 (2023). https://doi.org/10.1007/978-3-031-29962-9_6
41. Hafez niya, M.: Introduction to research in humanities, Organization of study and compilation of humanities textbooks of universities, Tehran, (2015)
42. Wu, C.F.J.: Jackknife, bootstrap and other resampling methods in regression analysis. Ann. Stat. **14**, 1261–1295 (1986)
43. Bakbaki, A., Nabhani, N., Anvaripour, B., Shirali, G.: Probabilistic risk assessment using fuzzy fault tree analysis based on two types of failure possibility distributions in process industries, J. Occup. Hyg. Eng. 4 41–52 (2017). https://doi.org/10.21859/johe.4.2.41.
44. Mezei, M.: Adaptive umbrella sampling: Self-consistent determination of the non-Boltzmann bias. J. Comput. Phys. **68**, 237–248 (1987)
45. Saari, D.G., Merlin, V.R.: The Copeland method. Econ. Theory. **8**, 51–76 (1996). https://doi.org/10.1007/BF01212012
46. Car traffic to enter Tehran Shomal Freeway, Iran. Students' News Agency. (2022)
47. Introduction and Overview of Phase One, Tehran Shomal Free. Co. (2019)
48. Bartels, M., Karplus, C.: Probability distributions for complex systems: adaptive umbrella sampling of the potential energy, J. Phys. Chem. B. 102 865–880 (1998)
49. Nouri, M., Soltani, B. A.: Evaluating the effect of tourism marketing mix on buying holiday homes in Cyprus, Int. J. Bus. Adm. 6 63 (2015)
50. ISO 45001, Occupational health and safety, Geneva, (2018)
51. Yazdi, M.: Improving failure mode and effect analysis (FMEA) with consideration of uncertainty handling as an interactive approach. Int. J. Interact. Des. Manuf. **13**, 441–458 (2019). https://doi.org/10.1007/s12008-018-0496-2

Enhancing Crowd Safety Resilience Through Systematic Analysis

Mohammad Yazdi and Esmaeil Zarei

Abstract Crowd safety has become an increasingly critical concern in contemporary society due to the convergence of various domains such as sports, entertainment, politics, and more. Consequently, it is imperative to prioritize a systematic and robust approach to safety and reliability analysis. The primary objective of crowd safety is to safeguard the health and well-being of all stakeholders, including patrons, event organizers, safety and security officers, and others. By conceptualizing crowd safety as a performance functionality inherent within the system they are part of, the concept of resilience emerges as a pivotal feature that can be seamlessly integrated into crowd safety management. This perspective views the resilience of a crowd as determined by evaluating four key characteristics: disruption, adaptation, absorption, and restoration. Additionally, it can evolve and improve over time through continuous learning. To assess the resilience of a crowd, this book chapter introduces the concept of Bayesian networks as a valuable tool for analysis. It then demonstrates how Bayesian networks can be effectively integrated to examine and enhance the resilience of crowd safety. This holistic approach ensures that crowd safety remains adaptable and responsive in the face of evolving challenges, ultimately contributing to all involved's overall well-being and security.

Keywords Crowd safety · Resilience analysis · Stakeholder well-being · Systematic approach · Adaptation · Disruption mitigation · Absorption capacity · Restoration strategies · Continuous learning · Event management · Integrated safety measures · Risk assessment · Evolving challenges

---

M. Yazdi (✉)
School of Computing, Engineering & Physical Sciences, University of the West of Scotland (UWS), London, UK
e-mail: mohammad.yazdi@mq.edu.au

Faculty of Science and Engineering, Macquarie University, Sydney, Australia

E. Zarei
Department of Safety Science, College of Aviation, Embry-Riddle Aeronautical University, Prescott, AZ, USA

Robertson Safety Institute (RSI), Embry-Riddle Aeronautical University, Prescott, AZ, USA

M. Yazdi (ed.), *Progressive Decision-Making Tools and Applications in Project and Operation Management*, Studies in Systems, Decision and Control 518,
https://doi.org/10.1007/978-3-031-51719-8_14

1 Introduction

In our contemporary world, characterized by rapid societal evolution and a multifaceted convergence of activities spanning sports, entertainment, large-scale ceremonies, celebratory gatherings, exhibitions, public protests, and beyond, the issue of crowd safety has soared to unparalleled importance. This surge in significance arises from the intricate interplay among these diverse domains, intensifying the demand for a systematic and committed approach to safety and reliability analysis. At the heart of the crowd safety discourse lies an unswerving commitment to a singular and fundamental objective: preserving the health and well-being of various stakeholders. These stakeholders encompass a rich tapestry of individuals, including event patrons seeking enjoyment and communal experiences, meticulous event organizers orchestrating seamless affairs, and vigilant safety and security officers entrusted with ensuring the welfare of the multitude. Against this backdrop of complexity and interdependence, the imperative of crowd safety transcends mere significance—it becomes truly imperative.

The following incidents serve as solemn reminders of the importance of prioritizing crowd safety, proper infrastructure, emergency preparedness, and effective crowd management at various gatherings and events. Learning from past tragedies is crucial to prevent similar incidents and ensure the well-being of event attendees and participants.

- Hajj Accident (2015): This tragic incident occurred during the annual Hajj pilgrimage in Mecca, Saudi Arabia. It resulted in the loss of more than 2,400 lives and left over 1,000 individuals injured. The event highlighted the immense challenges of managing massive crowds during religious gatherings.
- Love Parade Disaster (2010): The Love Parade festival in Duisburg, Germany, turned into a scene of tragedy when a crowd crush occurred in the queues leading to the festival site. This incident claimed the lives of 21 people and left over 500 individuals injured, emphasizing the importance of effective crowd management at large-scale events.
- Kumbh Mela Stampede (2013): During the Hindu festival of Kumbh Mela in Allahabad, India, a tragic stampede occurred, claiming the lives of at least 36 people and injuring dozens more. This incident highlighted the need for enhanced crowd control measures and safety protocols during large religious gatherings.
- The Station Nightclub Fire (2003): In a nightclub in West Warwick, Rhode Island, a devastating fire broke out during a performance, leading to the loss of 100 lives and causing injuries to over 200 individuals. The tragedy was exacerbated by the narrow exit available to the patrons, emphasizing the critical importance of fire safety and emergency exits in public venues.
- Ellis Park Stadium Disaster (2001): A soccer match between South Africa and Zimbabwe at Ellis Park Stadium in Johannesburg, South Africa, turned tragic when 43 people lost their lives, and over 150 were injured due to crowd-related incidents. This event highlighted the need for improved crowd control strategies at sports events.

Embracing an innovative perspective, we recognize crowd safety as an integral performance element woven into larger systems' fabric. Within this dynamic context, resilience emerges as the linchpin, providing a versatile framework that harmoniously melds with crowd safety management. When applied to a crowd, resilience hinges on meticulously evaluating four pivotal attributes: disruption, adaptation, absorption, and restoration. These elements collectively constitute the barometer by which we gauge the crowd's capacity to endure and rebound robustly from unforeseen adversities. In essence, through this lens of resilience, it is vital to unlock the true potential of crowd safety, viewing it not as a static set of measures but as a dynamic force poised to navigate the complex and ever-changing landscape of crowd dynamics and unforeseen challenges.

Furthermore, it's crucial to recognize that the concept of crowd resilience is far from static; it possesses the remarkable capacity to evolve and continually enhance its efficacy through an ongoing process of learning and adaptation. This adaptability allows crowd safety to maintain agility and responsiveness when confronted with the ever-evolving challenges that characterize our dynamic world. Within the chapters of this book, we embark on a compelling journey into the realm of crowd resilience, seeking to unravel its intricacies and unlock its potential for bolstering safety in diverse settings. Our objective is to understand the concept of resilience and explore practical methods for assessing and reinforcing it. As we jump deeper into this exploration, we introduce a pivotal analytical tool that underpins our comprehensive approach to enhancing crowd safety resilience using Bayesian networks. These sophisticated networks play a central role in our strategy, enabling us to dissect and fortify the underlying mechanisms that contribute to the resilience of crowds in various contexts. This integration of Bayesian networks offers a powerful means of gaining insights, making informed decisions, and ultimately elevating the safety and security of gatherings and events of all scales and purposes.

Ultimately, this holistic methodology ensures that crowd safety is not merely a static set of measures but a dynamic and adaptable system, continually safeguarding the well-being and security of all involved, regardless of the unpredictable nature of the world in which we live.

Upon a comprehensive examination of the current state of the art, it becomes evident that the majority of published works in the field of crowd safety and risk assessment have predominantly centered on intrinsic and extrinsic factors that pose potential threats to crowd safety. These studies include, but are not limited to [1–5].

Verma et al. [6] addressed the challenges posed by large religious gatherings in India, due to increased mobility and emotional participants. The paper aimed to assess and improve crowd management measures through empirical studies and observations from Kumbh Mela 2016, offering guidelines for future events. Another paper [7] explores overcrowding and stampede risks at public gatherings and events by investigating the causes, monitoring methods, and risk assessment techniques, summarising related studies. It systematically explains the evolution of crowd-gathering risks, categorises monitoring methods into three types, outlines articles on risk evaluation methods, and identifies three main paradigms in this field. Zhang et al. [8] addresses the significance of crowd video monitoring and analysis in computer vision and public management. It was introduced a method for assessing crowd safety through fuzzy inference, considering both pedestrian numbers and distribution uniformity as attributes. It was then estimated pedestrian numbers by foreground pixel count and calculates distribution uniformity using distribution entropy.

Drawing upon our individual interests and substantial industrial experience as a Workplace Health and Safety (WHS) advisor within the context of event safety and security organizations, and informed by a comprehensive review of the current state of the field, we have formulated the following research questions that we aim to address:

- What are the fundamental principles and strategies for enhancing safety and resilience in diverse crowd scenarios, and how can these principles be applied across different contexts and settings?
- How can crowd safety and resilience be reimagined and redefined to address the evolving challenges of our rapidly changing world, and what are the core elements of such a redefined approach?
- What role does interdisciplinary collaboration play in advancing our understanding of crowd safety and resilience, and how can insights from psychology, sociology, engineering, and data science be integrated to develop more comprehensive and effective crowd management strategies?

2 Integration of Bayesian Network and Resilience into the Crowed Safety

In system safety, reliability analysis, and decision-making processes, Bayesian networks offer a highly effective approach for handling complex and uncertain scenarios. These networks are a robust framework for modelling and quantifying the intricate relationships between various factors influencing system safety and reliability. Moreover, they facilitate informed decision-making by providing a structured means of assessing risks and optimizing strategies.

Introduction to Bayesian Networks in System Safety and Reliability Analysis:

Bayesian networks emerge as a critical tool in system safety and reliability analysis. These networks are adept at capturing the interplay of factors affecting the performance and safety of complex systems, such as industrial processes, transportation systems, and critical infrastructure. By representing variables and their dependencies in a graphical manner, Bayesian networks enable a comprehensive understanding of the underlying dynamics.

In a Bayesian network applied to system safety and reliability, nodes represent variables of interest, including component failures, environmental conditions, operational parameters, and more. The directed edges connecting these nodes convey probabilistic relationships, highlighting how one variable's state influences another.

This modelling approach not only accounts for causality but also acknowledges the inherent uncertainty in the system, some example are [9–12].

Decision-Making Process

Bayesian networks also play a pivotal role in the decision-making process within system safety and reliability analysis. When confronted with complex risk mitigation, maintenance scheduling, or resource allocation choices, decision-makers can leverage Bayesian networks to inform their strategies, some example are [13–15].

Here's how Bayesian networks facilitate this decision-making process:

1. Risk Assessment: Bayesian networks enable the quantification of risks associated with different system states or scenarios. By incorporating historical data and expert knowledge, the network can assess the likelihood of various outcomes, helping decision-makers prioritize actions to mitigate high-risk situations.
2. Scenario Analysis: Decision-makers can explore "what-if" scenarios by manipulating variables and observing their impact within the Bayesian network. This allows for the evaluation of different strategies and interventions to enhance system safety and reliability.
3. Resource Allocation: Bayesian networks assist in optimizing the allocation of resources, such as maintenance budgets or personnel, by considering the expected outcomes of various decisions. This ensures that resources are deployed efficiently to minimize risks and maximize system reliability.
4. Real-Time Decision Support: Bayesian networks can continuously update probabilities based on incoming data in dynamic environments, offering real-time decision support. This adaptability is invaluable for responding to evolving conditions and making timely adjustments.

As the mathematical foundations of Bayesian networks have been comprehensively elucidated in the existing literature, this chapter refrains from jumping into those details. Instead, it focuses on practical applications and real-world scenarios, offering a reference list that includes relevant sources for readers seeking an in-depth understanding of the mathematical underpinnings [16, 17]. These sources encompass a range of seminal works and comprehensive references to cater to both newcomers and seasoned practitioners in the field.

Resilience in complex systems like crowd safety and crowd management, whether natural ecosystems, organizations, or infrastructure, can be characterized by four plus one key attributes: disruption, adaptation, absorption, restoration, and learning. Here's an explanation of each of these characteristics:

- Disruption: Disruption refers to the capacity of a system to withstand and endure disturbances or shocks without suffering catastrophic failures or irreparable damage. Resilient systems have built-in mechanisms to absorb and dampen the initial impact of disruptions, preventing them from cascading into widespread shortcomings.
- Absorption: Absorption refers to the capability of a system to absorb and dissipate the energy or effects of a disturbance, minimizing the extent of damage or harm.

Resilient systems are designed to have mechanisms that can absorb shocks and reduce their negative impact, preventing them from overwhelming the system.
- Adaptation: Adaptation is the ability of a system to adjust and evolve in response to changing conditions or unforeseen challenges. Resilient systems are not static; they can modify their structures, processes, or strategies to better cope with new circumstances. This adaptability ensures the system's continued functionality in the face of evolving threats.
- Restoration: Restoration is returning a system to a functional state after a disruption has occurred. Resilient systems have well-defined recovery mechanisms and plans that enable them to efficiently and effectively restore normal operations. This includes repairing damage, replacing components, and regaining lost functionality promptly.
- Learning: Learning is a crucial aspect of resilience. Resilient systems recover from disruptions and use the experience to improve their future responses. This involves analyzing the causes of disorders, identifying vulnerabilities, and implementing proactive measures to enhance resilience in the long term. Learning ensures that the system becomes more robust and adaptive over time.

These four plus one characteristics collectively define the ability of a system to not only survive disturbances but to thrive and improve in the face of adversity. Resilient systems are essential in various domains, from disaster management and critical infrastructure to business continuity and ecological conservation [18, 19]. They enable societies and organizations to better prepare for, respond to, and recover from disruptions, ultimately ensuring long-term sustainability and success.

In the following discussion, we will consider Crowd Safety as a pivotal performance metric for the system. With this focus in mind, we define the four primary resilience characteristics, augmented by an additional attribute, collectively shaping the system's ability to withstand and adapt to various challenges. These five resilience characteristics are as follows:

- <u>Disruption</u>

We have identified several noteworthy safety concerns associated with mass gatherings by conducting a comprehensive review of existing literature [3] and industrial practices related to crowd safety. It is essential to recognize that any of these concerns, individually or in combination, can disrupt the safety and well-being of crowds. These concerns encompass:

- Crowd Density: Elevated crowd density and pressure can result in physical harm and even fatalities, particularly in enclosed spaces or during sudden surges in crowd movement. Such situations are commonly encountered at concerts, sporting events, and religious gatherings.
- Cascading Risks: A minor incident can trigger a disproportionate reaction, potentially leading to stampedes, crushes, or infrastructure collapses.
- Infectious Diseases: Mass gatherings can facilitate the spread of contagious diseases, especially when enforcing social distancing and other preventive measures is challenging.

- Terrorism: Acts of terrorism and violence pose significant threats to crowd safety during mass gatherings. These threats may include bombings, shootings, vehicular attacks, or actions by activists seeking to disrupt events for their agenda. The crowd's response can also provoke stampedes or crushes. Furthermore, if law enforcement's response is not well-understood, it can inadvertently endanger the crowd.
- Violence: Violence in the context of mass gatherings refers to physical altercations, confrontations, or aggressive behaviors that can occur among attendees, between attendees and security personnel, or between different groups with conflicting interests or ideologies.
- Extreme Weather: Extreme weather conditions, such as heatwaves or thunderstorms, can jeopardize the safety of crowds. These conditions can lead to heat-related illnesses, dehydration, or electrocution. In sudden weather changes, crowds may rush to seek shelter, potentially resulting in crushes, overcrowding of temporary structures, and limited escape options.
- Stampedes: Stampede incidents may occur when crowds abruptly rush toward specific entry or exit points, leading to injuries and deaths. This risk is prevalent at religious gatherings, sporting events, or concerts, mainly when crowd management is inadequate.
- Engineering Design Issues: Weak or poorly designed structures, including bridges, balconies, and grandstands, may collapse under the weight of large crowds, causing injuries or fatalities. Sudden shifts in crowd movement or unexpected stress on specific structures can trigger such incidents. Events like sports events with temporary stands, music festivals, and religious gatherings are susceptible to such risks.
- Flammability and Explosibility Hazards: Large crowds involving pyrotechnics or flammable materials carry a substantial risk of fires and explosions.

• Absorption

Absorption of crowd accidents and ensuring the safety of attendees during mass gatherings is a complex task that involves careful planning, coordination, and implementation of various strategies. When confronted with a disruption, the absorption system capability of the system comes into play, serving as a vital component. It readies itself and adjusts proactively to avert the disruptive impact, thereby minimizing the undesirable consequences. This preparation and self-adjustment can manifest in two distinct scenarios:

- The system can proactively and effectively prepare itself, self-adjusting to prevent disruptions from occurring in the first place.
- In the event of a disruption, the system demonstrates its potential to mitigate the rate of functionality change, effectively minimizing the extent of functionality decrease.

Here are some measures that can help address the specific risks you mentioned:

- Capacity Limits: Setting a strict maximum number of attendees based on venue safety standards and design to prevent overcrowding and maintain safety.
- Crowd Monitoring: Utilizing technology like crowd density sensors and cameras to monitor and manage crowd movements in real-time continuously.
- Emergency Response Plan: Develop comprehensive plans that outline procedures for various emergency scenarios, including minor incidents, to ensure a swift and coordinated response.
- Infrastructure Safety: Regularly inspecting and maintaining venue infrastructure to identify and mitigate potential bottlenecks and hazards.
- Health Protocols: Enforcing strict health protocols like mask mandates, social distancing, and hand sanitizing stations to reduce the spread of contagious diseases.
- Contact Tracing: Implementing a contact tracing system to track potential outbreaks and notify potentially exposed individuals.
- Security Measures: Collaborating with law enforcement for thorough security assessments, deploying security personnel, and screening measures to deter and respond to potential threats.
- Public Awareness: Conducting public awareness campaigns and establishing reporting mechanisms to encourage attendees to report suspicious activities.
- Visible Security: Deploy visible security personnel throughout the venue to deter aggressive behaviour and maintain order.
- Conflict Resolution: Training security staff in conflict de-escalation techniques and providing designated areas for peaceful conflict resolution.
- Weather Monitoring: Establishing a dedicated monitoring team to track forecasts and provide real-time updates.
- Structural Integrity: Ensuring that temporary structures are designed to withstand expected weather conditions and have backup plans for sudden changes.
- Crowd Flow Management: Implementing a well-designed crowd flow plan with signage, barriers, and designated entry and exit points.
- Attendee Education: Educating attendees on following crowd management instructions and staying calm in crowded situations.
- Structural Inspections: Creating a regular inspection schedule for all venue structures and engaging certified engineers for safety assessments.
- Safety Standards: Ensuring compliance with local safety regulations and international standards for all temporary structures.
- Fire Safety: Limiting flammable materials and pyrotechnics and conducting fire safety drills.
- Safety Rehearsals: Running safety rehearsals focused on fire and explosion scenarios and ensuring the availability of fire suppression equipment.

It is worth noting that several of the actions as mentioned above exhibit similarities with other facets of crowd safety resilience, such as performance and functionality. These shared elements are carefully considered as integral components in the overall computational process. There are also several adaptation factors that can be considered to enhance safety at mass gatherings. These factors can vary depending on the specific nature of the event and its location, and in this chapter we might exclude in terms of computation process.

- Adaptation

- Local Regulations and Permits: Compliance with local regulations and obtaining necessary permits is critical to ensure the event is legal, safe, and well-managed.
- Medical Facilities: Adequate medical facilities and trained personnel are essential for timely assistance for injuries and illnesses.
- Accessibility: Ensuring accessibility for individuals with disabilities helps create an inclusive and safe environment for all attendees.
- Language and Communication: Effective communication through multilingual signage and instructions ensures everyone understands safety measures and procedures.
- Transportation and Parking: Well-planned transportation options and parking facilities help manage crowd flow and reduce congestion.
- Evacuation Plans: Clear and well-communicated evacuation plans are crucial to ensure all attendees' safety in emergencies.
- Volunteer and Staff Training: Properly trained volunteers and staff can assist attendees and manage crowd-related issues.
- Emergency Response Plan: Comprehensive plans for various emergency scenarios, including minor incidents, are essential for a swift and coordinated response.
- Security Measures: Collaborating with law enforcement and deploying security measures is vital to deter and respond to potential threats.
- Public Awareness: Conducting public awareness campaigns and establishing reporting mechanisms encourage attendees to report suspicious activities and stay vigilant.

- Restoration

Restoration of crowd safety, functionality, and performance after a disruption is critical to event management and emergency response. Ensuring that the event can resume smoothly and safely following a disorder is essential for minimizing the impact on attendees and maintaining overall event success. In the following, some restoration practices can be taken into account to restore crowed safety functionality to its prominent swift role:

- Assessment and Communication: Immediately after a disruption, thoroughly assess the situation. This includes evaluating injuries, structural damage, and crowd conditions. Maintain clear and consistent communication with event staff, emergency responders, and attendees to provide updates and instructions.
- Emergency Services Response: Coordinate with emergency services to ensure injured individuals receive prompt medical attention. Emergency responders should be well-trained in crowd management and triage procedures.

- Evacuation if Necessary: If the disruption poses ongoing safety risks, consider the orderly evacuation of the venue, following established evacuation plans and routes. Ensure that individuals with disabilities and vulnerable populations receive appropriate assistance.
- Crowd Control and Management: Implement crowd control measures to manage the movement of attendees, prevent panic, and maintain order. This may involve directing people to safe areas or exit points.
- Infrastructure Assessment: Conduct a rapid assessment of venue infrastructure and structures to identify any damage or safety hazards. Engage structural engineers to evaluate the integrity of affected areas.
- Clean-Up and Repairs: Once it is safe, clean up debris and repair damaged infrastructure. Ensure that repairs are completed to industry safety standards.
- Security Measures: Enhance security measures in response to the disruption, including increasing the presence of security personnel and implementing additional screening measures if necessary.
- Reentry Protocols: If the disruption requires evacuations, establish reentry protocols to allow attendees back into the venue in an organized and safe manner. This may involve rechecking tickets or conducting security screenings.
- Medical Facilities: Ensure that medical facilities are fully operational and well-staffed to address any ongoing medical needs of attendees.
- Public Relations and Communication: Maintain open and transparent communication with attendees and the public, providing updates on the situation, safety measures, and the resumption of the event.
- Monitoring and Surveillance: Continue monitoring crowd density and movement through technology such as crowd density sensors and surveillance cameras to prevent overcrowding and respond to emerging issues.
- Crowd Psychology and Behavior: Consider the psychological impact of the disruption on attendees and take measures to reassure and calm the crowd to prevent further panic or unrest.
- Legal and Regulatory Compliance: Ensure that all actions taken during the restoration process comply with local laws and regulations, including safety standards and permits.

Lastly, Learning from disruptions is crucial to improving future event safety and management. Here are some of the most critical learning factors to consider after a disruption at a mass gathering event:

- Learning

- Root Cause Analysis: Understanding the fundamental causes of the disruption is paramount. Identify what triggered the disruption, whether it was a structural issue, a security breach, a medical emergency, or another factor.
- Communication: Evaluate the effectiveness of communication during the disruption. Identify any breakdowns in communication channels, internal (among event staff and responders) and external (with attendees and the public).

- Emergency Response: Assess how well the emergency response plan was executed. Examine the timeliness and effectiveness of responses by emergency personnel, security teams, and medical staff.
- Crowd Behavior and Management: Analyze how the crowd reacted during the disruption. Understand the factors influencing crowd behaviour, including panic, and consider strategies for better crowd management and control.
- Infrastructure and Safety: Evaluate the structural integrity of the venue and the adequacy of safety measures in place. Determine if design flaws, maintenance issues, or safety protocol gaps need addressing.
- Security Protocols: Review the effectiveness of security protocols and measures. Examine how security personnel responded to the disruption and whether any vulnerabilities or areas for improvement were identified.
- Evacuation Procedures: Assess the clarity and execution of evacuation plans, including the efficiency of evacuation routes and the ability to safely guide attendees out of the venue.
- Training and Preparedness: Evaluate the training and preparedness of event staff, security personnel, and emergency responders. Identify any deficiencies in training or areas that require additional preparation.
- Technology and Tools: Examine the use of technology, such as surveillance systems, communication tools, and crowd monitoring devices, in managing and responding to disruption.
- Legal and Regulatory Compliance: Ensure all actions taken during and after the disruption comply with local laws and regulations. Address any compliance issues that may have arisen during the event.

The following section studies a practical illustration example designed to comprehensively evaluate crowd safety performance and functionality across various time intervals. This examination will offer insights into the dynamic nature of crowd management and provide a deeper understanding of how safety measures evolve and adapt over time.

3 An Illustration of Resilience a Crowed

In this section, we spotlight Vivid Sydney, an internationally renowned annual festival that transforms the city of Sydney, Australia, into a mesmerizing blend of light, music, and innovative ideas. Traditionally spanning multiple weeks during the winter season of May and June, Vivid Sydney stands out as one of the world's premier cultural events.

Notably, Vivid Sydney reached a historic milestone in 2023, solidifying its status as Australia's largest festival by attracting an astounding crowd of over 3.28 million attendees. A remarkable testament to its allure, the festival's 2023 centrepiece, "Written in the Stars," featured breathtaking drone shows that captivated the imaginations of more than 500,000 spectators across seven spellbinding performances

[20]. As can be seen from Fig. 1, the Light Walk map of Vivid Sydney 2023 was officially released by Destination NSW (https://www.vividsydney.com). There are some critical features of Vivid Sydney:

- Light Installations: One of the main attractions of Vivid Sydney is its spectacular light installations that transform the city's landmarks and buildings into a vibrant and colourful canvas. Iconic structures like the Sydney Opera House, Sydney Harbour Bridge, and the Royal Botanic Garden are illuminated with intricate and dynamic light displays.

Fig. 1 Light walk map of Vivid Sydney 2023

- Music and Performances: Vivid Sydney features a diverse program of live music performances, concerts, and DJ sets. The festival showcases local and international talent, covering various music genres, from electronic and pop to classical and experimental.
- Ideas Exchange: As part of the festival, Vivid Ideas hosts a series of talks, workshops, and discussions on various topics, including technology, innovation, art, and culture. It brings together creative thinkers, entrepreneurs, and thought leaders to explore and exchange ideas.
- Interactive Installations: Many of the light installations at Vivid Sydney are interactive, allowing festivalgoers to engage with the art. These installations often respond to movement, sound, or touch, creating an immersive experience for visitors.
- Harbor Cruises: A popular way to experience Vivid Sydney is by taking a harbour cruise. These cruises offer breathtaking views of the illuminated landmarks along Sydney Harbour, providing a unique perspective on the festival.
- Food and Night Markets: The festival also features food stalls and night markets, where attendees can sample various culinary delights and shop for unique arts and crafts.
- Family-Friendly: Vivid Sydney is a family-friendly event with many activities and installations suitable for children. It's an excellent opportunity for families to enjoy the city's transformation together.
- Photography Opportunities: The harbour's stunning light displays and reflections make Vivid Sydney a paradise for photographers and Instagram enthusiasts. It's a fantastic time to capture the beauty of Sydney at night.

It should be added that Vivid Sydney has grown in popularity since its inception and has become a significant tourism drawcard for the city. It celebrates creativity and innovation and showcases the beauty of Sydney's landmarks in a unique and captivating way. The festival continues to evolve each year, offering new and exciting experiences for locals and visitors alike.

Building upon the previously mentioned overview of Vivid Sydney, it is evident that crowd safety performance functionality is vulnerable to disruptions, which are unwelcome and potentially detrimental. We have drawn from the official Vivid website's available information and insights from experienced risk assessors involved in Vivid events to examine this aspect.

In our pursuit of a comprehensive understanding, we constructed a Bayesian Network encompassing all potential contributing factors influencing crowd safety resilience. It's important to note that this study represents a partial exploration, focusing primarily on three discrete states for each node—High, Medium, and Low. These subject terms demonstrate the extent to which contributing factors influence their respective parental nodes. The conditional probability tables associated with these nodes were constructed subjectively, taking into account the evidence at hand, which was gathered during on-site assessments and field experiences.

Furthermore, we developed a dynamic Bayesian network to illustrate the evolving nature of crowd safety performance functionality over time. This temporal approach

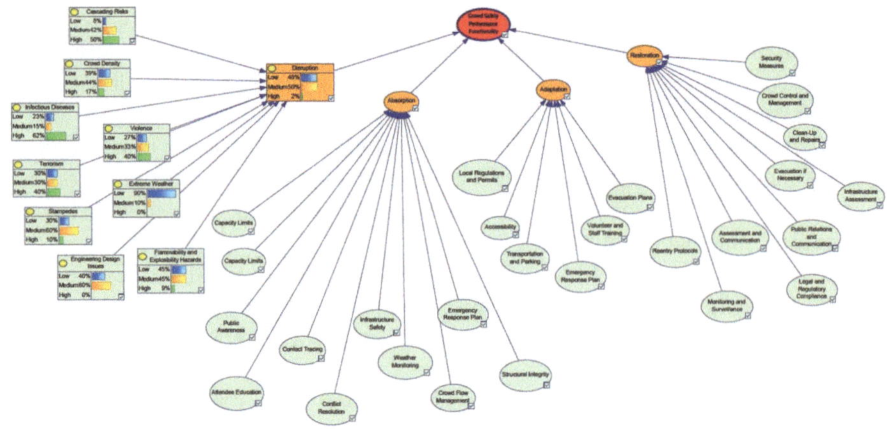

Fig. 2 The initial influence diagram for crowd safety performance functionality

allows us to observe how crowd safety performance behaves across multiple time slices, providing insights into its resilience and adaptability in changing conditions and disruptions. Figure 2 illustrates the Bayesian network as the initial influence diagram for crowd safety performance functionality.

In this scenario, we assumed that once the "High" state rate in the Disruption node reaches or exceeds 75%, it triggers disruptions in crowd safety performance functionality. Subsequently, a dynamic Bayesian model is employed to assess the resilience of crowd safety performance functionality across 20 different time intervals. Figure 3 illustrates the structural dynamic Bayesian network within these 20-time intervals, where each interval can represent various units of time, such as months, seconds, minutes, hours, days, weeks, or years.

The results indicate that at time $t = 0$, the characteristics disruption, absorption, adaptation, and restoration have values of 0.75, 0.79, 0.72, and 0.89, respectively. To evaluate resilience from $t = 1$ to $t = 20$, the disruption node is set to the "High" state. As depicted in Fig. 4, system resilience gradually declines, reaching its lowest point at $t = 5$, where it equals 0.21.

Based on the information from Fig. 3, it can be inferred that achieving a 90% recovery of performance loss from this lowest point corresponds to a value of 0.75 (calculated as $0.90 \times (0.81 - 0.21) + 0.21$) within 20-time intervals is not possible and we might event to reach that point as a matter of the fact. Additionally, it is worth noting that a lower absorption ability in the system results in a more significant reduction in system performance (resilience) and that a greater absorption ability leads to higher performance loss. After the rapid performance loss decline, the system's adaptation and restoration abilities gradually recover, 0.72.

Furthermore, an important observation is that the system continues to improve even after achieving a 90% recovery of performance loss, ultimately stabilizing at $t > 20$ and beyond.

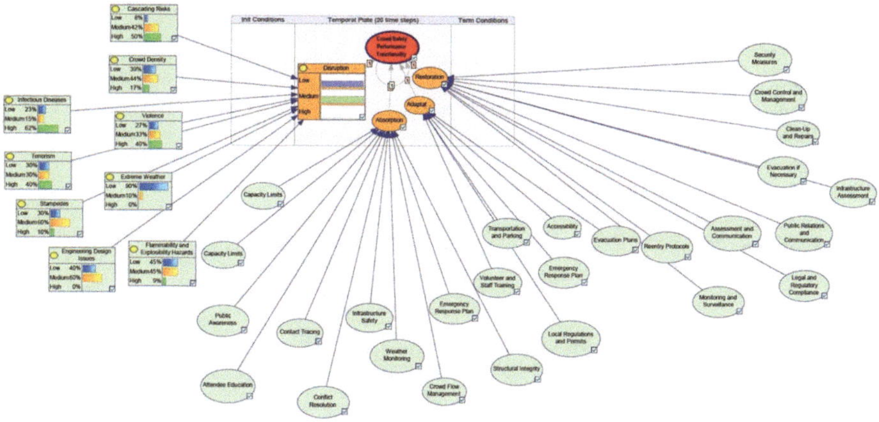

Fig. 3 The structural dynamic Bayesian network for assessing the resilience of crowed safety over time

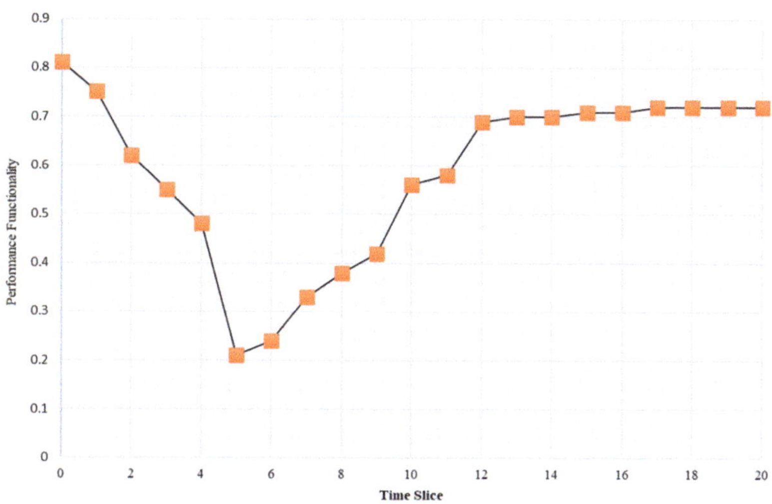

Fig. 4 The resilience assessment of crowed safety over time

As previously discussed, node Learning plays a crucial role in our dynamic Bayesian network model, serving as an external factor that leverages past experiences and knowledge to predict and mitigate future crowd safety disruptions. This addition of the Learning node is illustrated in Fig. 5, with the node having three distinct states: "High," "Moderate," and "Low."

The results of our analysis, depicted in Fig. 6, reveal that crowd safety performance functionality benefits significantly from higher learning ability, resulting in a more rapid recovery rate. Specifically, when the Learning node is in the "High" state, the system's resilience improves substantially compared to when it's in the "Low" state.

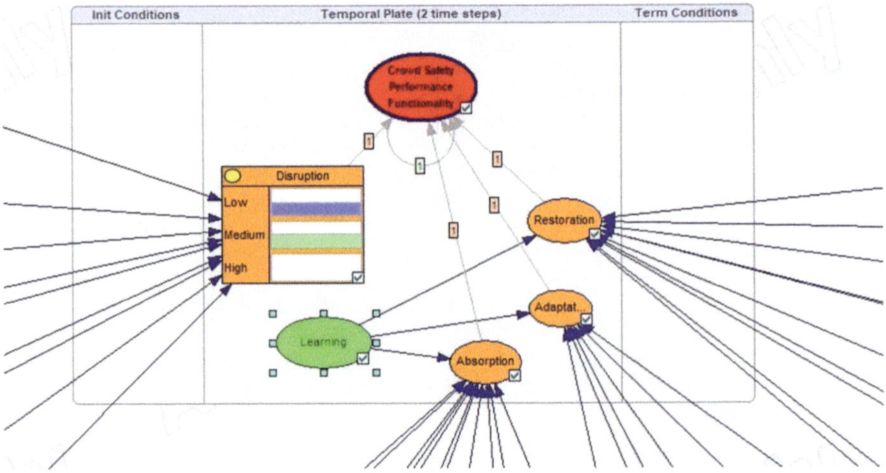

Fig. 5 Performing functionality of crowed safety analysis by adding the node learning to the dynamic Bayesian network

Notably, achieving a 90% recovery from the lowest performance point corresponds to a resilience value of 0.75, occurring at different times depending on the Learning state—20 time slice or more.

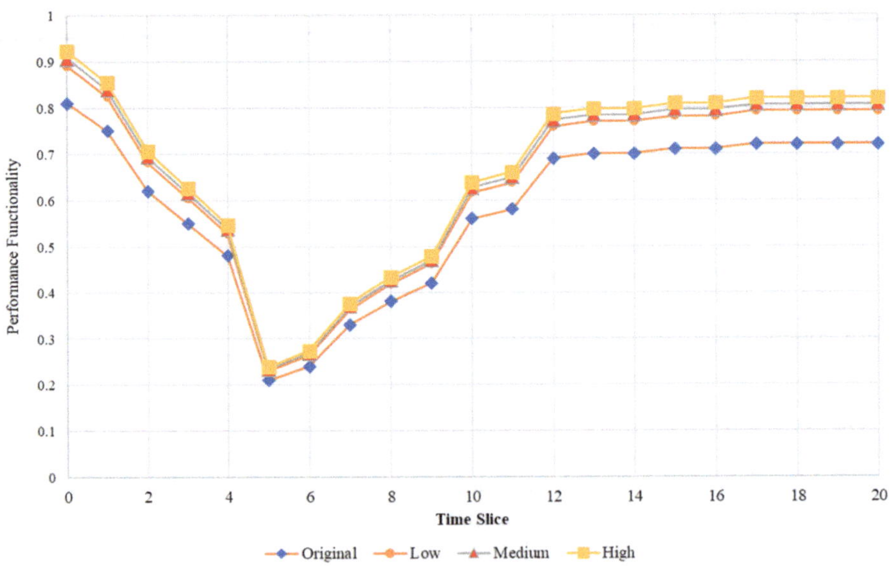

Fig. 6 Resilience assessment for crowed safety considering node Learning with different states

Learning from past experiences empowers decision-makers to enhance the system's resilience attributes in preparation for potential future disruptions. Furthermore, a robust learning ability facilitates the generation of valuable insights from past disruptions, aiding in developing new knowledge that can be applied to rectify inadequate technical guidelines. This knowledge also assists specialists in predicting and mitigating undesirable disruptions while allowing for timely adjustments in the subsea pipeline system.

Integrating the Learning node into our dynamic Bayesian network model enhances crowd safety performance functionality and fosters a proactive approach to addressing and learning from disruptions, ultimately contributing to a more robust and resilient system.

Computer simulation models are pivotal in enhancing crowd safety and their performance functionality. They empower planners and designers to assess and refine crowd management strategies effectively. This includes:

Predictive Analysis: Simulation models are invaluable tools for predicting crowd behaviour in diverse scenarios. Using these models, planners can proactively identify potential risks and formulate appropriate mitigation strategies.

Optimizing Crowd Management Strategies and Design: Simulation models facilitate the testing of various crowd management strategies. These strategies encompass elements like the positioning of entry and exit points, the configuration of barriers, and the deployment of security personnel. Through simulations, planners can fine-tune these strategies and ascertain the most efficient and practical approaches for crowd management across different scenarios.

In essence, computer simulation models offer a dynamic platform for experimentation and analysis, improving crowd safety measures and ensuring the well-being of event attendees and participants.

4 Conclusion

This chapter has explored the multifaceted and critically important crowd safety and resilience analysis domain. It has underscored the profound significance of this field in our modern society, where various events and gatherings draw together diverse and sizable crowds. The primary focus has been establishing a systematic and robust approach to crowd safety, with an unwavering commitment to safeguarding the health and well-being of all participants, event attendees, organisers, and security personnel.

Throughout the chapter, the concept of resilience has emerged as a central and transformative feature within the domain of crowd safety management. Viewing crowd safety through the resilience lens has allowed a deeper understanding of how crowds can effectively navigate unforeseen adversities. The four key characteristics of resilience—disruption, adaptation, absorption, and restoration—have provided a framework for evaluating the capacity of crowds to endure and rebound from troubles.

Integrating Bayesian networks as a tool for analysis has been introduced as a valuable means of assessing and enhancing crowd resilience. By employing Bayesian networks, stakeholders can gain insights, make informed decisions, and ultimately elevate the safety and security of gatherings and events of various sizes and purposes.

The chapter has also emphasised the dynamic nature of crowd safety, particularly in light of past incidents that tragically underscore the consequences of inadequate safety measures. These incidents are solemn reminders of prioritising safety and risk assessment in mass gatherings.

The measures and strategies discussed in the chapter, from addressing crowd density to managing infectious diseases and responding to acts of terrorism, provide a comprehensive guide for event organisers and safety professionals. The focus on absorption, adaptation, and restoration during and after disruptions underscores the commitment to maintaining crowd safety under challenging circumstances.

Additionally, the chapter has stressed the vital role of continuous learning and root cause analysis in enhancing crowd safety resilience. Learning from past disruptions and analysing their causes are essential to improve safety measures and better prepare for future challenges. As a direction for future study, we must expand our understanding of crowd safety and its performance functionality by employing various causality analysis tools. In this regard, two promising avenues for exploration are the advanced and Dynamic Decision-Making Trial and Evaluation Laboratory (DEMATEL) [21–23] and Fuzzy Cognitive Maps (FCM) [24, 25]. Incorporating Dynamic DEMATEL and FCMs into future studies on crowd safety can provide a more slight and comprehensive understanding of the factors influencing safety and resilience in mass gatherings. These tools can help researchers and practitioners make data-driven decisions, improve safety protocols, and enhance event attendees' overall well-being and security.

References

1. Feliciani, C., Corbetta, A., Haghani, M., Nishinari, K.: Trends in crowd accidents based on an analysis of press reports. Saf. Sci. **164**, 106174 (2023). https://doi.org/10.1016/j.ssci.2023.106174
2. Liao, X.-C., Chen, W.-N., Guo, X.-Q., Zhong, J., Hu, X.-M.: Crowd management through optimal layout of fences: an ant colony approach based on crowd simulation. IEEE Trans. Intell. Transp. Syst. **24**, 9137–9149 (2023). https://doi.org/10.1109/TITS.2023.3272318
3. Haghani, M., Coughlan, M., Crabb, B., Dierickx, A., Feliciani, C., van Gelder, R., Geoerg, P., Hocaoglu, N., Laws, S., Lovreglio, R., Miles, Z., Nicolas, A., O'Toole, W.J., Schaap, S., Semmens, T., Shahhoseini, Z., Spaaij, R., Tatrai, A., Webster, J., Wilson, A.: A roadmap for the future of crowd safety research and practice: introducing the Swiss Cheese Model of crowd safety and the imperative of a vision zero target. Saf. Sci. **168**, 106292 (2023). https://doi.org/10.1016/j.ssci.2023.106292
4. Bahamid, A., Mohd Ibrahim, A.: A review on crowd analysis of evacuation and abnormality detection based on machine learning systems. Neural Comput. Appl. **34**, 21641–21655 (2022). https://doi.org/10.1007/s00521-022-07758-5
5. Harihara Subramanian, G., Verma, A.: Crowd risk prediction in a spiritually motivated crowd. Saf. Sci. **155**, 105877 (2022). https://doi.org/10.1016/j.ssci.2022.105877
6. Verma, A., Gayathri, H.S., Karthika, P.S., Choubey, N., Khandelwal, T.: Crowd management guidelines for mass religious gatherings. In: Maurya, A.K., Vanajakshi, L.D., Arkatkar, S.S.,

Sahu, P.K. (eds.) Transportation Research in India: Practices and Future Directions, pp. 151–165. Springer Singapore, Singapore (2022). https://doi.org/10.1007/978-981-16-9636-7_8

7. Weng, W., Wang, J., Shen, L., Song, Y.: Review of analyses on crowd-gathering risk and its evaluation methods. J. Saf. Sci. Resil. **4**, 93–107 (2023). https://doi.org/10.1016/j.jnlssr.2022.10.004
8. Zhang, X., Yu, Q., Wang, Y.: Fuzzy evaluation of crowd safety based on pedestrians' number and distribution entropy. Entropy **22** (2020). https://doi.org/10.3390/e22080832
9. Li, H., Peng, W., Adumene, S., Yazdi, M.: Operations management of critical energy infrastructure: a sustainable approach. In: Li, H., Peng, W., Adumene, S., Yazdi, M. (eds.) Intelligent Reliability and Maintainability of Energy Infrastructure Assets, pp. 39–52. Springer Nature Switzerland, Cham (2023). https://doi.org/10.1007/978-3-031-29962-9_3
10. Wang, X., Zhang, C., Gao, W.: Risk assessment of hydrogen leakage in diesel hydrogenation process. Int. J. Hydrogen Energy **47**, 6955–6964 (2022). https://doi.org/10.1016/j.ijhydene.2021.12.027
11. Li, H., Yazdi, M.: Integration of the Bayesian network approach and interval type-2 fuzzy sets for developing sustainable hydrogen storage technology in large metropolitan areas. In: Li, H., Yazdi, M. (eds.) Advanced Decision-Making Methods and Applications in System Safety and Reliability Problem, pp. 69–85. Springer International Publishing, Cham (2022). https://doi.org/10.1007/978-3-031-07430-1_5
12. Yazdi, M., Khan, F., Abbassi, R., Rusli, R.: Improved DEMATEL methodology for effective safety management decision-making. Saf. Sci. **127**, 104705 (2020). https://doi.org/10.1016/j.ssci.2020.104705
13. Li, H., Yazdi, M., Huang, H.-Z., Huang, C.-G., Peng, W., Nedjati, A., Adesina, K.A.: A fuzzy rough copula Bayesian network model for solving complex hospital service quality assessment. Complex Intell. Syst. (2023). https://doi.org/10.1007/s40747-023-01002-w
14. Yazdi, M., Adumene, S., Zarei, E.: Introducing a probabilistic-based hybrid model (fuzzy-BWM-Bayesian network) to assess the quality index of a medical service. In: Yazdi, M. (ed.) Linguistic Methods Under Fuzzy Information in System Safety and Reliability Analysis, pp. 171–183. Springer International Publishing, Cham (2022). https://doi.org/10.1007/978-3-030-93352-4_8
15. Nedjati, A., Yazdi, M., Abbassi, R.: A Sustainable Perspective of Optimal Site Selection of Giant Air-Purifiers in Large Metropolitan Areas. Springer Netherlands (2021). https://doi.org/10.1007/s10668-021-01807-0
16. Zarei, E., Yazdi, M., Abbassi, R., Khan, F.: A hybrid model for human factor analysis in process accidents: FBN-HFACS. J. Loss Prev. Process Ind. 142–155 (2019). https://doi.org/10.1016/j.jlp.2018.11.015
17. Adumene, S., Nwaoha, T.C.: Dynamic cost-based integrity assessment of oil and gas pipeline suffering microbial induced stochastic degradation. J. Nat. Gas Sci. Eng. **96**, 104319 (2021). https://doi.org/10.1016/j.jngse.2021.104319
18. Yazdi, M., Khan, F., Abbassi, R., Quddus, N.: Resilience assessment of a subsea pipeline using dynamic Bayesian network. J. Pipeline Sci. Eng. **2**, 100053 (2022). https://doi.org/10.1016/j.jpse.2022.100053
19. Zubieta, L.L.: Resilience Framework for Critical Infrastructures (2013)
20. Vivid Sydney biggest on record, Minist. Jobs, Minist. Music Night-Time Econ. Minist. Tour. (n.d.). Australia's biggest festival, of more than 3.28 million. https://www.nsw.gov.au/media-releases/vivid-sydney-biggest-on-record#:~:text=VividSydney%2C
21. Li, H., Yazdi, M.: Dynamic decision-making trial and evaluation laboratory (DEMATEL): improving safety management system. In: Li, H., Yazdi, M. (eds.) Advanced Decision-Making Methods and Applications in System Safety and Reliability Problems: Approaches, Case Studies, Multi-criteria Decision-Making, Multi-objective Decision-Making, Fuzzy Risk-Based Models, pp. 1–14. Springer International Publishing, Cham (2022). https://doi.org/10.1007/978-3-031-07430-1_1
22. Li, H., Yazdi, M.: What are the critical well-drilling blowouts barriers? A progressive DEMATEL-game theory. In: Li, H., Yazdi, M. (eds.) Advanced Decision-Making Methods

and Applications in System Safety and Reliability Problems: Approaches, Case Studies, Multi-criteria Decision-Making, Multi-objective Decision-Making, Fuzzy Risk-Based Models, pp. 29–46. Springer International Publishing, Cham (2022). https://doi.org/10.1007/978-3-031-07430-1_3

23. Yazdi, M., Moradi, R., Pirbalouti, R.G., Zarei, E., Li, H.: Enabling safe and sustainable hydrogen mobility: circular economy-driven management of hydrogen vehicle safety. Processes **11** (2023). https://doi.org/10.3390/pr11092730
24. Li, H., Yazdi, M.: How to deal with toxic people using a fuzzy cognitive map: improving the health and wellbeing of the human system. In: Li, H., Yazdi, M. (eds.) Advanced Decision-Making Methods and Applications in System Safety and Reliability Problems: Approaches, Case Studies, Multi-criteria Decision-Making, Multi-objective Decision-Making, Fuzzy Risk-Based Models, pp. 87–107. Springer International Publishing, Cham (2022). https://doi.org/10.1007/978-3-031-07430-1_6
25. Jetter, A.J., Kok, K.: Fuzzy cognitive maps for futures studies—a methodological assessment of concepts and methods. Futures **61**, 45–57 (2014). https://doi.org/10.1016/j.futures.2014.05.002

Decision-Making During Crisis in Megaprojects

Rehab Iftikhar

Abstract Crisis and its management have always remained the subject of interest in project studies but not much studied from crisis and its management viewpoint. The importance of crisis as low probability-high impact event increased when setting is high scale (in terms of budget and reputation), multi-organizational, complex, and temporary such as megaproject. The impact of crisis in megaproject is tremendous which could jeopardize the projects and organizations; in the worst-case scenario, it causes reputational damages to participatory organizations, it took ages for them to recover from crisis and some might never enjoy the same reputation. This chapter will uncover steps for decision-making and decision-making styles in crises, and present decision-making process model.

Keywords Crisis · Decision-making · Megaproject

1 Introduction

Crisis is an unexpected, unforeseen, and unpredictable event [51]. In this chapter, crisis is defined as "a low probability and high impact event" [44, p. 60]. Risk is frequently used interchangeably with crisis, but they are different [24]. Risk is considered as a foreseen event that can result in predictable negative effects [21]. Risk is manageable as it is a known-unknown event, which means it is identifiable, but it is not possible to find out when it will occur [32, 40]. However, crisis on other hand is an unforeseen event, comprising an element of surprise with no contingency plan [24, 25, 44]. To avoid confusion, we only use the term crisis in this chapter and no other relevant terms such as uncertainty, issues, deviations etc.

Megaprojects are large-scale and complex infrastructure and investment projects with a long duration and a high budget that involve multiple stakeholders [18, 35].

R. Iftikhar (✉)
School of Computing, Engineering & Physical Sciences, University of the West of Scotland (UWS), London, UK
e-mail: Rehab.Iftikhar@uws.ac.uk

M. Yazdi (ed.), *Progressive Decision-Making Tools and Applications in Project and Operation Management*, Studies in Systems, Decision and Control 518,
https://doi.org/10.1007/978-3-031-51719-8_15

Megaprojects frequently encounter performance problems, crisis, and failure [15, 19]. Even a well-managed project can face a crisis. For example, Berlin International Airport is an illustration of a high-profile failure due to its delay of over four years and overrun of at least 70%. Opening dates for the airport have been moved four times, originally from 2011 to 2012, early 2013 and then late 2013. Given the complexity of the technical issues faced and due to poor planning, the project has suffered continuous delays, which led to its opening in 2020. As a result, all parties involved, such as designers, contractors, politicians, and Germany as a whole, have experienced severe reputational damage [14]. This example illustrates that a project can face crises which could cause severe damages.

Prior studies have considered a crisis in a wide range of contexts, focused on one specific crisis event such as the Mann Gulch fire [61], the Challenger disaster [57] and the Utrecht terrorist attack [66]. These are highly symbolic crisis events resulting in extensive damage to human life or the environment [53]. Moreover, there are a few studies on crisis and its management from a single organization perspective [9, 62]. Prior research scholars have focused on high-reliability organizations (such as fire department, military, NASA, emergency response organizations, crisis organizations and SWAT teams) which have developed the capability to manage unexpected events and seeking to avoid crises [5, 29, 65]. However, contemporary research rarely considers crises triggered in a performance-based environment comprised of a large group of diverse organizations especially in real-time ongoing crisis. Many crises facing organizations do not necessarily result in major damages or life-threatening events. According to Kornberger et al. [33], crises also include "normal accidents", as described by Perrow [46] which do not require a life-threatening event to precipitate a crisis. Simple organizational breakdowns and misuse of organizational information can also greatly impact organizations viability [44, 45, 52].

What is neglected in the existing literature is about making decisions during crises in megaproject, where different and multiple organizations work together in complex temporary settings [2, 62]. Prior research highlights the urgency of decision-making in a crisis [3, 44], however its importance is emphasized during megaprojects because (i) multiple stakeholders are involved to accomplish a complex task; and (ii) project temporariness put extra pressure on organizations to finish tasks on-time and make prompt decisions in crisis. Decision-making is integral to the management of projects, but decisions related to projects are complex and multifaceted [23, 26]. The focus of this chapter is not a particular crisis or a type of crisis but on crises in general. Due to their temporariness, complexity and large scale, megaprojects are distinct in their demands for decision-making in crisis compared to organizational decision-making in normal situations. The primary purpose of this chapter is to investigate and to understand these different steps and styles during the decision-making process in crises when multiple organizations are involved in megaprojects.

2 Crisis

Crises are unpredictable [43, 44], unexpected, and potentially disruptive events that which undermine the goals of an organization and have profound implications for its relationships with its stakeholders [5, 25]. It is, however, difficult for researchers to agree on a common definition of a crisis. For example, Stam et al. [55, p. 2861] define crises as "events that are associated with intense feelings of negative affect regarding the current situation and strong feelings of uncertainty regarding the future". For Weick [60, p. 305], a crisis is a "low probability-high consequence event that threatens the most fundamental goals of the organization". These definitions are broad and often fail to consider the fundamental elements and characteristics of crisis. In agreement with recent studies [43, 63], the most widely cited and well-recognized definition, proposed by Pearson and Clair [44] is adopted in this chapter. Pearson and Clair define a crisis as "a low probability, high impact event that threatens the viability of the organization and is characterized by ambiguity of cause, effect, and means of resolution, as well as by a belief that decisions must be made swiftly" [44, p. 60].

According to this definition, a crisis can be defined as a major, unpredictable event that interferes with the normal conduct of business and threatens survival. They have a low probability of occurring, are unpredictable, and are characterized by time pressures and the need for quick action to minimize their impact [3, 68]. Hence, a crisis threatens organizations' high-priority values, is restricted to the time available to respond to it and is unexpected or unanticipated by the organization [22]. Examining the above definition, the following characteristics are highlighted. A crisis is an unexpected event that has the potential of wrecking the structure of an organization, it may affect not only the employees and other members internal to the organization, but also key stakeholders external to the organization. A crisis may occur in any organization of any size and in any industry [8, 31]. Finally, a crisis may affect the legitimacy of an organization [48].

3 Decision-Making

A decision is a "cognitive phenomenon and conceptualized as the goal or endpoint for a more or less complex process of deliberation which includes an assessment of consequences and uncertainties" [42, p. 76]. According to Smart and Vertinsky [54], decision quality depends upon three factors: (a) the quality of information (effective information flows, thus preventing overloads and reducing noise in communication channels, noise depends upon the distance between units in the organization), (b) the fidelity of objective articulation and trade-off evaluation (input: cognitive abilities and group think output quality decision) and (c) cognitive abilities of the decision group (the abilities of the decision unit to interpret information, generate options and calculate and make choices between alternative courses of action). A decision process consists of the articulation of objectives, the generation of alternate courses of action, an appraisal of their feasibility, an evaluation of the consequences of the given alternatives, and a choice of the alternative which contributes most to the attainment of organizational objectives [54]. In conventional terms, the task of decision-making can be decomposed into five subtasks: (1) identifying the relevant goals, (2) searching for alternative courses of action; (3) predicting the consequences of following each alternative; (4) evaluating each alternative in terms of its consequences for goal achievement; and (5) selecting the best alternative for achieving the goal [1, 54].

During a crisis, decision-making is critical to make accurate and timely decisions [36] which are pressed by perceived time constraints and colored by cognitive limitations [44]. The consequences of a crisis are high, as it is a low probability, high severity event; however, its impact can be reduced by rapid and accurate decision-making [28, 37]. Decisions must be made quickly despite the uncertainty, time pressure, and high stakes associated with a crisis [44]. Sawle [50] described the importance of decisions in a crisis as follows: "the worst decision is no decision, and the second worst decision is a late one". It is critical to integrate crises in the decision-making process. The objective is to make the right decisions and execute them effectively. Moreover, the novelty, magnitude, and frequency of decisions demanded by a crisis suggest that no organization will respond in a manner that is completely effective or completely ineffective [44]. Decision-making is complex, and at times of crisis it is more complicated [64], as a crisis is an unexpected, unusual, and abnormal event [34], the subtle quality of decisions can easily get lost in the crisis events [20]. The core elements that define a crisis—ambiguity, urgency, and high stakes—are also severe constraints on the ability of individuals to make decisions effectively [44].

4 Megaproject

Megaprojects are large-scale, complex ventures that typically cost from several hundred million to billions of dollars, take many years to develop and build, involve multiple public and private stakeholders, are transformational, and impact millions of people [16, 35]. The strategic nature and high costs of these undertakings usually imply that public institutions hold a significant role in the project-financing phase [35]. Megaprojects are temporary endeavors (i.e., projects) characterized high commitment in terms of investment made, extreme complexity, and long-lasting impacts on the economy, the environment, technological development, and society [4, 69]. Megaprojects often involve the building of physical infrastructures, which may be roads, railways, airports, bridges, hydropower plants, etc. [47]. Each megaproject is characterized by a distinct governance structure, and a temporary production system [41].

As a standalone temporary organization, megaprojects can be led by a client, contractor, or some form of temporary joint venture, or coalition of multiple parties (sponsors, clients, contractors, suppliers, and other stakeholders) that work jointly on a shared activity for a limited period of time in an uncertain environment [27, 39]. Each megaproject is usually decomposed into many smaller inter-related projects and organized as a program. A large organization such as the client or prime contractor is established to coordinate and integrate the efforts of numerous subgroups and suppliers involved in project activities [11, 10, 39]. Megaprojects usually play an important role in the sustainable development of public infrastructures [59]. Megaprojects are commissioned by the collaboration of public and private sectors [7]. Organizations responsible for carrying out megaprojects face a performance paradox—a significant gap occurs between what is expected from the huge investment of resources and what is obtained from the project investment [17, 67]. Megaprojects are uncertain projects that usually suffer from cost overruns, delays, and reduced functionality of the delivered outcome [15, 19]. If not managed properly, any crisis during megaprojects could threaten its viability [30]. With respect to the critical role of megaprojects to the national economy and social development [33], so importance to identify crises and coping strategies multifold when diverse stakeholders of megaprojects are involved.

5 Decision-Making During Crisis in Megaprojects

Megaprojects are highly uncertain projects that usually suffer from cost overruns, delays, and reduced functionality of the delivered outcome [15, 19]. If not managed properly, the crisis during megaprojects could threaten project viability [30]. A crisis requires a rapid decision, and its importance is increased in megaprojects as it requires interacting with a variety of stakeholders such as clients, consultants, contractors, subcontractors [33]. With respect to the critical role of megaprojects in the national economy and social development [67], it is important to study decision-making to resolve crises when diverse actors/stakeholders of megaprojects are involved.

The conception of crisis is defined as a decisional situation with three characteristics: (1) identification of the origin of the event—whether external or internal for the decision makers; (2) the decision time available for response—whether short, intermediate, or long, and (3) the relative importance of the values at stake to the participants—whether high or low [22]. Decision-making in crises is characterized by a high level of uncertainty, urgency to act, a narrowing of options and high-stakes implications for organizational survival. At the time of crisis, the challenge for any organization is to make decisions quickly and accurately [3] as individuals make decisions based on their perceptions [58]. During a crisis, one must secure a high-quality decision-making process. Figure 1 is graphical representation step-wise approach of the decision-making process in crisis and decision-making styles in crisis. However, Fig. 1 was derived from International Islamabad—Airport project, which was a megaproject in Pakistan. This project began construction in April 2008 and became operational on May 3, 2018, with a cost of Rs. 92,865 billion rupees (US$887 million; [23]). The examples provided for each decision-making step and style are from the International Islamabad Airport project.

There are three steps of decision-making in crisis: (i) understanding the crisis, (ii) information gathering, and (iii) evaluating alternatives and selecting the best

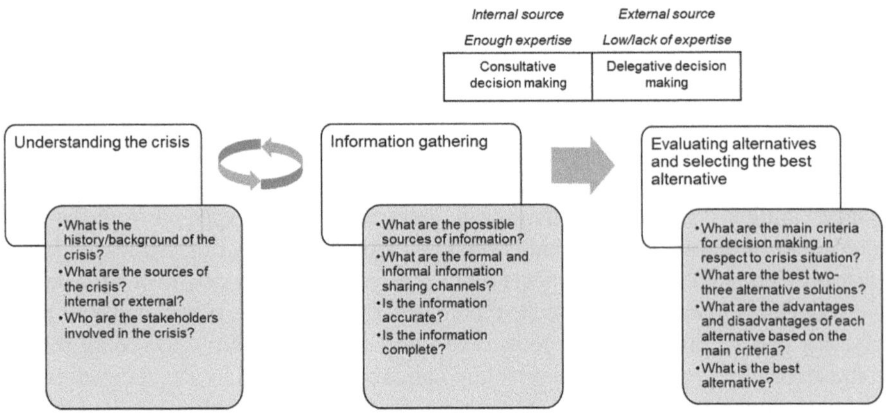

Fig. 1 Framework for the decision-making process during crisis in megaproject (Adopted from [23])

alternatives. Due to the high number of actors and activities involved in megaprojects [30], it is crucial to understand the crisis. Therefore, understanding of the crisis is defined as a separate step of decision-making to shed light on the process by which multiple organizations are involved and negotiate and interpret information. Generally, an understanding of the crisis defines where the crisis lies and what the sources and characteristics of the crisis are, meaning whether the crisis lies internally or externally, such as within an organization or other organizations such as the client, contractor, consultant, and designer involved in the project, the reason for the crisis and whether the crisis is technical, economic, or social in nature. Therefore, understanding the crisis requires gathering various types of information about the background of the crisis [12]. This information helps decision makers to not only identify the sources of the crisis but to recognize the organizations and stakeholders who are involved with impacts and actions. As stated by the project management officer of contractor 7: *The most important thing in decision-making is to identify the problem. Once the problem is identified then you have to coordinate with the stakeholders. If the problem is because of you or due to the client, the stakeholders or any third party, you just have to identify that problem maker... The first thing is identification of the problem and identification of the stakeholders.* By interpreting the information, decision makers obtain an understanding (meaning) of the crisis [49].

The next step is information gathering, which is the process of monitoring the environment and providing data to managers. The inter-organizational aspects of megaprojects force decision makers to seek information from internal and external individuals, teams of different organizations, such as clients, designers, contractors, etc. [56]. Internal sources refer to information collected from the project team members and the staff of the respective organization, whereas external sources include other organizations, such as the client, contractor, consultant, designer, and suppliers. As stated by the project director of contractor 2: *We can gather information from our own staff, but also, we can gather information from the project management consultant or some other stakeholders.* In a time of crisis, decision makers may receive several pieces of information from different sources either formally such as meetings, documents, and contracts, or informally such as email, WhatsApp, SMS, etc. As illustrated by integration manager from contactor 4: *In this is the era of communication, we have provided our employees show us the issue by taking pictures and transferred them to WhatsApp simultaneously, so we get information about the issue... We get information and instructions from the client by writing a letter or in meetings.*

Moreover, decision-making in a crisis requires accurate and complete information. The quality of decisions largely depends on the quality of information which is measured by accuracy, timeliness, completeness, and relevance. To gather quality information, decision makers try to define what type of information they need, where the information is and how to get the information. As stated, earlier accuracy and completeness are important to the quality of information. Accordingly, the accuracy and completeness of the information are concerned, project actors attempt to obtain as much information as possible. As stated by the project director of the client: *The information has to be correct; accuracy depends upon the quality of the information. It has to come on time. It should be complete.* If there is incomplete information, then decision-making in crisis also depends on the judgment and experience of decision makers in evaluating the quality of information to be utilized for decision-making [13]. As illustrated by the project director from contractor 9: *You can't have 100% information, you can't wait till your file is completed by 100%, you cannot make decision, you also go to middle course and make a decision with the information you have.* Figure 1 reveals that understanding the crisis and information gathering steps have a bidirectional relationship. In one direction, gathered information from different sources helps to obtain a better understanding of the crisis. In the other direction, by understanding the crisis, decision makers can identify whether the crisis has internal or external sources, or the crisis has technical, economic, or social causes. In turn, this understanding leads to deeper information gathering from relevant sources.

Due to the unfavorable high impacts of a crisis on a project, decision makers try to develop multiple alternatives to ensure that they can manage the impacts of a crisis; however, the number of alternatives is usually limited to create a balance between making an accurate decision and minimizing the efforts of decision-making [42]. Decision makers have certain criteria to evaluate different alternatives to make sure that pros and cons of actions are well-understood [6]. Using the iron triangle (cost, time, quality) does not always provide sufficient criteria [38]. The decision-making process in crisis identified several criteria, including time, financial implications, resource availability, favorability for the company and the project, project and control specification and legality, needed to select the best alternative. The quote from the project director of client, illustrated the evaluation and selection of alternatives: *The most important step is to develop some options—2–3 at least... Then we do evaluate these options and after the evaluation, we select the best option... It must be in the interest of the project, not to cause loss of revenue, it must be legal, it must be viable, with no irregularities and wasteful expenditures... We keep in mind the option that gives the maximum benefit to a project.*

There are two decision-making styles presented in Fig. 1, i.e., consultative and delegative decision-making. Consultative decision-making involves a process in which one person makes the decision with input from other members. For consultative decision-making, a project manager asks his/her team for information that would be helpful for decision-making, but the final decision is made by the manager. Therefore, the purpose is to consult and to gather as much information as possible from a team before making a decision. In delegative decision-making managers are not making decisions, but they receive the decisions. To make the right decisions, sometimes the responsibility of decision-making is transferred to other organizations, i.e., client, consultant, designer and the third party. The reasons to transfer the decision-making responsibility are (a) the organization does not have expertise and enough knowledge and (b) the crisis source is external, produced by a contractor, client, or consultant, but affects an organization, which is incapable of decision-making, so the power of decision-making lies with some other organization [23]. As the project manager from contractor 4 stated that: *If the decision is related to project, then we inform client and project management consultant. We also get major decision from client.... If something is internal within the organization, decision is taken by project manager if he/she is aware of situation and have knowledge... but sometime if we don't have knowledge, we cannot make decisions. So, then we consult with our head office or with senior, then took any decision.*

The quote and Fig. 1 show that once the organization identifies the sources of crisis (internal vs. external sources) and the level of internal expertise, the organization can determine the suitable decision-making style for a crisis and can continue evaluating alternatives accordingly. In other words, we can say that when the source of crisis is internal and an organization has specific knowledge and skills in house for decision-making, consultative decision-making style has been adopted. On the contrary, when the source of the crisis is external to the organization and there is a need for specific knowledge and skills, decision-making could be delegated to another organization with suitable expertise.

6 Conclusion

This chapter is about the decision-making process during crisis in megaprojects. The megaproject of International Islamabad Airport has been used as an example. There is conventional decision-making process, however this chapter is providing decision-making process during crises in multi-organizational complex environment i.e., megaproject. Moreover, the decision-making steps suggested by Fig. 1 are (i) understanding the crisis, (ii) information gathering, and (iii) evaluating alternatives and selecting the best alternatives. For decision-making style, Fig. 1 provides evidence for consultative and delegative decision-making. This chapter has contributed in several ways: (i) provide deep insights about decision-making process during crises in megaprojects, which can affect the way organizations make decisions, (ii) it also provides a roadmap for which will enhance the successful management of projects.

References

1. Anderson, P.A.: Decision making by objection and the Cuban missile crisis. Adm. Sci. Q. **28**(2), 201–222 (1983)
2. Bakker, R.M., DeFillippi, R.J., Schwab, A., Sydow, J.: Temporary organizing: promises, processes, problems. Organ. Stud. **37**(12), 1703–1719 (2016)
3. Bonn, I., Rundle-Thiele, S.: Do or die—strategic decision-making following a shock event. Tour. Manag. **28**(2), 615–620 (2007)
4. Brookes, N.J., Locatelli, G.: Power plants as megaprojects: using empirics to shape policy, planning, and construction management. Util. Policy **36**, 57–66 (2015)
5. Bundy, J., Pfarrer, M.D., Short, C.E., Coombs, W.T.: Crises and crisis management: integration, interpretation, and research development. J. Manag. **43**(6), 1661–1692 (2017)
6. Chapman, R.J.: A framework for examining the dimensions and characteristics of complexity inherent within rail megaprojects. Int. J. Proj. Manag. **34**(6), 937–956 (2016)
7. Clegg, S., Pitsis, T., Rura-Polley, T., Maroosszeky, M.: Governmentality matters: designing an alliance culture of interorganizational collaboration for managing projects. Organ. Stud. **23**(3), 317–338 (2002)
8. Coleman, L.: The frequency and cost of corporate crises. J. Conting. Crisis Manag. **12**(1), 2–13 (2004)
9. Cornelissen, J.: Sensemaking under pressure: the influence of professional roles and social accountability on the creation of sense. Organ. Sci. **23**(1), 118–137 (2012)
10. Davies, A., Gann, D., Douglas, T.: Innovation in megaprojects: systems integration at London Heathrow terminal 5. Calif. Manag. Rev. **51**(2), 101–125 (2009)
11. Davies, A., Mackenzie, I.: Project complexity and systems integration: constructing the London 2012 Olympics and Paralympics games. Int. J. Proj. Manag. **32**(5), 773–790 (2014)
12. De Bruijn, H., Leijten, M.: Megaprojects and contested information. Transp. Plan. Technol. **30**(1), 49–69 (2007)
13. Eweje, J., Turner, R., Müller, R.: Maximizing strategic value from megaprojects: the influence of information-feed on decision-making by the project manager. Int. J. Proj. Manag. **30**(6), 639–651 (2012)
14. Fiedler, J., Wendle, A.: Public Infrastructure Project Planning in Germany: The Case of the BER Airport in Berlin-Brandenburg. Hertie School of Governance, Berlin (2015)
15. Flyvbjerg, B.: What you should know about megaprojects and why: an overview. Proj. Manag. J. **45**(2), 6–19 (2014)
16. Flyvbjerg, B.: Introduction: the iron law of megaproject management. In: Flyvbjerg, B. (ed.) The Oxford Handbook of Megaproject Management, pp. 1–18. Oxford University Press, Oxford (2017)
17. Flyvbjerg, B., Holm, M.K.S., Buhl, S.: What causes cost overrun in transport infrastructure projects? Transp. Rev. **24**(1), 3–18 (2004)
18. Flyvbjerg, Miller, R., Lessard, D.R.: The Strategic Management of Large Engineering Projects: Shaping Institutions, Risks, and Governance. MIT Press, Cambridge (2000)
19. Górecki, J., Díaz-Madroñero, M.: Who risks and wins?—Simulated cost variance in sustainable construction projects. Sustainability **12**(8), 3370 (2020)
20. Hart, P., Rosenthal, U., Kouzmin, A.: Crisis decision making: the centralization thesis revisited. Admin. Soc. **25**(1), 12–44 (1993)
21. Heath, R.L., Lee, J., Ni, L.: Crisis and risk approaches to emergency management planning and communication: the role of similarity and sensitivity. J. Public Relat. Res. **21**(2), 123–141 (2009)

22. Hermann, C.F.: Some consequences of crisis which limit the viability of organizations. Adm. Sci. Q. **8**(1), 61–82 (1963)
23. Iftikhar, R., Momeni, K., Ahola, T.: Decision-making in crisis during megaprojects. J. Mod. Proj. Manag. **9**(3), 33–49 (2022)
24. Iftikhar, R., Müller, R.: Taxonomy among triplets: opening the black box. Int. J. Manag. **10**(2), 63–85 (2019)
25. Iftikhar, R., Müller, R., Ahola, T.: Crises and coping strategies in megaprojects: the case of the Islamabad–Rawalpindi Metro Bus Project in Pakistan. Proj. Manag. J. **52**(4), 394–409 (2021)
26. Iftikhar, R., Majeed, M., Drouin, N.: Crisis management process for project-based organizations. Int. J. Manag. Proj. Bus. **16**(8):100–125 (2023)
27. Jones, C., Lichtenstein, B.: Temporary inter-organizational projects: how temporal and social embeddedness enhance coordination and manage uncertainty. In: Cropper, S., Ebers, M., Huxman, C., Smith Ring, P. (eds.) The Oxford Handbook of Inter-organizational Relations, pp. 231–255. Oxford University Press (2008)
28. Kahn, W.A., Barton, M.A., Fellows, S.: Organizational crises and the disturbance of relational systems. Acad. Manag. Rev. **38**(3), 377–396 (2013)
29. Kalkman, J.P.: Boundary spanners in crisis management. Int. J. Emerg. Serv. **9**(2), 233–244 (2020)
30. Kardes, I., Ozturk, A., Cavusgil, S.T., Cavusgil, E.: Managing global megaprojects: complexity and risk management. Int. Bus. Rev. **22**(6), 905–917 (2013)
31. Keeffe, M.J., Darling, J.R.: Transformational crisis management of organization development: the case of talent loss at Microsoft. Organ. Dev. J. **26**(4), 43–58 (2008)
32. Knight, F.H.: Risk, Uncertainty and Profit. Hart, Schaffner & Marx; Houghton Mifflin, Boston, MA (1921)
33. Kornberger, M., Leixnering, S., Meyer, R.E.: The logic of tact: how decisions happen in situations of crisis. Organ. Stud. **40**(2), 239–266 (2019)
34. Lacombe, P.: Agronomic research and research into crisis situation. J. Conting. Crisis Manag. **10**(4), 205–206 (2002)
35. Locatelli, G., Mancini, M.: Risk management in a mega-project: the universal EXPO 2015 case. Int. J. Proj. Org. Manag. **2**(3), 236–253 (2010)
36. Loosemore, M.: The three ironies of crisis management in construction projects. Int. J. Proj. Manag. **16**(3), 139–144 (1998)
37. Mallak, L.A., Kurstedt, H.S., Jr.: Planning for crises in project management. Proj. Manag. J. **28**(2), 14–24 (1997)
38. Marques, G., Gourc, D., Lauras, M.: Multi-criteria performance analysis for decision making in project management. Int. J. Proj. Manag. **29**(8), 1057–1069 (2011)
39. Merrow, E.W.: Industrial Megaprojects: Concepts, Strategies, and Practices for Success. Wiley (2011)
40. Meyer, D.A., Loch, H.C., Pich, T.M.: Managing project uncertainty: from variation to chaos. MIT Sloan Manag. Rev. **43**(2), 60–67 (2002)
41. Miller, R., Hobbs, B.: Governance regimes for large projects. Proj. Manag. J. **36**(3), 42–51 (2005)
42. Müller, R., Spang, K., Ozcan, S.: Cultural differences in decision making in project teams. Int. J. Manag. Proj. Bus. **2**(1), 70–93 (2009)
43. Oh, O., Agrawal, M., Rao, H.R.: Community intelligence and social media services: a rumor theoretical analysis of tweets during social crisis. MIS Q. **37**(2), 407–426 (2013)
44. Pearson, C.M., Clair, J.A.: Reframing crisis management. Acad. Manag. Rev. **23**(1), 59–76 (1998)
45. Pearson, C.M., Mitroff, I.I.: From crisis prone to crisis prepared: a framework for crisis management. Acad. Manag. Exec. **7**(1), 48–59 (1993)
46. Perrow, C.: Normal Accidents: Living with High Risk Technologies. Basic Books, New York (1984)
47. Priemus, H.: Decision-making on mega-projects: drifting on political discontinuity and market dynamics. Eur. J. Transp. Infrastruct. Res. **10**(1), 19–29 (2010)

48. Ray, S.J.: Strategic Communication in Crisis Management: Lessons from the Airline Industry. Quorum Books, London (1999)
49. Saunders, F.C., Townsend, E.A.: Delivering new nuclear projects: a megaprojects perspective. Int. J. Manag. Proj. Bus. **12**(1), 144–160 (2019)
50. Sawle, C.: Concerns of project managers: crisis project management. PM Netw. **5**(1), 25–29 (1991)
51. Seeger, M.W.: Chaos and crisis: proposition for general theory of crisis communication. Public Relat. Rev. **28**(4), 329–337 (2002)
52. Shrivastava, P., Mitroff, I.I.: Strategic management of corporate crises. Columbia J. World Bus. **22**(1), 5–11 (1987)
53. Shrivastava, P., Mitroff, I.I., Miller, D., Miglani, A.: Understanding industrial crises. J. Manag. Stud. **24**(4), 285–303 (1988)
54. Smart, C., Vertinsky, I.: Designs for crisis decision units. Adm. Sci. Q. **22**(4), 640–657 (1977)
55. Stam, D., van Knippenberg, D., Wisse, B., Pieterse, A.N.: Motivation in words: promotion and prevention-oriented leader communication in times of crisis. J. Manag. **44**(7), 2859–2887 (2018)
56. Turner, B.: The organizational and interorganizational development of disasters. Adm. Sci. Q. **21**(3), 378–397 (1976)
57. Vaughan, D.: The Challenger Launch Decision. University of Chicago Press, Chicago (1996)
58. Wang, A., Pitsis, T.S.: Identifying the antecedents of megaproject crises in China. Int. J. Proj. Manag. **38**(6), 327–339 (2020)
59. Wang, Y., Wang, Y., Wu, X., Li, J.: Exploring the risk factors of infrastructure PPP projects for sustainable delivery: a social network perspective. Sustainability **12**(10), 4152 (2020)
60. Weick, K.E.: Enacted sensemaking in crisis situations. J. Manag. Stud. **25**(4), 305–317 (1988)
61. Weick, K.E.: The collapse of sensemaking: the Mann Gulch disaster. Adm. Sci. Q. **38**(4), 628–652 (1993)
62. Weick, K.E.: Sensemaking in Organizations. Sage Publications, Thousand Oaks, CA (1995)
63. Wilding, R., Paraskevas, A.: Crisis management or crisis response system? A complexity science approach to organizational crises. Manag. Decis. **44**(7), 892–907 (2006)
64. Wilson, T.C.: Risk management lessons learned from the financial crisis: one CRO's view. J. Risk Manag. Financ. Inst. **6**(2), 167–177 (2013)
65. Wolbers, J., Boersma, K.: The common operational picture as collective sensemaking. J. Conting. Crisis Manag. **21**(4), 186–199 (2013)
66. Wolbers, J.: Understanding distributed sensemaking in crisis management: the case of the Utrecht terrorist attack. J. Conting. Crisis Manag. 1–11 (2021)
67. Wu, G., Zhao, X., Zuo, J., Zillante, G.: Effects of contractual flexibility on conflict and project success in megaprojects. Int. J. Confl. Manag. **29**(2), 253–278 (2017)
68. Yang, X., Wang, L., Zhu, F., Müller, R.: Prior and governed stakeholder relationships: the key to resilience of inter-organizational projects. Int. J. Proj. Manag. **40**(1), 64–75 (2022)
69. Zhai, L., Xin, Y., Cheng, C.: Understanding the value of project management from a stakeholder's perspective: case study of megaproject management. Proj. Manag. J. **40**(1), 99–109 (2009)

Harnessing AI for Project Risk Management: A Paradigm Shift

He Li, Mohammad Yazdi, Arman Nedjati, Rosita Moradi, Sidum Adumene, Uyen Dao, Amirhossein Moradi, Aida Haghighi, Francis Eemmanuel Obeng, Cheng-Geng Huang, Hooi Siang Kang, Reza Ghasemi Pirbalouti, Esmaeil Zarei, Majeed Dehghan, Mahlagha Darvishmotevali, Peiman Ghasemi, Payam Shayan Fard, and Harish Garg

Abstract Project risk management is pivotal in ensuring the success of endeavours within contemporary organizations. As businesses navigate an intricate and competitive landscape, integrating artificial intelligence (AI) introduces a paradigm shift in how project risks are recognized, evaluated, and addressed. This chapter go

H. Li
Centre for Marine Technology and Ocean Engineering (CENTEC), Instituto Superior Técnico, Universidade de Lisboa, Lisbon, Portugal

M. Yazdi (✉)
School of Computing, Engineering & Physical Sciences, University of the West of Scotland (UWS), London, UK
e-mail: mohammad.yazdi@mq.edu.au

M. Yazdi · M. Dehghan
Faculty of Science and Engineering, Macquarie University, Sydney, Australia

A. Nedjati
Industrial Engineering Department, Quchan University of Technology, Quchan, Iran

R. Moradi
University of Wollongong, Northfields Avenue, Wollongong, NSW 2522, Australia

S. Adumene
Department of Marine Engineering, Rivers State University, Port Harcourt, Nigeria

U. Dao · A. Moradi · F. E. Obeng
Faculty of Engineering & Applied Science, Memorial University, St. John's, NL, Canada

U. Dao
The Faculty of Petroleum and Energy, Hanoi University of Mining and Geology, Hanoi, Vietnam

A. Moradi
Department of Environment, Health and Safety, Conagra Brands, Vancouver, BC, Canada

A. Haghighi
School of Occupational and Public Health, Toronto Metropolitan University, Toronto, ON, Canada

F. E. Obeng
Department of Marine Engineering, Regional Maritime University, Accra, Ghana

M. Yazdi (ed.), *Progressive Decision-Making Tools and Applications in Project and Operation Management*, Studies in Systems, Decision and Control 518,
https://doi.org/10.1007/978-3-031-51719-8_16

through into the multifaceted impact of AI on project risk management, reshaping strategies and enhancing decision-making processes to elevate project outcomes. By leveraging predictive analytics and machine learning, AI uncovers hidden patterns in historical project data, enabling proactive risk anticipation and simulation-based mitigation strategies. AI's cognitive capabilities expedite risk assessment by aggregating diverse data sources, fostering agility in response strategies through real-time communication and collaboration among project teams. Beyond operational facets, AI-driven algorithms continuously adapt to evolving risk landscapes, refining risk management strategies and resource allocation. Ethical considerations arise with the expanding role of AI, emphasizing the importance of transparency and accountability in algorithmic decision-making. This chapter envisions a dynamic convergence of AI and project risk management, fostering a future where innovation harmonizes with human understanding, propelling projects towards excellence, resilience, and growth.

Keywords Project risk management · Artificial intelligence (AI) · Predictive analytics · Machine learning · Hidden patterns · Proactive risk anticipation ·

C.-G. Huang
School of Automation Engineering, University of Electronic Science and Technology of China, Chengdu 611731, People's Republic of China

H. S. Kang
Faculty of Mechanical Engineering, Marine Technology Center, Institute for Vehicle Systems and Engineering, Universiti Teknologi Malaysia, Johor Bahru, Johor, Malaysia

R. G. Pirbalouti
Department of Mathematical and Industrial Engineering, Polytechnique Montréal, Montreal, Canada

E. Zarei
Department of Safety Science, College of Aviation, Embry-Riddle Aeronautical University, Prescott, AZ, USA

Robertson Safety Institute (RSI), Embry-Riddle Aeronautical University, Prescott, AZ, USA

M. Darvishmotevali
School of Hotel and Tourism Management, The Hong Kong Polytechnic University, Kowloon, Hong Kong

P. Ghasemi
Department of Business Decisions and Analytics, University of Vienna, Kolingasse 14-16, 1090 Vienna, Austria

P. S. Fard
Department of Fire Safety Engineering, Faculty of Engineering and Science, Western Norway University of Applied Sciences, Haugesund, Norway

Technical Safety, Department of Engineering and Systems, Aibel AS, Haugesund, Norway

H. Garg
School of Mathematics, Thapar Institute of Engineering & Technology (Deemed University), Patiala 147004, Punjab, India

Simulation-based mitigation · Cognitive capabilities · Data aggregation · Response strategies · Collaboration · Decision-making · Ethical considerations · Transparency · Accountability · Algorithmic decision-making · Innovation · Growth

1 Introduction

Within project management, the ability to recognize, evaluate, and counteract potential risks holds the utmost significance in guaranteeing success in endeavours [1]. As contemporary organizations endeavour to navigate through an ever-more intricate and cutthroat business domain, the fusion of artificial intelligence (AI) stands out as a revolutionary influence, reshaping the essence of approaching project risk management [2].

In this chapter, we explore the myriad dimensions through which AI technology can impact project risk management. This technological integration not only prompts a reformulation of prevailing strategies but also empowers decision-making processes, culminating in an overall enhancement of project outcomes that stand head and shoulders above previous achievements.

AI's introduction into project risk management heralds an era where predictive analytics and machine learning algorithms collaboratively scrutinize historical project data, unearthing hidden patterns and correlations that human analysis might overlook [3, 4]. This proactive approach and AI's capacity to forecast potential risks affords project managers a unique vantage point to anticipate challenges even before they manifest. Moreover, AI-powered simulations enable the construction of virtual risk scenarios, fostering a space for testing various mitigation strategies before implementation [5, 6].

Furthermore, AI's cognitive prowess accelerates risk assessment by swiftly sifting through vast information repositories, aggregating diverse data sources, from market trends to unforeseen disruptions. This amalgamation of insights equips decision-makers with a comprehensive comprehension of potential threats and opportunities, fostering agility in response strategies. Collaborative AI-driven platforms facilitate real-time communication among project teams, ensuring that risk-related information flows seamlessly and is acted upon promptly [7].

The transformational impact of AI extends beyond the operational facets. Strategic planning becomes more nuanced as AI algorithms continuously adapt to evolving risk landscapes. By analyzing internal and external dynamics, AI aids in refining risk management strategies, prompting timely adjustments to project blueprints and resource allocation. Some examples are:

AI has been explored through four perspectives based on thinking and acting dimensions, with two philosophies: human-centred and rationalist. AI has diverse applications like healthcare, security, finance, and more [8–13].

The AI landscape includes various techniques: (a) Machine Learning (ML) [14, 15]: It uses training data to classify and predict without explicit programming, employing algorithms like Random Forest, Support Vector Machine (SVM), Decision Trees, and K-means, (b) Deep Learning (DL) [16]: It handles complex raw data, building models with multiple layers for abstract data representation. (c) Neural Networks (NNs) [17]: Inspired by the human brain, NNs consist of nodes emulating neurons and synapses for information transfer and learning tasks without explicit programming. (d) Natural Language Processing (NLP) [18, 19]: It aims to enable AI to communicate with humans and extract information from text through tasks like classification, retrieval, and extraction. (e) Fuzzy Approaches: Fuzzy logic represents uncertainty in decision-making, used notably in expert systems [20–23] with fuzzy membership functions and rules for data reasoning [24–26].

As the boundaries of project risk management expand, ethical considerations emerge alongside technological advancements. The transparency and accountability of AI algorithms become paramount, necessitating careful oversight to prevent biases and ensure fair risk evaluations. Some key challenges Presented provided in Fig. 1.

This chapter highlights a dynamic narrative where AI and project risk management converge to yield unparalleled insights and strategies. It uncovers a future wherein technological innovation harmonizes with human understanding, propelling projects toward success in an environment of complexity and competition. Through AI's transformative potential, organizations fortify their resilience and catalyze a new era of project management excellence, where risks are not merely managed but harnessed to foster innovation and growth.

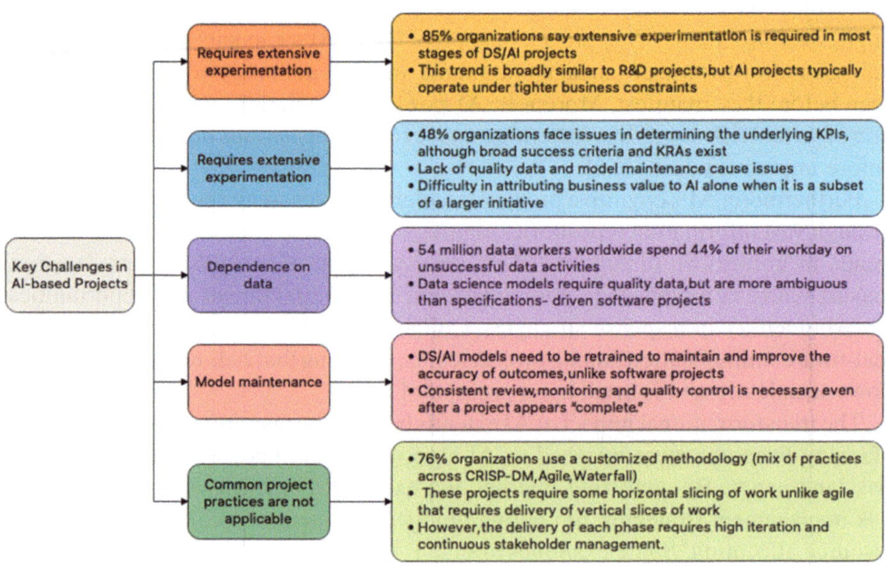

Fig. 1 Summery of challenges (Redrawn after [27])

2 The Evolution of Project Risk Management

Throughout history, project risk management has predominantly leaned on subjective judgment, retrospective data analysis, and the insights of experts [1, 28]. Although these approaches hold intrinsic value, they frequently need to be revised when confronting contemporary projects' intricate and unforeseeable characteristics. Enter the era of AI, which heralds an innovative paradigm by infusing data-driven revelations and anticipatory prowess, enriching and complementing the traditional risk management methodologies.

In times past, project risk management was often a product of human intuition and historical precedent [29]. While these human-centred perspectives provided crucial guidance, they struggled to grapple with modern projects' multifaceted uncertainties. The rapid pace of change, coupled with the intricacies of global interconnectedness, has rendered such conventional strategies increasingly insufficient.

With the advent of AI, project risk management experienced a metamorphosis. AI, powered by data analytics and machine learning, extracts profound insights from vast troves of historical data, enabling it to recognize intricate patterns and correlations that might elude human cognition. As a result, risk assessment transcends the limitations of human intuition, offering a more comprehensive and nuanced perspective on potential threats and opportunities [30–32].

Moreover, introducing predictive capabilities amplifies the potential of risk management strategies. AI's algorithms are adept at forecasting potential risks by extrapolating from historical data and discerning emerging trends. This foresight empowers project managers to proactively address challenges before they materialize, thereby altering the dynamics of risk mitigation from reactive to proactive.

Nevertheless, the true strength of AI lies in its symbiotic relationship with traditional methodologies. Rather than replacing existing approaches, AI harmoniously supplements them. Expert opinions and human insights remain pivotal, but now they are bolstered by AI-driven analytics' quantitative precision and scalability. The result is a synergy that equips decision-makers with a holistic perspective, helping them navigate the intricacies of risk with more informed and decisive actions.

As the evolution continues, it becomes increasingly clear that AI is not merely a disruptive force but an evolutionary leap forward. This technological integration elevates project risk management from a retrospective and subjective exercise to a forward-looking, data-enriched discipline. It empowers organizations to harness their historical experiences, enriches their capacity to anticipate the future, and ultimately fosters a new era of project success that thrives amidst the uncertainty of our modern world.

In this unfolding narrative, the role of AI extends beyond individual projects. It contributes to the collective wisdom of organizations by accumulating insights from various projects and sectors, enabling organizations to cultivate a repository of best practices and lessons learned. This shared intelligence enhances an organization's risk management approach and fosters a culture of continuous improvement and adaptability.

However, integrating AI into project risk management is full of challenges. Ethical considerations, data privacy concerns, and the potential for algorithmic bias demand careful attention. Maintaining transparency, ensuring accountability, and mitigating the risk of unintended consequences are essential facets of harnessing AI's potential responsibly.

In conclusion, the evolution of project risk management spurred by AI underscores the dynamic nature of progress itself. The journey from subjective intuition to data-driven foresight is marked by a willingness to embrace innovation while retaining the core tenets of human judgment and expertise. As organizations navigate the complexities of an ever-changing landscape, AI is a reliable partner, enabling them to chart a course toward improved decision-making, heightened resilience, and, ultimately, triumphant project outcomes.

3 AI-Driven Risk Identification

At the forefront of modern project management, AI-driven risk identification stands as a transformative capability. Leveraging advanced algorithms, AI possesses the remarkable capacity to meticulously comb through vast and heterogeneous datasets, unearthing potential risks that might have remained concealed using conventional methods. In this new era of risk management, the integration of NLP emerges as a pivotal catalyst, enabling the analysis of unstructured data extracted from project documentation, emails, and reports. This cutting-edge approach transcends the confines of structured data analysis, facilitating the identification of subtle nuances, latent patterns, and even the most cryptic early warning signs that foreshadow emerging risks.

Traditionally, risk identification often hinged upon structured data, such as historical records and quantitative metrics. While valuable, this methodology inherently had its limitations, particularly when grappling with the complexity of modern projects characterized by multifaceted variables and intertwined uncertainties. AI fundamentally revolutionizes this landscape by its ability to ingest, process, and comprehend various data types, bridging the gap between structured and unstructured information.

Natural language processing, a cornerstone of AI-driven analysis, equips the technology with a remarkable linguistic prowess. By dissecting the textual fabric of project documentation, emails, and reports, NLP uncovers the implicit narratives that traditional data analysis methods might overlook. It deciphers the implicit tones, sentiments, and contextual cues woven within the written word, enriching risk identification with a layer of understanding that resonates with human comprehension.

Incorporating AI-driven risk identification into project management ushers in an era of unparalleled proactive vigilance. AI algorithms tirelessly scrutinise every word and phrase as project-related communications and documents flow through the organizational ecosystem. By recognizing linguistic subtleties indicative of potential risks, AI becomes an invaluable collaborator in the quest to foresee challenges before they materialize. This real-time assessment empowers project teams with an informed edge, enabling them to pivot strategies and allocate resources to mitigate risks in their infancy.

However, the infusion of AI introduces a new dimension of responsibility. Ensuring the integrity of AI's risk identification process necessitates continuous validation, fine-tuning, and ethical scrutiny. The algorithms must be trained on high-quality, diverse datasets to avoid reinforcing biases and skewed assessments. Transparent governance is paramount to build trust and ensure accountability for the outcomes derived from AI-driven insights.

As the adoption of AI for risk identification matures, organizations can derive additional benefits beyond immediate risk mitigation. By analyzing patterns in historical data and recognizing recurrent risk factors, AI can offer insights that enable organizations to fine-tune their overall project strategies. Moreover, AI's ability to learn from new data and adapt to evolving risk landscapes empowers it to refine its risk identification capabilities over time.

In summation, AI-driven risk identification symbolises the dawn of a data-enriched era in project management. This technological evolution transcends the limitations of human-centric analysis, enabling the recognition of subtle indications that herald potential risks. By deciphering unstructured data with linguistic finesse, AI unlocks insights that augment traditional methodologies and empower organizations to navigate an intricate risk landscape with sagacity and foresight. As AI continues to mature, its role in risk identification evolves from a novelty to an indispensable tool for those who aspire to master the art of risk management in the modern age. In embracing this evolution, organizations embark on a transformative journey toward proactive risk intelligence and heightened project success.

4 Predictive Analytics for Risk Assessment

At the forefront of modern project management, AI-driven risk identification stands as a transformative capability. Leveraging advanced algorithms, AI possesses the remarkable capacity to meticulously comb through vast and heterogeneous datasets, unearthing potential risks that might have remained concealed using conventional methods. In this new era of risk management, the integration of NLP emerges as a pivotal catalyst, enabling the analysis of unstructured data extracted from project documentation, emails, and reports. This cutting-edge approach transcends the confines of structured data analysis, facilitating the identification of subtle nuances, latent patterns, and even the most cryptic early warning signs that foreshadow emerging risks.

Traditionally, risk identification often hinged upon structured data, such as historical records and quantitative metrics. While valuable, this methodology inherently had its limitations, particularly when grappling with the complexity of modern projects characterised by multifaceted variables and intertwined uncertainties. AI fundamentally revolutionises this landscape by its ability to ingest, process, and comprehend various data types, bridging the gap between structured and unstructured information.

Natural language processing (NLP), a cornerstone of AI-driven analysis, equips the technology with a remarkable linguistic prowess. By dissecting the textual fabric of project documentation, emails, and reports, NLP uncovers the implicit narratives that traditional data analysis methods might overlook. It deciphers the implicit tones, sentiments, and contextual cues woven within the written word, enriching risk identification with a layer of understanding that resonates with human comprehension.

Incorporating AI-driven risk identification into project management ushers in an era of unparalleled proactive vigilance. AI algorithms tirelessly scrutinise every word and phrase as project-related communications and documents flow through the organisational ecosystem. By recognising linguistic subtleties indicative of potential risks, AI becomes an invaluable collaborator in the quest to foresee challenges before they materialise. This real-time assessment empowers project teams with an informed edge, enabling them to pivot strategies and allocate resources to mitigate risks in their infancy.

However, the infusion of AI introduces a new dimension of responsibility. Ensuring the integrity of AI's risk identification process necessitates continuous validation, fine-tuning, and ethical scrutiny. The algorithms must be trained on high-quality, diverse datasets to avoid reinforcing biases and skewed assessments. Transparent governance is paramount to build trust and ensure accountability for the outcomes derived from AI-driven insights.

As the adoption of AI for risk identification matures, organisations can derive additional benefits beyond immediate risk mitigation. By analysing patterns in historical data and recognising recurrent risk factors, AI can offer insights that enable organisations to fine-tune their overall project strategies. Moreover, AI's ability to learn from new data and adapt to evolving risk landscapes empowers it to refine its risk identification capabilities over time.

AI-driven risk identification symbolises the dawn of a data-enriched era in project management. This technological evolution transcends the limitations of human-centric analysis, enabling the recognition of subtle indications that herald potential risks. By deciphering unstructured data with linguistic finesse, AI unlocks insights that augment traditional methodologies and empower organisations to navigate an intricate risk landscape with sagacity and foresight. As AI continues to mature, its role in risk identification evolves from a novelty to an indispensable tool for those who aspire to master the art of risk management in the modern age. In embracing this evolution, organisations embark on a transformative journey toward proactive risk intelligence and heightened project success.

5 Real-Time Risk Monitoring

In the ever-evolving area of project risk management, the emergence of AI as a steadfast partner brings forth an innovation that transcends boundaries. At the heart of this transformation lies the monumental capacity of real-time monitoring—an advancement that redefines how project risks are perceived, assessed, and acted upon. Through an intricate interplay of AI-powered sensors, Internet of Things (IoT) devices, and dynamic data streams, project managers ascend to an echelon of heightened vigilance, equipped with the tools to monitor project parameters and external influences in real-time seamlessly. This symbiotic fusion of technology empowers proactive decision-making and swift responses to burgeoning risk scenarios, epitomizing the synergy between human understanding and AI's analytical precision.

Traditionally, risk management encompassed periodic assessments and manual updates that relied heavily on historical and scheduled data. While this approach had its merits, it often needed to capture the dynamic and rapid changes that characterize modern project environments. Fuelled by AI, the introduction of real-time monitoring marks a paradigm shift that redefines risk awareness from a static snapshot to a dynamic tapestry woven by many interconnected threads.

Incorporating AI-powered sensors and IoT devices, projects transform into living ecosystems of data generation. These sensors incessantly harvest data from diverse aspects of the project, ranging from environmental conditions to machinery performance, thereby offering a comprehensive understanding of the project's health. Moreover, integrating external data streams, such as market trends and geopolitical factors, elevates the monitoring spectrum to encompass influences that transcend the project's immediate scope.

The treasure trove of real-time data is seamlessly ingested by AI algorithms that can decipher patterns, anomalies, and trends near-instantaneously. This dynamic analysis yields a multifaceted understanding of ongoing risks, enabling project managers to anticipate challenges as they unfold rather than reacting after the fact. The real-time awareness provided by AI-driven monitoring aligns risk management with the agile and rapidly changing nature of contemporary projects.

This new frontier of risk vigilance is a catalyst for proactive decision-making. Armed with real-time insights, project managers possess the agility to swiftly recalibrate strategies, allocate resources, and deploy contingency plans with precision. The result is an organization empowered to navigate an ever-shifting risk landscape with unprecedented skill and efficacy.

Yet, this technological evolution also comes with its share of considerations. Security, privacy, and data integrity form a trifecta of paramount concerns. Ensuring a seamless data flow without compromising sensitive information is an intricate challenge that demands robust protocols and encrypted communication pathways.

The journey into real-time risk monitoring is a testament to the adaptive capacity of organizations. It calls for embracing a culture of continuous learning and adjustment, where AI's insights serve as guiding beacons in steering the project ship through tumultuous waters. Collaboration between AI and human experts is a symphony of intellects, where data-driven intelligence combines with intuitive discernment to make informed choices that shape project destinies.

The synergy between AI and real-time monitoring revolutionizes project risk management by immersing it in a continuous stream of data-driven insights. This partnership empowers project managers to transcend traditional limitations and go through into a new area of anticipatory decision-making. As AI algorithms unravel the threads of real-time data, they weave a tapestry of risk awareness that empowers organizations to remain resilient, agile, and attuned to the ever-evolving nuances of modern projects [33]. Through real-time risk monitoring, AI infuses projects with a living pulse that harmonizes with the rhythm of change and advances the boundaries of success. With AI as a vigilant co-pilot, organizations embark on a journey where the future is navigated with poise, adaptability, and the unwavering pursuit of excellence.

6 Automating Routine Tasks

In the rapidly evolving landscape of project management, the integration of AI emerges as a transformative force that not only reshapes strategies but also streamlines operations at their very core. A profound testament to this transformation lies in task automation, where AI's capabilities are harnessed to alleviate project managers from the burden of routine risk management duties. By entrusting AI with responsibilities such as data collection, report generation, and trend analysis, project managers embark on a journey that liberates their bandwidth, enabling them to pivot their focus toward high-level strategic decision-making and orchestrating meticulous risk mitigation strategies.

Traditionally, the contours of risk management were intertwined with an array of repetitive tasks that spanned data collection, documentation, and the generation of insightful reports. While these tasks are essential, they often consume excessive time and divert attention away from the pivotal strategic aspects of risk management. The advent of AI-driven automation introduces a crucial shift, where mundane tasks are relegated to technology, and project managers are unburdened to operate in a area where their expertise can flourish.

AI algorithms, equipped with machine learning, can learn from patterns and historical data. They can adeptly navigate complex data landscapes, curating the information required for risk analysis. This autonomous data collection ensures accuracy and consistency and accelerates the process, freeing project managers from the shackles of manual data aggregation.

Report generation, a cornerstone of risk communication, is another arena where AI-driven automation shines. Drawing from the amassed data, AI algorithms can craft comprehensive reports infused with insights and trends. AI algorithms enhances the speed of report creation and imbues these documents with a layer of analytical depth that augments decision-making.

Furthermore, the analytical prowess of AI extends to trend analysis. By dissecting historical data, AI algorithms unveil hidden trends and correlations, which are the foundation for informed risk assessments. This analytical agility empowers project managers to anticipate potential pitfalls and design strategies underpinned by quantifiable insights.

The transformative effect of AI-driven automation in risk management extends beyond efficiency alone. As the administrative load lightens, project managers can engage in strategic thinking and proactive risk mitigation. Freed from the minutiae, they can devote their intellectual capital to discerning patterns, analyzing critical decision points, and charting courses that minimize vulnerabilities.

However, the adoption of AI automation also poses considerations. Striking a balance between machine-generated insights and human intervention is crucial. Human judgment and context cannot be replaced but should be augmented by AI's analytical power. The ethical implications of automation, including transparency and accountability, also deserve diligent attention.

In summation, the fusion of AI and task automation reframes the landscape of project risk management. The burden of routine tasks is transferred from project managers to technology, liberating their cognitive bandwidth for strategic pursuits. By capitalizing on AI's analytical precision, project managers can direct their energies toward envisioning futures, designing robust mitigation strategies, and navigating the ever-shifting terrain of risk with astute clarity. As organizations embrace this transformation, the area of risk management metamorphoses into a tapestry woven by collaboration between human insight and AI-driven efficiency.

7 Enhanced Decision-Making Through Data-Driven Insights

In the evolution of project management, the ascent of AI as a strategic ally has engendered a paradigm shift in decision-making dynamics. Anchoring this transformation is the monumental capacity of AI to generate actionable insights by dissecting vast datasets and distilling pertinent information. As decision-making becomes increasingly data-driven, project managers find themselves empowered to wield these insights as formidable tools, guiding them towards astute choices, effective risk mitigation strategies, and the meticulous optimization of project plans.

Historically, decision-making within the project management domain often leaned on instinct, experience, and an amalgamation of qualitative inputs. While these time-honoured practices hold their merit, they occasionally need to catch up in the face of complexity and multifaceted risk landscapes. AI emerges as a potent catalyst, leveraging its analytical prowess to sift through colossal volumes of data that humans could scarcely manage. By unravelling patterns, correlations, and subtle nuances buried within these data repositories, AI illuminates insights that resonate with a clarity unattainable through traditional approaches.

AI's value lies not solely in its data processing capabilities but in its ability to distil this raw information into actionable insights. These insights serve as a compass, guiding project managers through the intricate maze of decisions that dictate project trajectories. Armed with AI-generated insights, decision-makers can decipher trends, anticipate potential challenges, and project the consequences of different courses of action.

Integrating AI's insights within the decision-making framework is a transformative endeavour. Once informed primarily by historical experiences, risk mitigation strategies are now fortified by predictive insights derived from AI's historical and real-time data analysis. This augmentation enhances the precision of the method, fostering a proactive approach that champions risk containment before it transforms into adversity.

Moreover, AI's data-driven insights exert their influence on project plans. Project managers can recalibrate strategies, allocate resources more astutely, and navigate projects through turbulence with incredible foresight by evaluating the patterns and behaviours extracted from data. This orchestration of plans in alignment with data-driven intelligence sets the stage for optimized project outcomes.

As project managers leverage AI insights, it's imperative to remember that the collaboration between human intuition and AI-generated insights enhances the decision-making journey. While AI furnishes quantitative data, human expertise contributes to the qualitative aspects—context, judgment, and the synthesis of multidimensional considerations. Furthermore, AI facilitates the complex decision-making process, often encompassing uncertainties, various inputs, and subjective interpretations from decision-makers. By generating comprehensive reports, data-driven insights and trends through AI algorithms, AI contributes to enhancing objectivity and confidence within the complex decision-making process.

In the tapestry of progress, ethical considerations emerge. The ethical use of data and algorithms, guarding against biases, and ensuring transparency in decision-making algorithms are paramount. An honest AI partnership provides the fruits of technological innovation while upholding the values and principles that underpin responsible decision-making.

In summation, the fusion of AI-generated insights and human acumen ushers in a new era of informed decision-making within project management. The marriage of quantitative precision and qualitative judgment paves the way for strategies that are agile, adaptive, and poised for success. By harnessing the capabilities of AI, project managers transcend the limitations of traditional approaches, propelling projects toward horizons illuminated by data-driven insights. In this partnership, the symphony of human intellect and AI's analytical might harmonize to orchestrate a future where data and wisdom guide decisions.

As AI evolves, its potential to uncover subtle trends and correlations deepens. The iterative nature of machine learning ensures that AI adapts to changing circumstances, continuously refining its insights and strengthening its contributions to decision-making. Moreover, integrating AI insights extends beyond individual projects, fostering a collective organizational wisdom that informs strategic planning and fosters a learning culture.

The profound alliance between AI-powered data insights and project management heralds an era of decision-making excellence. This transformative synergy empowers project managers to transcend the boundaries of intuition and venture into a area of precision and foresight. As organizations embrace this evolutionary path, they chart a course toward project success illuminated by AI's analytical brilliance, navigated by the wisdom of human judgment, and defined by a commitment to innovation and excellence.

8 Empowering Risk Navigation Through AI-Powered Simulations

In the dynamic landscape of project management, where uncertainty and complexity are constants, the infusion of AI-powered simulations marks an epochal advancement. At the nucleus of this transformation lies the ability to model an array of risk scenarios, dissect their potential ramifications, and illuminate the path toward strategic success. By harnessing the analytical might of AI, project teams embark on a journey that transcends the area of conjecture, yielding insights that foster a more profound comprehension of risks while equipping them with the understanding to orchestrate strategies that navigate uncertainties with finesse.

Historically, comprehending and anticipating risk scenarios was fraught with ambiguity, often reliant on conjecture and past experiences [34, 35]. The advent of AI-powered simulations revolutionizes this landscape by providing a tangible platform to explore myriad hypothetical risk situations. These simulations, fuelled by AI's expansive data processing capacities, allow project teams to meticulously scrutinize the interplay of variables, external influences, and internal dynamics that shape potential risk outcomes.

AI simulations, rooted in machine learning and predictive modelling, empower project managers to orchestrate a dance of variables, crafting a spectrum of risk scenarios that reflect the multifaceted intricacies of real-world projects. Project teams obtain a comprehensive panorama of potential outcomes by running these simulations through AI algorithms, from best-case schemes to the most daunting contingencies.

The beauty of AI-powered simulations lies not solely in their ability to unveil the spectrum of potential risk trajectories but in their capacity to quantify impacts. By distilling the complex web of factors into quantifiable metrics, simulations lay the foundation for informed decision-making. This numerical insight empowers project managers to discern which risks bear the most significant potential for disruption and which might be absorbed with lesser consequences.

Furthermore, the practice of simulating risk scenarios becomes an incubator for innovation. It encourages project teams to envision strategies that might have remained undiscovered without the guidance of AI-powered simulations. By dissecting the impacts of different decisions and manoeuvres in a virtual environment, project managers cultivate a space to experiment, iterate, and fine-tune strategies before implementing them in the real world.

However, the potency of AI simulations is amplified when combined with the creative flair and contextual understanding of human expertise. AI is a powerful tool that provides data-driven insights. Still, human intuition and judgment provide the essential layer of wisdom that contextualizes these insights within the broader organizational objectives and cultural nuances.

Ethical considerations also loom large in this domain. Ensuring the simulations accurately represent reality, avoiding algorithmic biases, and maintaining transparency in modelling assumptions are all pivotal facets of ethical AI-powered simulations.

In the grand tapestry of progress, AI-powered simulations resonate as a cornerstone of the future. As AI algorithms evolve and datasets expand, these simulations' accuracy and predictive prowess will continue to deepen. This evolution will transform risk management practices and extend its influence on strategy formulation, resource allocation, and even exploring uncharted project territories.

The interplay between AI-powered simulations and human expertise is not a mere juxtaposition but a harmonious convergence that enriches both areas. While AI unravels the quantitative facets of risks, human insight offers qualitative understanding and contextualization. Together, they weave a narrative of performance and strategy that propels projects toward success.

The synergy between AI and simulations ushers in an era where risks are no longer enigmatic adversaries but manageable components of the project landscape. This transformative partnership empowers project teams to go through into the spectrum of possibilities, anticipate challenges, and devise strategies with the foresight needed to navigate uncertainties. As organizations embrace this evolution, they assume a future defined by innovation, resilience and the collective wisdom of AI-powered simulations harmonizing with human intellect. This alliance heralds an era where the uncharted territories of project management become fertile grounds for exploration and triumph. Figure 2 (modified after [27]) presents Resources and best-practices-based toolkit for AI-based projects.

9 Challenges and Considerations

Integrating AI into the fabric of project risk management is undoubtedly a transformative journey, ushering in a area of possibilities that extend far beyond conventional methodologies. However, as with any technological evolution, this integration is full of challenges. Ethical quandaries, data privacy dilemmas, the elusive nature of model interpretability, the imperative for human oversight, scalability considerations, and the complex interplay of AI and organizational culture emerge as critical facets that demand careful consideration to ensure a responsible, equitable, and effective AI-driven risk management paradigm.

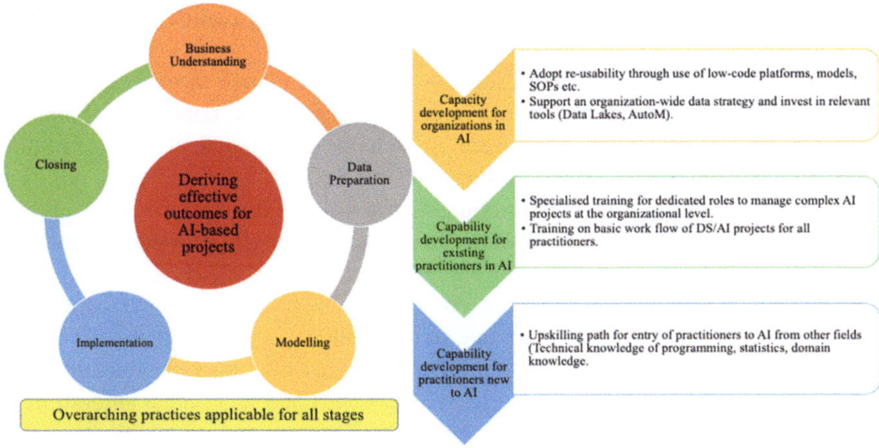

Fig. 2 Resources and best-practices-based toolkit for AI-based projects

- Ethical and Bias Concerns: As AI algorithms glean insights from diverse datasets, the potential for hidden biases to be perpetuated is a pressing concern. Tendencies present in historical data might be replicated, further exacerbating existing inequalities. It is essential to proactively address overt and subtle biases to ensure AI-driven risk management remains fair, inclusive, and representative of diverse perspectives. Developing methodologies for bias detection and mitigation becomes pivotal in upholding ethical standards.
- Data Privacy and Security: The sheer volume of data harnessed by AI-driven risk management systems raises questions about privacy and security. Safeguarding sensitive information while allowing AI to perform its analysis necessitates robust encryption, secure storage, and stringent access controls. Striking a balance between data utilization and protecting individual privacy is paramount. Adhering to evolving data protection regulations and ensuring transparent data handling practices are crucial in building stakeholder trust.
- Model Interpretability: AI algorithms, particularly those based on deep learning, often operate as black boxes, rendering their decision-making processes inscrutable. For risk management to be effective, it's imperative to bridge the gap between the complexity of AI algorithms and the need for understandable insights. Developing methods to explain the rationale behind AI-generated predictions is essential for building trust and facilitating informed decision-making. Techniques such as explainable AI and model visualizations can enhance interpretability.
- Human Oversight and Judgment: While AI possesses remarkable analytical capabilities, it is not a replacement for human judgment and contextual understanding. The collaboration between AI insights and human expertise is pivotal for striking a harmonious balance between quantitative analysis and qualitative sense. Establishing clear lines of responsibility and maintaining human oversight is crucial to ensure compliance with AI-generated recommendations. Human intervention

becomes crucial in interpreting AI outputs and making contextually informed decisions.

- Continual Learning and Adaptation: AI algorithms evolve through continuous learning from data, but the environment in which they operate is dynamic. Ensuring that AI-driven risk management systems can adapt to changing circumstances and developing risks requires a well-defined framework for ongoing training, validation, and recalibration. Constant monitoring of model performance, updates to account for evolving threats, and adapting to new data patterns are vital aspects of maintaining AI efficacy.
- Resistance to Change: Introducing AI into established risk management practices might encounter resistance due to unfamiliarity, perceived threats to job roles, or scepticism about AI's capabilities. A culture of openness, education, and collaboration is vital to overcome these challenges and facilitate a smooth transition. Clear communication about AI's role, benefits, and its augmentation of human capabilities is essential in addressing concerns and fostering buy-in from stakeholders.
- Resource Investment: Integrating AI into risk management necessitates resource allocation for technology adoption, data infrastructure enhancement, and talent acquisition. Organizations must weigh the potential benefits against the associated costs and align their strategies with their objectives. Adequate investment in training, technology infrastructure, and talent acquisition is required to ensure that AI-driven risk management is effective and sustainable.
- Organizational Culture Alignment: Integrating AI into risk management also requires alignment with the existing corporate culture. The integration of AI may involve overcoming resistance to change, fostering a culture of data-driven decision-making, and ensuring that AI complements the values and mission of the organization. An organization's readiness to embrace AI and its ability to adapt to a more data-driven approach are critical factors that can influence the success of AI implementation.
- Accountability and Responsibility: Integrating AI into project risk management introduces new dimensions of accountability and responsibility. While AI can assist in generating insights and recommendations, the ultimate decisions and actions remain under human jurisdiction. Defining clear lines of accountability becomes essential to ensure that AI-driven recommendations are correctly followed with human review and validation. In case of adverse outcomes, it is crucial to establish a framework for attributing responsibility and understanding AI's role in the decision-making process. This accountability framework helps balance leveraging AI's capabilities and upholding human oversight and responsibility.
- Governance and Regulation: The need for comprehensive control and regulation becomes evident as AI becomes more embedded in risk management processes. Establishing guidelines, standards, and frameworks for AI's ethical and responsible use is critical. Regulatory bodies and industry associations are increasingly

focusing on developing policies addressing AI transparency, fairness, accountability, and safety. Organizations must align their AI practices with these evolving regulations to ensure compliance and maintain ethical standards.

- Continuous Learning and Skill Development: The successful integration of AI-driven risk management requires a workforce with the necessary skills to work with AI systems effectively. Organizations must invest in continuous learning and skill development programs to ensure employees understand AI capabilities, limitations, and best practices. Continuous learning and skill development programs empower teams to collaborate with AI systems, interpret their outputs, and make contextually informed decisions.
- Transparency and Explainability: The clarity and explainability of AI-driven risk management systems are crucial for building trust and understanding among stakeholders. As AI algorithms make complex decisions, explaining how these decisions are reached is essential. Ensuring that stakeholders, including project managers, team members, executives, and even external parties, can understand the rationale behind AI-generated insights enhances the credibility and acceptance of AI-driven risk management. Transparent communication about the data sources, methodologies, and assumptions used in AI models fosters a sense of confidence and encourages informed decision-making.
- Change Management and Cultural Shift: Integrating AI-driven risk management represents more than a technological shift; it necessitates a cultural transformation within an organization. Change management becomes pivotal as teams adapt to new workflows, processes, and decision-making paradigms. Ensuring that employees understand the benefits of AI, are equipped with the skills to collaborate effectively with AI systems and are supported through the transition is essential for successful implementation. The cultural shift also fosters an innovation culture, encouraging experimentation and learning from AI-driven insights. Embracing a mindset that values continual improvement and adaptation to changing risk landscapes can lead to the effective assimilation of AI into risk management practices. Organizations should invest in comprehensive change management strategies, including training, communication, support mechanisms, and a clear roadmap for AI integration. By nurturing a culture that welcomes AI as an enabler rather than a disruptor, organizations can harness its capabilities to elevate risk management processes and outcomes.

Incorporating these additional considerations rounds out the holistic view of challenges and considerations in AI-driven project risk management. The multifaceted landscape demands a comprehensive approach that addresses ethical, technical, organizational, and regulatory dimensions. By acknowledging and proactively addressing these facets, organizations can navigate the complexities of AI integration while upholding moral integrity, fostering innovation, and driving positive outcomes in risk management.

10 Conclusion

In conclusion, integrating AI into project risk management represents a monumental paradigm shift, fundamentally altering how organizations approach risk mitigation. The infusion of AI's advanced capabilities empowers enterprises to adopt a proactive and meticulously data-driven stance towards managing risks within projects. This transformative journey encompasses the entire spectrum of risk management, from the early stages of risk identification and assessment to ongoing monitoring and sophisticated decision-making processes.

The marriage of AI with project risk management equips project managers with unprecedented agility and confidence when navigating the intricate terrains of today's complex business environment. AI's ability to identify potential risks, assess their potential impact, and predict their likelihood of occurrence revolutionizes how projects are conceived, planned, and executed swiftly and accurately. By harnessing the power of AI, organizations can anticipate challenges, devise effective mitigation strategies, and allocate resources with precision, ultimately driving projects towards successful outcomes.

As AI technologies continue their rapid evolution, the trajectory of their influence on project risk management appears destined for even more significant transformation. The symbiotic relationship between AI and risk management promises to redefine further established methodologies and best practices. As algorithms become more sophisticated and machine learning models more adept at understanding nuanced patterns, the potential for accurate risk forecasting and adaptive risk responses becomes increasingly evident.

In the coming years, we can anticipate a landscape wherein AI enhances the efficiency and effectiveness of risk management processes and catalyzes innovation and creativity in project management strategies. Organizations that embrace AI's potential will be positioned to navigate uncertainty with heightened resilience, responding to dynamic changes in real time and capitalizing on emerging opportunities. The fusion of human expertise with AI's computational prowess will give rise to a new era of project risk management characterized by informed decision-making, optimized resource allocation, and the consistent delivery of successful projects.

The infusion of AI into project risk management marks a profound evolution, enabling organizations to transcend conventional limitations and embark on a proactive, agile, and data-infused risk mitigation journey. As we stand at the cusp of this transformative era, it is evident that AI's trajectory within project risk management will continue to shape, refine, and ultimately revolutionize the way projects are envisioned, executed, and triumphantly delivered.

Declaration of generative AI and AI-assisted technologies in the writing process

During the preparation of this work, the author(s) utilized generative AI and AI-assisted technologies, including the Bing AI, to enhance the writing process. These tools were employed to improve language clarity and improving language fluency. Following the utilization of this tool/service, the author(s) thoroughly reviewed and edited the content, ensuring its coherence and accuracy in accordance with the intended message. The author(s) hold(s) complete responsibility for the final content and scholarly integrity of the publication.

References

1. PMI: Project Management Body of Knowledge (PMBOK® Guide). Proj. Manag. Institute, Inc. Newt. Square, PA (2017)
2. Russell, S.J.: Artificial Intelligence a Modern Approach. Pearson Education, Inc. (2010)
3. Xu, F., Lin, S.-P.: Theoretical framework of Fuzzy-AI model in quantitative project management. J. Intell. Fuzzy Syst. **30**, 509–521 (2016)
4. Vaishya, R., Javaid, M., Khan, I.H., Haleem, A.: Artificial Intelligence (AI) applications for COVID-19 pandemic. Diabetes Metab. Syndr. Clin. Res. Rev. **14**, 337–339 (2020)
5. Aljebory, K.M., QaisIssam, M.: Developing AI based scheme for project planning by expert merging Revit and Primavera software. In: 2019 16th International Multi-conference on Systems, Signals & Devices, pp. 404–412. IEEE (2019)
6. Dam, H.K., Tran, T., Grundy, J., Ghose, A., Kamei, Y.: Towards effective AI-powered agile project management. In: 2019 IEEE/ACM 41st International Conference on Software Engineering: New Ideas and Emerging Results, pp. 41–44. IEEE (2019)
7. Jallow, H., Renukappa, S., Suresh, S.: Knowledge management and artificial intelligence (AI). In: ECKM 2020 21st European Conference on Knowledge Management, p. 363. Academic Conferences International Limited (2020)
8. Frazer, H.M.L., Qin, A.K., Pan, H., Brotchie, P.: Evaluation of deep learning-based artificial intelligence techniques for breast cancer detection on mammograms: results from a retrospective study using a BreastScreen Victoria dataset. J. Med. Imaging Radiat. Oncol. **65**, 529–537 (2021)
9. Agarwala, N., Chaudhary, R.D.: Artificial intelligence and international security. In: Towards an International Political Economy of Artificial Intelligence, pp. 241–254. Springer (2021)
10. Warin, T., Stojkov, A.: Machine learning in finance: a metadata-based systematic review of the literature. J. Risk Financ. Manag. **14**, 302 (2021)
11. Thakkar, A., Lohiya, R.: A survey on intrusion detection system: feature selection, model, performance measures, application perspective, challenges, and future research directions. Artif. Intell. Rev. **55**, 453–563 (2022)
12. Yazdi, M., Khan, F., Abbassi, R.: A dynamic model for microbiologically influenced corrosion (MIC) integrity risk management of subsea pipelines. Ocean Eng. **269**, 113515 (2023). https://doi.org/10.1016/j.oceaneng.2022.113515
13. Taboada, I., Daneshpajouh, A., Toledo, N., de Vass, T.: Artificial intelligence enabled project management: a systematic literature review. Appl. Sci. **13** (2023). https://doi.org/10.3390/app13085014
14. Zhang, X.-D.: A Matrix Algebra Approach to Artificial Intelligence (2020)
15. Huang, C.-G., Men, C., Yazdi, M., Han, Y., Peng, W.: Transfer fault prognostic for rolling bearings across different working conditions: a domain adversarial perspective. Int. J. Adv. Manuf. Technol. (2022). https://doi.org/10.1007/s00170-022-09452-1
16. LeCun, Y., Bengio, Y., Hinton, G.: Deep learning. Nature **521**, 436–444 (2015)
17. Yegnanarayana, B.: Artificial Neural Networks. PHI Learning Pvt. Ltd. (2009)
18. Yazdi, M., Golilarz, N.A., Nedjati, A., Adesina, K.A.: An improved lasso regression model for evaluating the efficiency of intervention actions in a system reliability analysis. Neural Comput. Appl. (2021). https://doi.org/10.1007/s00521-020-05537-8
19. Liddy, E.D.: Natural Language Processing (2001)
20. Jackson, P.: Introduction to Expert Systems (1986)

21. Li, H., Yazdi, M.: Advanced Decision-Making Methods and Applications in System Safety and Reliability Problems. Springer, Cham (2022). https://link.springer.com/book/9783031074295
22. Li, H., Guo, J.-Y., Yazdi, M., Nedjati, A., Adesina, K.A.: Supportive emergency decision-making model towards sustainable development with fuzzy expert system. Neural Comput. Appl. **33**, 15619–15637 (2021). https://doi.org/10.1007/s00521-021-06183-4
23. Adesina, K.A., Yazdi, M., Omidvar, M.: Emergency decision making fuzzy-expert aided disaster management system. In: Yazdi, M. (ed.) Linguistic Methods Under Fuzzy Information in System Safety and Reliability Analysis, pp. 139–150. Springer International Publishing, Cham (2022). https://doi.org/10.1007/978-3-030-93352-4_6
24. Chen, T.Y.: New Chebyshev distance measures for Pythagorean fuzzy sets with applications to multiple criteria decision analysis using an extended ELECTRE approach. Expert Syst. Appl. **147**, 113164 (2020). https://doi.org/10.1016/j.eswa.2019.113164
25. Li, H., Yazdi, M.: An advanced TOPSIS-PFS method to improve human system reliability. In: Li, H., Yazdi, M. (eds.) Advanced Decision-Making Methods and Applications in System Safety and Reliability Problems: Approaches, Case Studies, Multi-criteria Decision-Making, Multi-objective Decision-Making, pp. 109–125. Springer International Publishing, Cham (2022). https://doi.org/10.1007/978-3-031-07430-1_7
26. Jiang, G.-J., Huang, C.-G., Nedjati, A., Yazdi, M.: Discovering the sustainable challenges of biomass energy: a case study of Tehran metropolitan. Environ. Dev. Sustain. (2023). https://doi.org/10.1007/s10668-022-02865-8
27. Playbook for Project Management in Data Science and Artificial Intelligence Projects, PMI South Asia NASSCOM (2020)
28. van Rooij, S.W.: Scaffolding project-based learning with the project management body of knowledge (PMBOK®). Comput. Educ. **52**, 210–219 (2009)
29. Richardson, G.L.: Project Management Theory and Practice. CRC Press (2010)
30. Yazdi, M., Mohammadpour, J., Li, H., Huang, H.-Z., Zarei, E., Pirbalouti, R.G., Adumene, S.: Fault tree analysis improvements: a bibliometric analysis and literature review. Qual. Reliab. Eng. Int. (2023). https://doi.org/10.1002/qre.3271
31. Li, H., Yazdi, M.: A holistic question: is it correct that decision-makers neglect the probability in the risk assessment method? In: Li, H., Yazdi, M. (eds.) Advanced Decision-Making Methods and Applications in System Safety and Reliability Problems: Approaches, Case Studies, Multi-criteria Decision-Making, Multi-objective Decision-Making, Fuzzy Risk-Based Models. Springer International Publishing, Cham (2022). https://doi.org/10.1007/978-3-031-07430-1_10
32. Li, H., Peng, W., Adumene, S., Yazdi, M.: Intelligent Reliability and Maintainability of Energy Infrastructure Assets. Springer Nature Switzerland (2023)
33. Yazdi, M., Zarei, E., Pirbalouti, R.G., Li, H.: A comprehensive resilience assessment framework for hydrogen energy infrastructure development. Int. J. Hydrogen Energy (2023). https://doi.org/10.1016/j.ijhydene.2023.06.271
34. Li, H., Yazdi, M., Huang, H.-Z., Huang, C.-G., Peng, W., Nedjati, A., Adesina, K.A.: A fuzzy rough copula Bayesian network model for solving complex hospital service quality assessment. Complex Intell. Syst. (2023). https://doi.org/10.1007/s40747-023-01002-w
35. Li, H., Yazdi, M.: Advanced decision-making neutrosophic fuzzy evidence-based best–worst method. In: Li, H., Yazdi, M. (eds.) Advanced Decision-Making Methods and Applications in System Safety and Reliability Problems: Approaches, Case Studies, Multi-criteria Decision-Making, Multi-objective Decision-Making, Fuzzy Risk-Based Models, pp. 153–184. Springer International Publishing, Cham (2022). https://doi.org/10.1007/978-3-031-07430-1_9

MIX
Papier aus verantwortungsvollen Quellen
Paper from responsible sources
FSC® C105338

If you have any concerns about our products,
you can contact us on
ProductSafety@springernature.com

In case Publisher is established outside the EU,
the EU authorized representative is:
Springer Nature Customer Service Center GmbH
Europaplatz 3, 69115 Heidelberg, Germany

Printed by Libri Plureos GmbH
in Hamburg, Germany